Groundwater Pollution, Aquifer Recharge and Vulnerability

Geological Society Special Publications
Series Editor A. J. FLEET

It is recommended that reference to all or part of this book should be made in one of the following ways:

ROBINS, N. S. (ed.) 1998. *Groundwater Pollution, Aquifer Recharge and Vulnerability*. Geological Society, London, Special Publications, **130**.

ADAMS, B. & MACDONALD, A. M. 1998. Aquifer susceptibility to side-effects of groundwater exploitation. *In:* ROBINS, N. S. (ed.) *Groundwater Pollution, Aquifer Recharge and Vulnerability*. Geological Society, London, Special Publications, **130**, 71–76.

GEOLOGICAL SOCIETY SPECIAL PUBLICATION NO. 130

Groundwater Pollution, Aquifer Recharge and Vulnerability

EDITED BY

N. S. ROBINS
British Geological Survey
Wallingford
UK

1998

Published by

The Geological Society

London

THE GEOLOGICAL SOCIETY

The Society was founded in 1807 as The Geological Society of London and is the oldest geological society in the world. It received its Royal Charter in 1825 for the purpose of 'investigating the mineral structure of the Earth'. The Society is Britain's national society for geology with a membership of around 8000. It has countrywide coverage and approximately 1000 members reside overseas. The Society is responsible for all aspects of the geological sciences including professional matters. The Society has its own publishing house, which produces the Society's international journals, books and maps, and which acts as the European distributor for publications of the American Association of Petroleum Geologists, SEPM and the Geological Society of America.

Fellowship is open to those holding a recognized honours degree in geology or cognate subject and who have at least two years' relevant postgraduate experience, or who have not less than six years' relevant experience in geology or a cognate subject. A Fellow who has not less than five years' relevant postgraduate experience in the practice of geology may apply for validation and, subject to approval, may be able to use the designatory letters C Geol (Chartered Geologist).

Further information about the Society is available from the Membership Manager, The Geological Society, Burlington House, Piccadilly, London W1V 0JU, UK. The Society is a Registered Charity, No. 210161.

Published by The Geological Society from:
The Geological Society Publishing House
Unit 7, Brassmill Enterprise Centre
Brassmill Lane
Bath BA1 3JN
UK
(*Orders*: Tel. 01225 445046
Fax 01225 442836)

First published 1998

British Library Cataloguing in Publication Data
A catalogue record for this book is available from the British Library.

ISBN 1-897799-98-5
ISSN 0305-8719

Typeset by The Alden Press, Osney Mead, Oxford, UK.

Printed by The Alden Press, Osney Mead, Oxford, UK.

Distributors

USA
AAPG Bookstore
PO Box 979
Tulsa
OK 74101-0979
USA
(*Orders*: Tel. (918) 584-2555
Fax (918) 560-2652)

Australia
Australian Mineral Foundation
63 Conyngham Street
Glenside
South Australia 5065
Australia
(*Orders*: Tel. (08) 379-0444
Fax (08) 379-4634)

India
Affiliated East-West Press PVT Ltd
G-1/16 Ansari Road
New Delhi 110 002
India
(*Orders:* Tel. (11) 327-9113
Fax (11) 326-0538)

Japan
Kanda Book Trading Co.
Tanikawa Building
3-2 Kanda Surugadai
Chiyoda-Ku
Tokyo 101
Japan
(*Orders*: Tel. (03) 3255-3497
Fax (03) 3255-3495)

Contents

Referees

The Editor is grateful to the following people for their assistance with the reviewing of papers submitted to this Special Publication.

Dr P. J. Aldous	Thames Water Utilities, Reading
Mr D. J. Allen	British Geological Survey, Wallingford
Dr T. C. Atkinson	School of Environmental Sciences, University of East Anglia
Dr W. Back	US Geological Survey, Reston, VA., USA
Mr D. Banks	Norges Geologiske Undersokelse, Trondheim, Norway
Professor J. A. Barker	Department of Geological Sciences, University College London
Mr J. R. P. Bennett	Hydrogeological & Environmental Services, Belfast
Dr W. G. Burgess	Department of Geological Sciences, University College London
Mr D. Burgess	Environment Agency, Anglian Region, Peterborough
Mr K. Cullen	K T Cullen & Co, Dublin, Republic of Ireland
Mr I. R. Davey	Environment Agency, Thames Region, Reading
Dr R. A. Downing	Twyford, Reading
Mr J. Ferry	Geraghty & Miller, Cambridge
Dr A. J. Fleet	Department of Mineralogy, The Natural History Museum
Mr S. W. Fletcher	Environment Agency, Midlands Region, Solihull
Dr T. S. Haines	Entec Ltd, Shrewsbury
Dr J. Heathcote	Entec UK, Shrewsbury
Dr T. H. E. Heaton	NIGL, Keyworth
Dr K. M. Hiscock	School of Environmental Sciences, University of East Anglia
Dr D. G. Kinniburgh	British Geological Survey, Wallingford
Professor D. Lerner	Department of Civil Engineering, University of Bradford
Dr A. Lilly	The Macaulay Land Use and Research Institute, Aberdeen
Mr B. D. R. Misstear	Trinity College, Dublin, Republic of Ireland
Mr M. J. Morrey	Aspinwall & Co, Shrewsbury
Mr B. L. Morris	British Geological Survey, Wallingford
Professor R. N. Mortimore	Department of Civil Engineering, University of Brighton
Professor J. Mulready	Department of Civil Engineering, University of Leeds
Ms H. Nash	Wardell Armstrong, Newcastle-under-Lyme
Dr M. Owen	Environmnent Agency, Thames Region, Reading
Mr R. J. Peart	British Geological Survey, Keyworth
Mr M. Price	PRIS, University of Reading
Dr M. Robinson	Institute of Hydrology, Wallingford
Dr J. C. Rodda	Brightwell-cum-Sotwell, Oxon
Professor K. R. Rushton	School of Civil Engineering, University of Birmingham
Dr A. C. Skinner	Environment Agency, Midlands Region, Solihull
Dr P. L. Smedley	British Geological Survey, Wallingford
Mr D. C. Watkins	Camborne School of Mines, Redruth
Mr G. M. Williams	British Geological Survey, Keyworth
Dr E. P. Wright	Aston Tirrold, Oxon
Dr P. L. Younger	Department of Civil Engineering, University of Newcastle

Recharge: the key to groundwater pollution and aquifer vulnerability

N. S. ROBINS

British Geological Survey, Maclean Building, Crowmarsh Gifford, Wallingford, Oxfordshire, OX10 8BB, UK

Abstract: Recharge is pivotal to understanding the processes by which groundwater pollution can occur. It is implicit in the classification of aquifer units according to their vulnerability to pollution. The management of both groundwater resources and of individual groundwater sources cannot sensibly be undertaken without some knowledge of recharge: its quantity, its seasonality and, above all, the different routes through the sub-soil and the unsaturated zone by which it can occur. However, current estimates of recharge, other than on a research site basis, may be poor, both in the UK and overseas. This volume provides a review of current research into these issues; this introductory paper attempts to highlight the thread throughout all of this work which collectively provides the basic information in support of the current and future management of groundwater resources and sources.

This volume serves to highlight current developments in the assessment of groundwater pollution, aquifer recharge and vulnerability, but more particularly it illustrates the breadth of scientific enquiry that is currently being applied to this complex, but topical, field of study. Moreover, the volume demonstrates that the key to understanding many aspects of groundwater pollution lies alongside the reliable estimation of recharge, and also implicitly in the definition of aquifer vulnerability. To this end, a great deal of effort is now being put into investigating the processes that control recharge and to the application of contemporary findings.

Various groundwater management tools have been, and are being put into place by the Environment Agency in the UK. Many of these were first described in *Policy and Practice for the Protection of Groundwater* (NRA 1992), a document which has since been emulated by Scotland and will shortly also be followed by Northern Ireland. This policy document clearly recognizes the important role of recharge and acknowledges potential areas in which knowledge and data are currently inadequate for the rigid application of the strategy. Data are least available in areas of upland Britain, including Scotland and Northern Ireland, where information has not been collected so assiduously by the developers of the lesser aquifers as it has been of necessity by workers on the major aquifers such as the Chalk, the Triassic sandstones and the Jurassic limestones of England. Lack of detailed knowledge and understanding, particularly of relevant processes, is apparent from much of the work described in this volume; a great deal of research has yet to be undertaken, on both a generic and a site-specific basis.

This volume concentrates on the key role of recharge in groundwater management. This introductory paper attempts to highlight some of the issues pertaining to recharge estimation, and to the measurement and modelling of recharge; to the complex relationship between recharge, groundwater pollution and aquifer vulnerability; and to the importance of the recharge estimate.

Recharge estimation: the pitfalls

Traditional estimates of recharge have depended on the closure of the water balance for a given catchment or hydrogeological unit. Measurement of rainfall, evapotranspiration and run-off provides an indirect estimate of infiltration by remainder. However, history has shown that each new method of determining infiltration and recharge based on secondary observations (e.g. climatic and hydrological data) falls ultimately into part disfavour.

Calculation of evaporation from open water using an evaporation pan, and the determination of the same from meteorological data (Penman 1948) was overtaken by the estimation of transpiration for particular types of vegetation (Penman 1963). The general expression for transpiration depends on the sum of the available energy and an aerodynamic term, the latter a function of wind-speed and humidity. Nowadays this technique, modified by Monteith (1965), forms the basis of the Morecs 40 by 40 km block averages for the UK. In the absence of site-specific information, these are generally the only available data.

Estimates of run-off using low surface water flow data (Gustard *et al.* 1992), although widely available, provide little insight into the nature of the baseflow contribution that is actually maintaining the low flow. They do, however,

Robins, N. S. 1998. Recharge: the key to groundwater pollution and aquifer vulnerability. *In*: Robins, N. S. (ed.) *Groundwater Pollution, Aquifer Recharge and Vulnerability*. Geological Society, London, Special Publications, **130**, 1–5

allow regional estimates of recharge to be made based on geolines for aquifers and non-aquifers, and low flow indices (Robins 1990). The hydraulic relationship between surface and groundwater at any given point in a stream or river bed has first to be defined before low flow data can be used rationally. That being so, data are available for large parts of the UK.

Even a simple parameter like rainfall is not easily assimilated into a recharge calculation. The rainfall event, its intensity and duration, are as critical as the existing soil moisture status. Furthermore, microclimatic variation may provide significant difficulties in rational extrapolation of data even across a small catchment area. Clearly, accurate determination of recharge must be done on a site-specific basis using a direct measurement such as a lysimeter (see Jones & Cooper 1998) or using soil moisture measurement and event modelling (see Blackie *et al.* 1998). These techniques are, however, expensive. It may be that the way forward is through regional digital hydrogeological models which will assist in establishing overall renewable resource values for a given aquifer unit rather than point estimates of actual recharge.

A further significant problem arises in Britain through the existence of glacial till, which overlies bedrock over a significant part of upland Britain and much of lowland Britain north of the Severn and the Wash, and the Clay-with-Flint that occurs on some of the Chalk outcrop. It is often considered that the existence of till reduces potential recharge to an underlying aquifer by 30%, and such a rule of thumb has served, for the most part, reasonably satisfactorily. But what of the nature of the till itself? There is a considerable difference between the thick clay till cover of the Fylde, and the often sandy till with pebble lenses that occurs in many areas elsewhere. The thick till deposits of North Yorkshire and Humberside clearly inhibit all recharge, but how effective are the thin veneers of till and Clay-with-Flint?

Mapping geologists have always concentrated on bedrock mapping; at best till is depicted as a pale blue wash, with little regard for its integrity or thickness. Recent mapping of superficial strata, such as that reported by Browne & McMillan (1991) for the Glasgow area and elsewhere in Scotland, has attempted to provide the detail with which recharge distribution can be attempted, but areas so far covered are few. For the most part, site-specific investigations are necessary in areas where low-conductivity till is believed to be present. This is clearly advised on the aquifer vulnerability maps of the Environment Agency, Scottish Office and the Environment Service of Northern Ireland. Besides, the role of till in inhibiting recharge is locally zero wherever it is removed or effectively removed, for example in a quarry or a railway cutting, or indeed where the source is below the protective cover, such as a leaking sewer set at the base of the till. The published vulnerability maps identify 5 m as the critical thickness of till as a protecting cover to an aquifer, but acknowledge that till integrity is crucial but generally unknown (Palmer *et al.* 1995).

The actual mechanism by which recharge occurs is also very important. Price *et al.* (1993) argue that the Chalk has three sets of porosity: microfissures which allow rapid egress of storm water; microfissures which promote storage and transport; and the intergranular matrix of the Chalk blocks. Jones & Cooper (1998) demonstrate that 30% of the potential recharge within the Fleam Dyke lysimeter in Cambridgeshire takes place within fractures. Many other aquifers offer dual porosity, the Triassic sandstones are largely fractured and may also contain dilated bedding planes sufficient to offer rapid egress to percolating water. At the extreme end of the scale, karstic limestone provides selective, but potentially more rapid egress, transport and discharge, in this case between sinks and risings, but with almost no storage (Daly & Warren 1998).

The role of the soil zone is critical in the infiltration process. The texture and grain size together control the ability of a soil to retain or to drain soil moisture, both saturated and unsaturated. Palmer & Lewis (1998) explain in some detail how the structure and density of a soil influence the progress of permeating fluids, and how the presence of clay minerals and organic matter influence the so-called leaching potential of a soil (the ability of a soil to transmit non-conservative pollutants). Cation exchange also assists degradation and attenuation of some pollutants within the soil zone.

Estimation of recharge is, therefore, fraught with difficulty, a theme which is expanded by Simmers (1998). Recharge can be measured directly and accurately on a site-specific basis, but extrapolation of such data, either in time at the measured site or spatially away from the measured site, is difficult, not least because the prevailing conditions do not remain constant and because the principal recharge mechanism may differ from one location to another.

There is additional uncertainty in estimating recharge in a semi-arid climate where recharge events need not necessarily occur on an annual basis and may even be complicated by urbanization and man-made lagoons (Alderwish & Dottridge 1998).

Groundwater pollution and recharge

The obvious difficulties of measuring recharge and even of understanding the prevailing recharge processes, detract considerably from studies of groundwater pollution in the unsaturated zone. Recharge has two direct effects on dispersed pollutants: the greater the recharge the greater is the potential for dilution, but the more recharge the more efficient the transport of pollutant from surface to water-table. In the former case, the occurrence of nitrate in unconfined aquifers correlates directly to rainfall: nitrate concentrations in groundwater are least because of dilution in the higher rainfall areas of Northern Ireland and west England than in the low rainfall areas of eastern England, notably East Anglia, Scotland, and even Jersey.

Whereas calculations of recharge derived from regional models calibrated with long-term rainfall and hydrograph responses enable a renewable resource estimate to be derived, these techniques have little to offer pollution studies. The question of whether our understanding of recharge mechanisms is adequate to characterize flow rates and pathways in the phreatic aquifer recharge process and in transfer to semi-confined aquifers is taken up by Foster (1998). This question has obvious implications for groundwater pollution protection policies in the UK and elsewhere. The recharge mechanism is also critical; the role of by-pass routes and the occurrence of fractures and other preferential flow paths clearly influence pollutant transport potential, although the specific issues pertaining to fractured rocks are not described in the policy document for England and Wales (NRA 1992).

Management of potential sources of pollution also forms an important component of groundwater protection strategy. This too hinges on understanding the water balance at a particular site (Sears 1998). Conversely, the identification of the source of pollution is critical to remediation. Green *et al.* (1998) used nitrogen isotopes to prove that elevated nitrate concentrations in Jersey groundwater derive principally from agricultural fertilizer rather than septic tanks and soakaways.

Aquifer vulnerability and recharge

In Britain, aquifer vulnerability is determined according to the nature of the overlying soil cover; the presence and nature of the drift; the nature of strata; and the thickness of the unsaturated zone (NRA 1992). Collectively, these four variables provide a function which is implicitly related to recharge, or more strictly to the accessibility of the aquifer to recharge, independent of effective rainfall. There are many different systems for deriving a working model and hence a map of the degrees of vulnerability (Vrba & Zoporozec 1994) and this is amply illustrated by comparing work in England and Wales (Palmer & Lewis 1998) and in the Republic of Ireland (Daly & Warren 1998).

The two national strategies start from quite different vantages. The former developed from a mapping programme designed to delineate the vulnerability of aquifers to pollution from agricultural fertilizer, i.e. from diffuse pollution caused by the leaching of excess nitrate applied to the ground (Robins *et al.* 1994). These maps have now given way to the current series of aquifer vulnerability maps, which provide for any reasonably conservative diffuse pollutant, including nitrate, but also for any point-source pollutant discharging at the surface to produce saturated conditions at or near the surface (Adams & Foster 1991). This system, therefore, places emphasis on the role of the topsoil as an inhibitor to the downward migration of diffuse sources of pollution. However, the main source of pollution in the Republic of Ireland is not diffuse or surface point sources, but rather the point-source pollution posed by numerous underground septic tanks. These are designed to overflow at a depth of between one and two metres below ground surface; consequently the topsoil does not feature in the equation, which in Ireland emphasizes permeability and thickness of unsaturated zone strata.

Other significant differences are that the UK mapping programme only classifies the main or regionally important aquifers (major and minor aquifers). Not all the aquifers in the Republic of Ireland fall into these categories and it would be an inappropriate method to apply, where all aquifers – regionally important, locally important and poor aquifers – are classified. The Irish system is thus a scheme to map the vulnerability of all groundwater and not just the more important aquifers. The other significant difference is that whereas the aquifer vulnerability classification is used separately from source protection zonation to define control measures in the UK, vulnerability and source protection together define control measures in a single matrix in the strategy recommended for the Republic of Ireland.

The problems of dealing with karstic limestones are actively being considered in Ireland. Recommended policy in the Republic of Ireland is that each sink be mapped with a 30 m radius circle which is designated 'extreme vulnerability', and this is extended to include the whole of the surface water catchment that drains into it. The karst is clearly an area that will tax groundwater managers for a long time to come; the Environment Agency is now recognizing the importance of karst in England and Wales, particularly with regard to Mendip and other similar areas, and contemporary literature (e.g. MacDonald *et al.* 1998) suggests that the English Chalk is in part karstic.

Groundwater protection

The key to a successful groundwater management policy is an understanding of recharge and recharge processes. In an English context this depends heavily on good quality Quaternary mapping, but this is not yet universally available. Burgess & Fletcher (1998) describe in detail the regulatory policy applied for the delineation of groundwater source protection zones. Together with aquifer vulnerability zoning, be it the English method or the Irish method, these zones enable suitable control measures to be applied in order to safeguard groundwater from pollution.

All groundwater protection policies must be reviewed from time to time. As technology advances, and as understanding of groundwater processes allows, the existing policies will be adapted to provide the best possible and most suitable policies for resource and source management. To this end, the discussion on karstification in the Chalk (MacDonald *et al.* 1998) and the discussion of whether the Kellaways Sand is an aquifer containing groundwater that should be controlled (Mather *et al.* 1998) provide lines of enquiry to be pursued by the policy makers. Perhaps the concept of mapping the susceptibility of aquifers to over-exploitation may become an important future management tool (Adams & MacDonald 1998) although it is most likely to be applicable in regions of new and rapid groundwater development.

In conclusion, the current policy and strategy for groundwater protection in the UK (NRA 1992), and the policy now recommended in the Republic of Ireland, represent the state of the art and provide the best possible and best practical means of managing both resources and sources at the moment.

This volume arises from a symposium on 'Groundwater Pollution, Aquifer Recharge and Vulnerability' which was held as part of the first biennial meeting of the Geological Society on 'Applied Geology' at Warwick University in April 1996, and from a workshop on 'Aquifer Vulnerability' which was presented at the meeting by the Environment Agency. This paper is published by permission of the Director, British Geological Survey (NERC).

References

ADAMS, B. & FOSTER, S. S. D. 1991. *National groundwater protection policy: hydrogeological criteria for division of the land surface area.* National Rivers Authority, R&D Note 6.

—— & MACDONALD, A. M. 1998. Aquifer susceptibility to side-effects of groundwater exploitation. *This volume.*

ALDERWISH, A. M. & DOTTRIDGE, J. 1998. Recharge components in a semi-arid area: the Sana'a Basin, Yemen. *This volume.*

BLACKIE, J. R., HOUGHTON-CARR, H. A., MCCARTNEY, M. P. & MOORES, J. P. 1998. Estimation of groundwater recharge on Jersey. *This volume.*

BROWNE, M. A. E.& MCMILLAN, A. A. 1991. British Geological Survey thematic geology maps of Quaternary deposits in Scotland. *In*: FORSTER, A., CULSHAW, M. G., CRIPPS, J. C., LITTLE, J. A. & MOON, C. F. (eds), *Quaternary Engineering Geology*. Geological Society, London, Special Publication, **7**, 511–518.

BURGESS, D. B. & FLETCHER, S. W. 1998. Methods used to delineate groundwater source protection zones in England and Wales. *This volume.*

DALY, D. & WARREN, W. P. 1998. Mapping groundwater vulnerability: the Irish perspective. *This volume.*

FOSTER, S. S. D. 1998. Groundwater recharge and pollution vulnerability of British aquifers: a critical overview. *This volume.*

GREEN, A. R., FEAST, N. A., HISCOCK, K. M. & DENNIS, P. F. 1998. Identification of the source and fate of nitrate contamination of the Jersey bedrock aquifer using stable nitrogen isotopes. *This volume.*

GUSTARD, A., BULLOCK, A. & DIXON, J. M. 1992. *Low flow estimation in the United Kingdom.* Institute of Hydrology Report No. 108.

JONES, H. K. & COOPER, J. D. 1998. Water transport through the unsaturated zone in the Middle Chalk: a case study from Fleam Dyke lysimeter. *This volume.*

MACDONALD, A. M., BREWERTON, L. J. & ALLEN, D. J. 1998. Evidence for rapid groundwater flow and karst-type behaviour in the Chalk of Southern England. *This volume.*

MATHER, J. D., HALLIDAY, D. & JOSEPH, J. B. 1998. Is all groundwater worth protecting? The example of the Kellaways Sand. *This volume.*

MONTEITH, J. L. 1965. Evaporation and environment. *In: The State and Movement of Water in Living Organisms.* Proceedings of the 19th Symposium of the Society of Experimental Biology, Swansea, 1964, **19**, 205–234.

NRA. 1992. *Policy and Practice for the Protection of Groundwater*. National Rivers Authority, Bristol.

PALMER, R. C. & LEWIS, M. A. 1998. Assessment of groundwater vulnerability in England and Wales. *This volume*.

——, HOLMAN, I. P., ROBINS, N. S. & LEWIS, M. A. 1995. *Guide to Groundwater Vulnerability Mapping in England and Wales*. HMSO, London.

PENMAN, H. L. 1948. Natural evaporation from open water, bare soil and grass. *Proceedings of the Royal Society of London (A)*, **193**, 120–145.

—— 1963. *Vegetation and Hydrology*. Commonwealth Agricultural Bureau, Soils, Technical Communication **53**.

PRICE, M., DOWNING, R. A. & EDMUNDS, W. M. 1993. The Chalk as an aquifer. *In:* DOWNING, R. A., PRICE, M. & JONES, G. P. (eds) *The Hydrogeology of the Chalk of North-West Europe*, Clarendon, Oxford, 35–58.

ROBINS, N. S. 1990. *The Hydrogeology of Scotland.* HMSO, London.

——, ADAMS, B., FOSTER, S. S. D. & Palmer, R. 1994. Groundwater vulnerability mapping: the British perspective. *Hydrogéologie*, **3**, 35–42.

SEARS, R. 1998. The British Nuclear Fuel's Drigg low-level waste site characterization programme. *This volume*.

SIMMERS, I. 1998. Groundwater recharge: an overview of estimation 'problems' and recent developments. *This volume*.

VRBA, J. & ZOPOROZEC, A. (eds) 1994. *Guidebook on Mapping Groundwater Vulnerability*. International Association of Hydrologists, Hannover, International Contributions to Hydrogeology, **16**.

Groundwater recharge and pollution vulnerability of British aquifers: a critical overview

S. S. D. FOSTER

British Geological Survey, Groundwater and Geotechnical Surveys Division, Nottingham, NG12 5GG, UK

Abstract: The concept of aquifer pollution vulnerability, and the way it has been developed and incorporated as a central element in British groundwater protection policies, is reviewed in detail. A related question is: do we understand enough about groundwater recharge mechanisms to assess pollution vulnerability realistically and to define necessary protection measures efficiently? Some key remaining areas of uncertainty are analysed in the light of recent research results.

Groundwater recharge and pollution vulnerability: the intimate link

Since the transport of most groundwater contaminants to saturated aquifers occurs in the aqueous phase as part of the recharge process, assessing aquifer pollution vulnerability is inextricably linked with understanding groundwater recharge mechanisms. The only exceptions are density-driven contamination by non-aqueous phase liquids (NAPLs) and saline effluent discharge, although even here the 'contaminant pathways' will still have comparability with natural recharge mechanisms.

The groundwater recharge mechanisms referred to include the soil infiltration process, the flow rates and pathways through the vadose (unsaturated) zone to phreatic aquifers and the vertical leakage through confining beds to semi-confined aquifers.

The need to understand recharge mechanisms is in marked contrast to the (earlier) requirements for water resources inventory, which were restricted to the estimation of approximate average recharge volumes, and their spatial and temporal variation in general terms.

Techniques of recharge evaluation

Traditional approaches to the evaluation of groundwater recharge (permeable catchment water balance and agrometeorological data analysis) provided no information on recharge mechanisms (Table 1). There has, however, been steady refinement of the latter method (Thompson *et al.* 1981), which is the normal 'starting point' for all studies of potential groundwater recharge.

It was the development of vadose zone environmental isotope (tritium and deuterium) profiling in the 1970s and 1980s that stimulated the debate on groundwater recharge and contaminant transport mechanisms, but there remain today significant limitations to the interpretation of these data in fractured porous media (Foster & Smith-Carington 1980). Artificial soil-injected tracers offer the possibility of studying the scale of heterogeneity and dispersive vadose zone flow in greater detail, at least on a representative site basis and provided that the bulk of the tracer mass can be accounted for, but to date they have not been used significantly in Britain.

Soil water physics monitoring (Table 1) also provides the possibility of studying vadose zone flow rates and recharge mechanisms (Gardner *et al.* 1992). In some types of Chalk they suggested upward flow and evapotranspiration in the summer months from depths in excess of 2 m, but these have the potential limitation of not recording high-volume preferential flow episodes of short duration. There also still remain significant differences to reconcile on the estimation of recharge volume, between soil water physics monitoring and the traditional approaches. Although the number of sites where the former have been undertaken are few, they appear to give significantly higher values of actual evapotranspiration (and lower values of deep infiltration to the Chalk) than indicated by the (areally more representative, but less precise) permeable catchment water balance method.

Importance of the vadose zone

The vadose zone merits special attention since it represents the first, and most important, natural

FOSTER, S. S. D. 1998. Groundwater recharge and pollution vulnerability of British aquifers: a critical overview. *In*: ROBINS, N. S. (ed.) *Groundwater Pollution, Aquifer Recharge and Vulnerability*. Geological Society, London, Special Publications, **130**, 7–22.

Table 1. *Summary of techniques for groundwater recharge evaluation*

Technique	Brief description		Information provided			
	Key elements	Main sources of error	Parameter determined	Measurement scale	Temporal resolution	Recharge mechanisms
Permeable catchment water balance	Recharge equated to discharge (determined by stream baseflow separation) after correction for aquifer storage changes and abstractions	Permeable catchment boundaries, stream gauge underflow, aquifer storativity	Water-table recharge (indirectly)	District–region	Annual	None
Agrometeorological data analysis	Recharge equated to excess rainfall determined by cumulative difference between gauged rainfall and actual evaporation (estimated from potential via soil moisture model)	Unrepresentativity/uncertainty of soil moisture model, intermittent surface run-off following some high-intensity rainfall events	Sub-soil infiltration (indirectly)	Site–district	Daily	None
Soil-water physics monitoring	Recharge equated to net change of moisture content (monitored by neutron probe) below zero-flux plane (ZFP) identified by tensionmeters	No ZFP present during winter so excess of gauged rainfall over potential evaporation used for this period; may not detect short-duration, high-volume preferential flow episodes	Downward water flux in matrix of vadose zone	Site	Weekly–monthly	Partial
Vadose zone environmental isotope profiling	Recharge determined as total soil moisture volume down to peak concentration/first appearance of isotope/tracer in profile, whose date is known	Highly dispersive isotope/tracer transport (but this at least provides information on mechanisms), vadose zone sampling difficulties and small-volume analytical problems	Part/most vadose zone downward water flux	Site	Annual or longer	Partial
Artificial soil-injected tracer studies				Site	Months to few years	Potentially good

defence against groundwater pollution. This is not only because of its strategic position between the land surface and the saturated aquifer, but also because its environment is generally more favourable for pollutant attenuation and elimination.

Vadose zone water movement is normally slow and restricted to the smaller pores with larger specific surface. The chemical condition is usually aerobic and frequently alkaline. In the vadose zone there is thus major potential for:

(a) interception, sorption and elimination of pathogenic bacteria and viruses;
(b) attenuation of heavy metals, and other inorganic chemicals, through precipitation (as carbonates or hydroxides), sorption or cation exchange;
(c) sorption and biodegradation of many hydrocarbon and synthetic organic compounds.

However, water movement in the vadose zone can be complex and its ability to attenuate pollutants difficult to predict. It is important to appreciate that marked changes in the behaviour of some pollutants can occur if the polluting activity has sufficient organic or acidic loading to bring about significant change in the Eh or pH of the zone. Moreover, in the case of persistent mobile pollutants the vadose zone merely introduces a large timelag before arrival at the water-table, without any beneficial attenuation.

In most instances the degree of contaminant attenuation will be largely dependent on vadose zone pollutant pathways and residence times. While natural flow rates for most porous formations do not exceed 0.2 m day^{-1} when averaged over long time periods, in the presence of artificial hydraulic surcharging, water flow and pollutant penetration rates may be more than an order of magnitude higher, especially in fractured formations. Thus fracturing is a key factor in vulnerability assessment, where microbial, biodegradable and more readily retarded contaminants are concerned.

Most of the processes causing pollutant elimination and attenuation in the deeper vadose zone occur at much higher rates in the biologically active soil zone, as a result of its higher clay mineral and organic contents, and very much larger microbial populations. The characteristics of the soil zone will also strongly influence the scale of nutrient and pesticide leaching from a given agricultural cultivation practice, and whether acidic aerial deposition is neutralized. However, in many point sources of contamination, the sub-surface contaminant load is applied below the soil zone at the base of excavations, such as pits, trenches, lagoons, soakaways and quarries, and the attenuation capacity of this part of the vadose zone does not contribute to reducing overall vulnerability.

Mapping of aquifer pollution vulnerability

Need for vulnerability mapping

Most groundwater originates as excess rainfall locally infiltrating the land surface, and thus activities at the land surface threaten groundwater quality. Simple and robust matrices need to be established, which indicate what activities are possible at an acceptable risk to groundwater (e.g. NRA 1992; Foster 1993). Instead of applying universal controls over land or soil use and effluent discharge to the ground, it is more cost-effective (and less prejudicial to economic development) to utilize the natural contaminant attenuation capacity of the strata overlying the saturated aquifer. This is the basic premise underlying the concept of aquifer pollution vulnerability and the need for related maps.

Over the past five years or so, the pros and contras of land-surface zoning have been argued by hydrogeologists. Some say that:

(a) hydrogeological conditions are so complex in detail that zoning schemes cannot encapsulate them;
(b) groundwater everywhere is vulnerable to surface-derived pollution and the only thing that differs geographically is the time-scale for effects to become manifest.

Whilst recognizing the technical basis for these arguments, there is an overriding case (Foster & Skinner 1995) for land-surface zoning as a general framework for the development and implementation of groundwater protection policy since:

(a) decisions will be made affecting groundwater in any event, and if land-use planners have no hydrogeological zoning this will mean less (not more) consultation;
(b) zoning helps set priorities for groundwater protection activities generally, and in particular for environmental audit of industrial premises and for targeting of soil leaching control within the agricultural advisory system.

Development of vulnerability concept

The term 'vulnerability' began to be used in hydrogeology from the 1970s in France (Albinet & Margat 1970) and more widely in the 1980s (Haertle 1983; Aller *et al.* 1987). While the implication was of relative susceptibility of aquifers to anthropogenic pollution, the term was used initially without any attempt at formal definition.

The expression began to mean different things to different people and Foster (1987) suggested that it was necessary and logical to define aquifer pollution vulnerability as 'the intrinsic characteristics of the strata separating the saturated aquifer from the immediately overlying land surface which determine its sensitivity to being adversely affected by a surface applied (anthropogenic) contaminant load'. In effect this is a combination of:

(a) the inaccessibility of the saturated zone, in a hydraulic sense, to the penetration of pollutants;
(b) the attenuation capacity of the strata overlying the saturated zone as a result of physicochemical retention or reaction of pollutants.

Groundwater pollution risk could then be defined as the probability that groundwater in the uppermost part of an aquifer will become contaminated to an unacceptable level by activities on the land surface, and will be the result of interaction between the aquifer vulnerability and the contaminant loading applied at the location concerned (Foster & Hirata 1988; Adams & Foster 1992). These authors considered that integrated vulnerability mapping was worth pursuing for land-use planning and environmental impact assessment (provided its definition was clear and its limitations understood), that inventories and audits of sub-surface contaminant load were of high priority and that both could be implemented on a GIS-compatible basis (Hiscock *et al.* 1995). It was necessary to resist the temptation, however, to set up too many mapped vulnerability zones, since this is not generally practicable and risks the loss of credibility.

Recently two major professional working groups have reviewed and pronounced upon the applicability of the vulnerability concept (NRC 1993; IAH 1994). Both came out in favour of its usefulness, but neither succeeded in producing a definition of vulnerability sufficiently specific for the term to gain uniformity of application amongst professional hydrogeologists. Some workers consider that a factor representing the natural mobility and persistence of pollutants in the saturated zone should be included, but this is considered illogical since it does not view vulnerability maps from the (correct) perspective of planning and controlling activities at the land surface. It is considered far preferable for vulnerability to be a statement about the intrinsic characteristics of the strata (unsaturated zone or confining beds) separating the saturated aquifer from the land surface, thus providing an indication of the impact of land-use decisions at that point on the immediately underlying groundwater.

Limitation of concept

Sub-surface water flow and contaminant transport are complex processes; in particular, the degree of attenuation will vary significantly with type of contaminant and pollution process in any given situation. Thus a 'general (integrated) vulnerability to a universal contaminant in a typical pollution scenario' has no validity in rigorous scientific terms (Foster & Hirata 1988).

For this reason Andersen & Gosk (1987) suggested that vulnerability mapping would be better carried out for individual contaminant groups in specific pollution scenarios, and such an approach had already been adopted in certain instances (Le Grand 1983). However, the implication would be an atlas of maps for any given area which would be difficult to handle for most applications, except the evaluation and control of diffuse agricultural pollution (Carter *et al.* 1987; Loague 1994).

The more practical way forward for general planning purposes, however, is to use integrated vulnerability maps, but make every effort to spell out adequate 'health warnings' against their misuse (Foster & Hirata 1988). Such warnings have been elegantly expressed in the recent US review (NRC 1993) in the form of three 'laws of groundwater vulnerability': first, 'all groundwater is to some degree vulnerable'; second, 'uncertainty is inherent in all vulnerability assessments'; third, 'there is risk that the obvious may be obscured and the subtle indistinguishable'. The latter is especially true of the more complex systems of vulnerability assessment.

The British approach

The aquifer pollution vulnerability systems currently adopted in Britain (Foster & Adams 1992; NRA 1992; DENI 1994, ADRIS 1995; Robins *et al.* 1995; Palmer *et al.* 1995) classify, by lithology and thickness, the predominant strata above the saturated aquifer as the principal indicator of pollution vulnerability. Only a few major classes are recognized (Fig. 1). In deciding the criteria for lithological classification there was concern that too much emphasis might inadvertently be placed on average vadose zone time-lag, and vulnerability would then become more a measure of when (as opposed to if and which) pollutants reach the aquifer (Foster & Skinner 1995). Thus greatest emphasis was put upon the likelihood of well-developed fracturing being present (rather than the thickness and porosity of the strata involved), since this is considered the most critical factor increasing vulnerability and reducing the chance of contaminant elimination, especially given that some hydraulic surcharging is associated with many pollution scenarios.

In terms of final vulnerability class, this scheme produces similar results to the earlier GOD scheme for vulnerability index estimation (Fig. 2) (Foster 1987; Foster & Hirata 1988), but on balance it was considered preferable not to use indexation, because of the danger of obscuring the basic hydrogeological data on which estimates are based.

A particularly important aspect of both schemes is clear interpretation of the classes of relative vulnerability established (Table 2). In this way it is possible to overcome many (if not

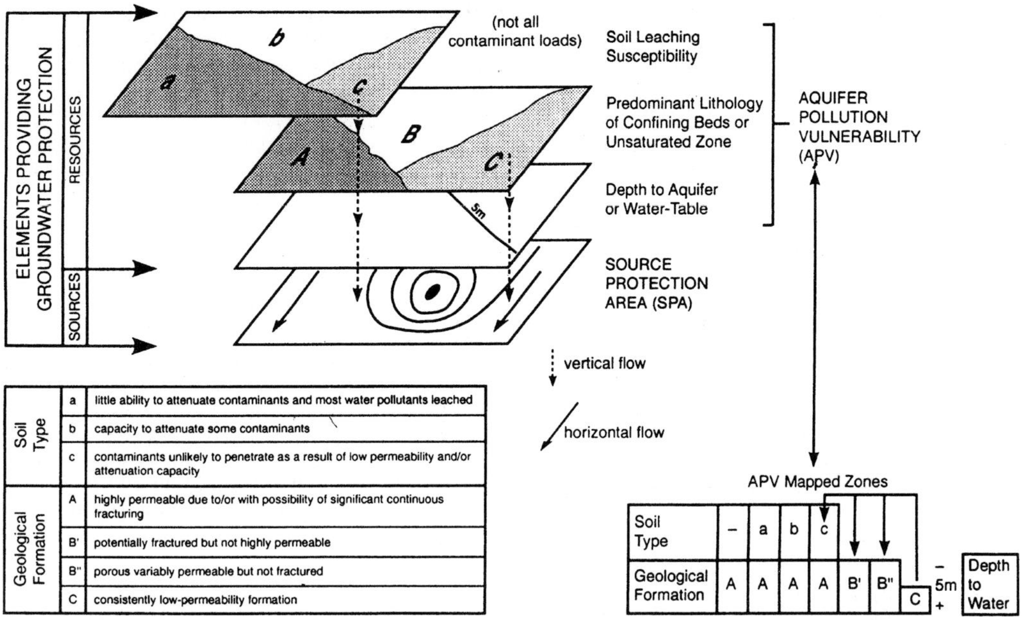

Fig. 1. Elements of British groundwater protection scheme and aquifer vulnerability classification (after Adams & Foster 1992).

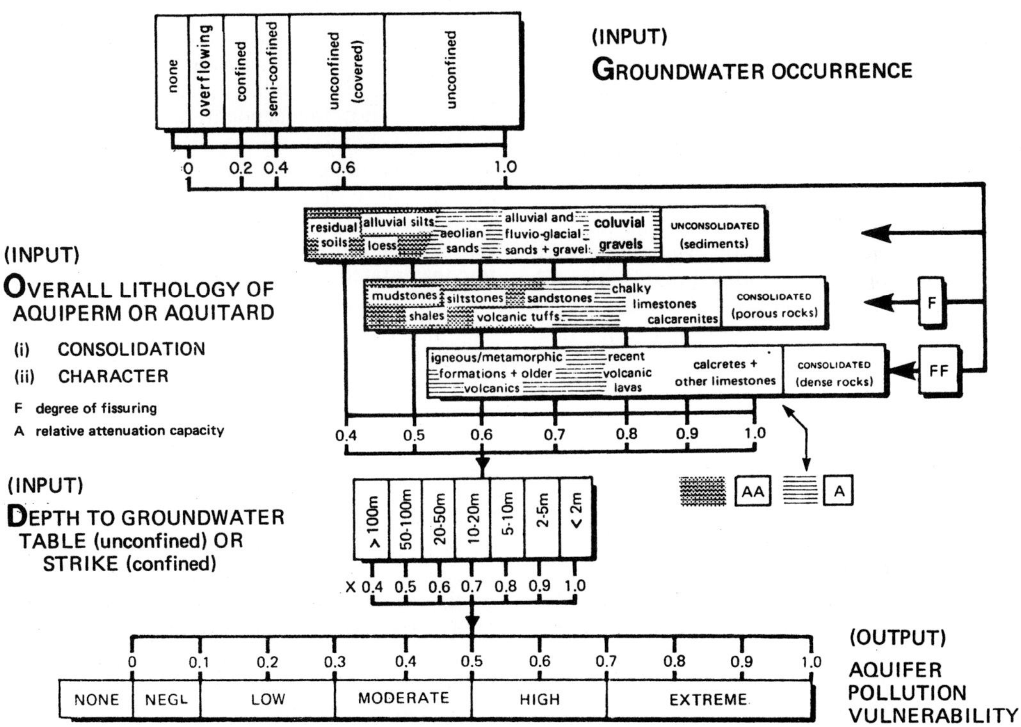

Fig. 2. GOD system for evaluation of aquifer pollution vulnerability index (after Foster 1987) (the term 'aquiperm' was coined as a shorthand for 'vadose (or unsaturated) zone of aquifers' and 'aquitard' is used to imply 'confining beds overlying a saturated aquifer').

Table 2. *Significance of relative classes of aquifer pollution vulnerability*

Vulnerability class	Practical definition
Extreme	Vulnerable to most water pollutants with relatively rapid impact in many pollution scenarios
High	Vulnerable to many pollutants, except those highly absorbed or readily transformed, in many scenarios
Moderate	Vulnerable to some pollutants, but only when continuously discharged/leached
Low	Only vulnerable to conservative pollutants in long term when continuously and widely discharged/leached
Negligible	Confining beds present with no significant vertical groundwater flow

all) the objections to the use of integrated vulnerability in groundwater protection policies.

A novel element of the new British approach is the way the soil zone is incorporated. Soils are classified into three categories by relative susceptibility to leaching and in areas of high hydrogeological vulnerability are taken as modifying the overall vulnerability ranking (Fig. 1). Since in urban areas, in particular, the soil is often removed during construction or the sub-surface pollutant load is applied below its base, the soil zone is assumed to be absent.

Comparison with DRASTIC methodology

The best known, and probably the most widely applied, scheme of vulnerability assessment was developed in the USA and is known as the DRASTIC methodology (Aller *et al.* 1987). It attempts to quantify relative vulnerability by the summation of weighted indices for the following seven hydrogeological variables: **D**epth to groundwater (X5), natural **R**echarge rates (X4), **A**quifer media (X3), **S**oil media (X2), **T**opographic aspect (X1), **I**mpact (effect) of vadose zone (X5) and hydraulic **C**onductivity (X3). The weighting for each variable is given in parentheses, but changes (especially for parameters S and T) if vulnerability to diffuse agricultural pollution alone is under consideration.

The method has been the subject of various evaluations, comparing DRASTIC vulnerabilities to the actual incidence of pollution: the pesticide alachlor state-wide in the USA (Holden *et al.* 1992), nitrate in NSW, Australia (Bates *et al.* 1993), volatile organic compounds in Nebraska, USA (Kalinski *et al.* 1994) and groundwater acidification in Nyköping, Sweden (Rosen 1994). All of these evaluations revealed benefits, but also numerous shortcomings, of the methodology. It should be noted, however, that the evaluations themselves may not be strictly objective unless the sub-surface contaminant load causing groundwater quality deterioration was distributed uniformly in the area under investigation.

While being totally sympathetic with the motivation of the architects of DRASTIC, the method appears to generate a vulnerability index whose meaning is rather obscure and whose significance is unclear, especially as a result of including a parameter reflecting contaminant mobility in the saturated zone. This is a consequence of the interaction of too many rather spuriously weighted parameters, some of which are not independent but quite strongly correlated. Similar indices can be obtained by a very different combination of circumstances which leads to the danger of making wrong interpretations. More specifically it under-estimates the vulnerability of fractured, compared to unconsolidated, aquifers (Rosen 1994).

Scientific uncertainties and misunderstandings

Vulnerability of dual-porosity formations

The most important British aquifers (the Permo-Triassic sandstones, the Cretaceous Chalk, the Jurassic limestones) are all (to varying degrees) of the dual-porosity type and all fall into the high vulnerability category in their outcrop/unconfined areas by virtue of extensive fracture networks.

In terms of definition of protection measures, however, the question arises as to just how vulnerable these unconfined aquifers are to particular types of pollution load scenario. This comes down to two factors:

(a) primarily, the probability and magnitude of (so-called) preferential vadose zone flow, which in effect rapidly by-passes the porous matrix and provides little possibility for contaminant attenuation;
(b) secondarily, the occurrence and effectiveness of contaminant attenuation processes occurring during porous matrix transport, whether direct or involving aqueous diffusion.

The presence of well-defined preferential flow paths is highly relevant in assessing aquifer vulnerability to faecally contaminated discharges (containing bacteria, viruses and even protozoa, such as the cryptosporidia) and to industrial solvent and hydrocarbon fuel spillages. They are also of major importance in relation to the transport of pesticides to aquifers, providing routes for deep penetration of any pesticide compounds leached from the soil. The rapid pollutant transport provides little opportunity for retardation through diffusion into matrix pore-water, and associated adsorption, reaction and degradation. However, even these routes offer some attenuation capacity for certain contaminants as a result of mineral coatings (such as manganese spots and iron-rich clays in the Chalk) on fracture walls (Shand & Bloomfield 1995). If preferential flow in fissures of large aperture occurs, groundwater pollution by less mobile pesticide compounds adsorbed on (and transported with) colloidal material is also a possibility.

In the case of the British aquifers, the potential for preferential flow is immediately evident from their fractured character and the existence of high fissure hydraulic conductivity, which has been proven in the case of the Chalk at various sites (Wellings & Cooper 1987; Allen & Price 1990). It can be caused by numerous factors (Beran & Germann 1982), and is often associated with instability of downward flow in situations where permeable deposits are overlain by a somewhat less permeable soil horizon (such that flow velocity increases with depth) and where surface drainage to the ground is concentrated by soakaways.

Preferential flow is difficult to prove, especially under natural rainfall conditions. Because of the 'pressure to model', it has almost certainly been under-estimated in past contaminant transport modelling in the vadose zone. Indirect evidence, however, comes from various lines of circumstantial evidence:

(a) intermittent groundwater contamination with faecal bacteria in areas of deep (10–25 m) Chalk water-table, despite the fact that profiles of environmental tritium indicate an average rate of vadose zone transport of non-reactive solutes of less than 1.0 $m\,a^{-1}$ (Foster & Smith-Carington 1980);
(b) the fact that it is not possible to reach satisfactory tritium profile mass balances, even at sites where the slow-flow component does not show significant dispersion effects, and realistically revised soil moisture models allow up to at least 20% of preferential flow (Table 3);
(c) the substantial number of vadose zone matrix pore-water profiles for environmental tritium showing marked dispersion with strong forward tailing of the 1963–65 peak concentrations (Fig. 3), which could only occur as a result of fissure flow in a dual porosity system where the velocity of water movement was high enough to prevent full interchange of solutes with the microporous matrix (Barker & Foster 1981; Geake & Foster 1989).

Various novel approaches are being introduced in order to understand further the signifi-

Table 3. *Analysis of environmental tritium (^3H) profiles from the Chalk vadose zone*

	Dorset (Sep. 1997)	Norfolk (Apr. 1977)
Mean rainfall ($mm\,a^{-1}$)*	880	700
Mean infiltration ($mm\,a^{-1}$)*	440	260
^3H in pore-water profile (TUm)	770†	480†
Estimated ^3H of infiltration (TUm)	720†	400†
Standard evaporation model (RC = 25–50 mm)		
Revised soil moisture model (RC = 25–50 mm/S_s = 100–300 mm)	890†	530†

Derived from Foster & Smith Carington (1980), where full definition of RC (root constant) and S_s (soil storage) can be obtained.
* For period April 1963 to date of profile
† Error band approximately ±40 TUm

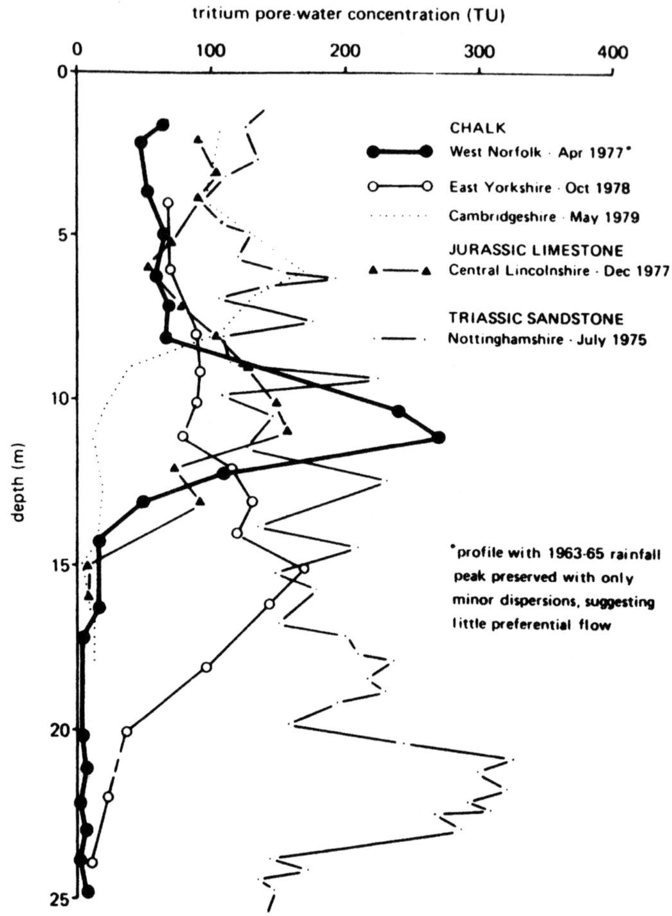

Fig. 3. Vadose zone profiles of environmental tritium providing evidence of dispersive (preferential) flow (after Foster *et al.* 1991).

cance and scale of recharge by preferential flow. In the case of pesticides this has involved use of a floating water-table sampling device, which has obtained evidence of higher pesticide concentrations with the onset of recharge at the water-table (Fig. 4). The concentrations recorded to date, however, are not as high as might be expected, bearing in mind that observed pesticide concentrations in soil water are sometimes in the range 50–500 $\mu g\,l^{-1}$ when pre-emergence (root-acting) herbicides are applied in late autumn/early winter.

Shallow-depth/high-resolution surface geophysical surveys, combined with modern data inversion technology, offer the possibility of mapping the sub-surface to locate potential preferential flow paths and horizons of lateral flow (Andrews *et al.* 1995).

Artificial soil-injected tracers have not yet been widely employed in Britain, but offer the possibility of studying in greater detail the scale of heterogeneity and dispersive vadose zone flow, at least at selected (and hopefully representative) sites. Two recent experiments on the Chalk have given (superficially) contradictory results:

(a) little non-dispersive flow was revealed by the use of a ^{2}H and Cl tracer at a site on the Upper Chalk of the Berkshire Downs over a period of 4 years, although only about 50% of the total porosity appeared to be active in solute transport (Barraclough *et al.* 1994);

(b) a Chalk vadose zone of 10 m thickness was traversed by a cocktail of tracers (lithium bromide, a bacteriophage and fluorescent

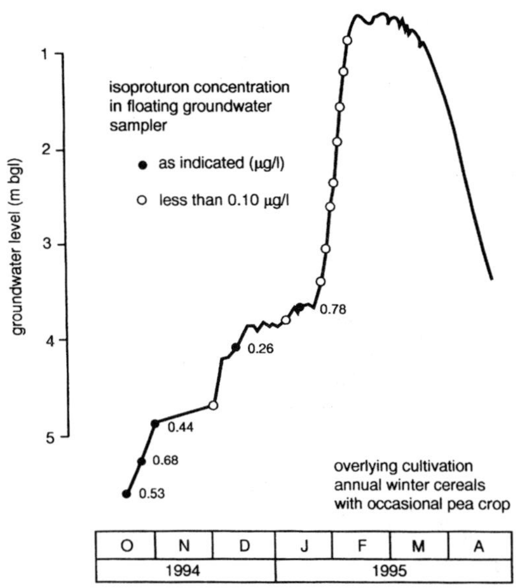

Fig. 4. Variation of isoproturon concentration at shallow Chalk water-table beneath winter cereal cultivation at a Hampshire site (after Chilton *et al.* 1995).

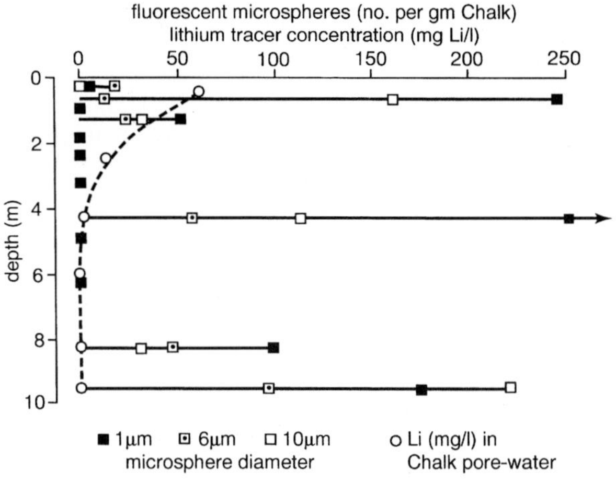

Fig. 5. Preferential flow in a Chalk vadose zone demonstrated by lithium bromide and fluorescent polystyrene microsphere tracers (after Ward *et al.* 1997).

polystyrene microspheres) in 0.6–3.0 days following 'light irrigation' (Ward *et al.* 1997), the fluorescent microspheres subsequently being detected on discernible fractures in drill cores (Fig. 5).

What is clear from the limited evidence available is that the important British aquifers present a relatively high probability of preferential vadose zone flow. It also appears likely that more localized parts of the Chalk and Jurassic limestone outcrop (as well as larger areas of the Carboniferous Limestone) could be defined as having extreme vulnerability (Table 2) in view of the frequent presence of karst-like solution features. In these aquifers, quality is likely to be intermittently affected by turbidity, and such groundwater sources need to be treated in the same way as 'lightly polluted' surface watercourses in quality terms. The identification of such areas and their protection against surface faecal pollution (especially by cryptosporidia) will be of paramount importance where they occur in production borehole capture zones.

Vadose zone contaminant breakdown

When considering the attenuation capacity of the 'matrix component' of vadose zone flow, the most relevant parameters in relation to many water contaminants are (a) the proportion of active organic matter available for contaminant adsorption, and (b) the population of viable indigenous bacteria capable of contaminant breakdown.

The proportion of organic matter in most British aquifers is low (Table 4). Organic carbon may be present in the aquifer matrix as a range of compounds, but not nearly enough is known about its precise specification, which normally falls into three main groups:

(a) fulvic and humic acids, which are immature polar materials, generally similar to much soil organic matter; adsorption from aqueous solution onto such materials is probably well represented by the equilibrium partition coefficient (K_{oc});
(b) bitumens and related materials, which are of lower molecular weight, but more mature, less soluble and related to crude oils; adsorption to these less polar materials would reduce, but should still be generally related to K_{oc};
(c) kerogen, which is a high molecular weight, insoluble organic material, having a graphitic character and unlikely to be very active in adsorption processes.

Amongst British aquifers, the Chalk contains (relatively) the largest amount of adsorbing species, which are disseminated widely within the matrix. The Jurassic limestones contain the highest overall proportion of organic carbon, but this is unevenly distributed and may include a much higher percentage of kerogen-type materials. The Permo-Triassic sandstones are believed to contain only low levels of mature refractory organic matter, consistent with their highly oxidized character.

It must also be borne in mind that the proportion of organic carbon and population of viable bacteria decrease rapidly with depth below the soil zone. Thus the published adsorption and degradation characteristics of pesticides, which relate to a (standard) fertile organic clayey soil, will not be representative of the sandy sub-soils often developed on aquifer outcrops and even less representative of deeper in the vadose zone (Table 5).

The low proportion of organic matter is a key factor in the persistence of oxygen and apparent absence of significant rates of natural denitrification in the vadose zone of the important British aquifers, a fact that increases their vulnerability to the leaching of agricultural nutrients (Foster 1989).

Effect of superficial clay cover

In general terms, a cover of superficial clay has traditionally been thought to offer significant protection to underlying aquifers. Two superficial deposits dominate the British hydrogeological map: the glacial tills (colloquially known as boulder clay), and the Clay-with-Flints deposit. On vulnerability maps their presence is recorded by stippling because their status requires further investigation. Recent research has begun to improve understanding of the hydrogeological behaviour of both these formations, although some caution over the interpretation of field hydraulic conductivity tests is necessary because of the problems of non-saturation, clay mineral reactions, test surface smearing, etc.

A typical site on the extensive and structurally little-disturbed Lowestoft till in Norfolk has been investigated in detail, including *in situ* field and laboratory core tests (triaxial testing being preferred to oedometers) (Klinck *et al.* 1996). It shows the following characteristics:

Table 4. *Typical characteristics of rock matrix of British aquifers*

Formation	Grain density (g/cm^3)	Total porosity	Specific retention*	Organic carbon content (% weight)
Chalk	2.72	0.30–0.45	0.29–0.43	0.05–0.2
Triassic sandstones	2.65	0.20–0.30	0.12–0.18	0.02–0.05
Jurassic limestones	2.72	0.10–0.25	0.08–0.22	0.2–2

After Foster *et al.* (1991)
* Corresponding to porosity range given and approximately equivalent to moisture content in unsaturated zone below zero flux plane

Table 5. *Variation in isoproturon half-life and partition coefficient, and related parameters, in Chalk vadose zone at a site in Hampshire*

Profile depth (m)	Isoproturon		Viable bacteria (cfu g^{-1})‡	Organic carbon (%) §
	Half-life (day)*	K_D (ml g^{-1})		
0.1	35	2.79	1×10^7	5.4
0.3	–	–	–	1.4
0.5	70	–	1×10^6	1.0
0.7	–	0.75	–	0.5
1.5	990	–	–	0.1
2.5	–	0.20	5×10^4	–
3.5	4200	–	–	–
8.5	–	0.02	9×10^2	–

* Determined in microcosm with 1 g core material/25 ml groundwater injected with 100 μg l^{-1} herbicide at 17°C
† On 3 g dried core material with 1000 μg l^{-1} herbicide in 3 ml 10 mM $CaCl_2$ for 1 day at 20°C
‡ By the TSAR agar-plate technique
§ Earlier data for same site

(a) the till is a silty overconsolidated clay, including about 15% clasts and containing some 25–40% clay-grade and 30–55% silt-grade sediment in the matrix (Fig. 6), most of which is rock-flour derived from the Chalk (and not clay minerals as such);
(b) the main clay mineral present is illite and overall there is only limited cation exchange capacity (10–20 meq/100 g);
(c) in the essentially unweathered state the till has a hydraulic conductivity of 10^{-8} to 10^{-10} m s^{-1}, but thick oxidized sequences are present at both its upper and lower contacts; the fracture density in this weathered till is higher and increases the hydraulic conductivity by 100–1000 times (Fig. 7);

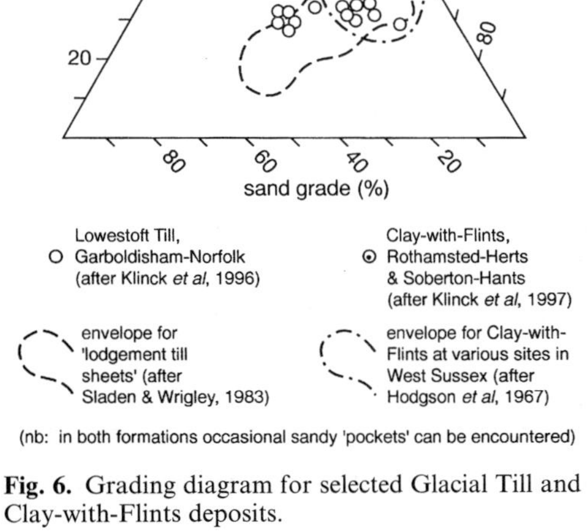

Fig. 6. Grading diagram for selected Glacial Till and Clay-with-Flints deposits.

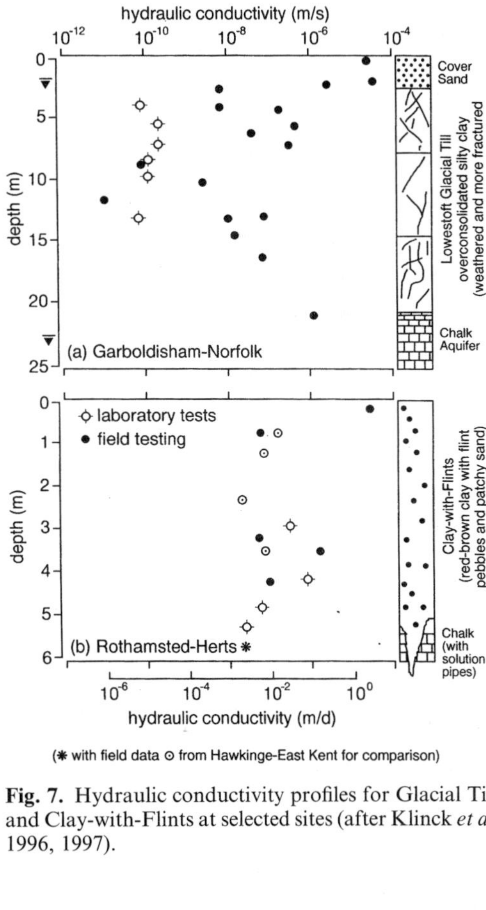

Fig. 7. Hydraulic conductivity profiles for Glacial Till and Clay-with-Flints at selected sites (after Klinck *et al.* 1996, 1997).

(d) the origin of fracturing in the over-consolidated clay may be variously glacial loading–unloading, freeze–thaw processes and/or glacial tectonics.

A conceptual model was developed of the division of groundwater recharge between flow to land drains in the cover sand/weathered till at the site and deep infiltration through the unweathered (and less fractured) till. Using reasonable parameter combinations the probable recharge through the till was estimated to be 20–40 mm a^{-1} (10–20% of the assumed excess rainfall), most of which was routed down fractures with rapid actual downward flow rates. However, any dissolved contaminants in this flow should be attenuated by aqueous matrix diffusion and/or clay mineral reactions.

Variations will clearly occur because some till sheets have either a higher clay mineral content or a larger proportion of silt and sand grade materials, as well as different fracture patterns. This conceptual model and numerical analysis, however, explain earlier work, which suggested from catchment hydrological balances that the recharge through glacial till in East Anglia varied from 25–100 mm a^{-1} (according to area) (Ineson & Downing 1967; Foster & Robertson 1977; Jackson & Rushton 1987). It also explains the deep penetration of thermonuclear tritium and agricultural nitrate at some locations of the same area (Lloyd *et al.* 1981; Parker *et al.* 1987), although alternative flow paths may be glacial sand-and-gravel bodies, highly disturbed during the advance of later ice sheets.

From the above it is evident that a superficial cover of glacial till does not offer complete protection to underlying aquifers, and that some pathways of potentially rapid downward flow exist. However, many aqueous-phase contaminants will still be strongly attenuated by matrix diffusion and clay mineral reactions, and perhaps the major concern is with dense non-aqueous phase liquids (DNAPLs) which are capable of migrating along, and perhaps even of enlarging, such pathways. There is also circumstantial evidence, from the relative penetration of environmental tritium and nitrate, that denitrification is more likely to occur in glacial tills (Fig. 8), and for that matter till-derived soils (Parker *et al.* 1987).

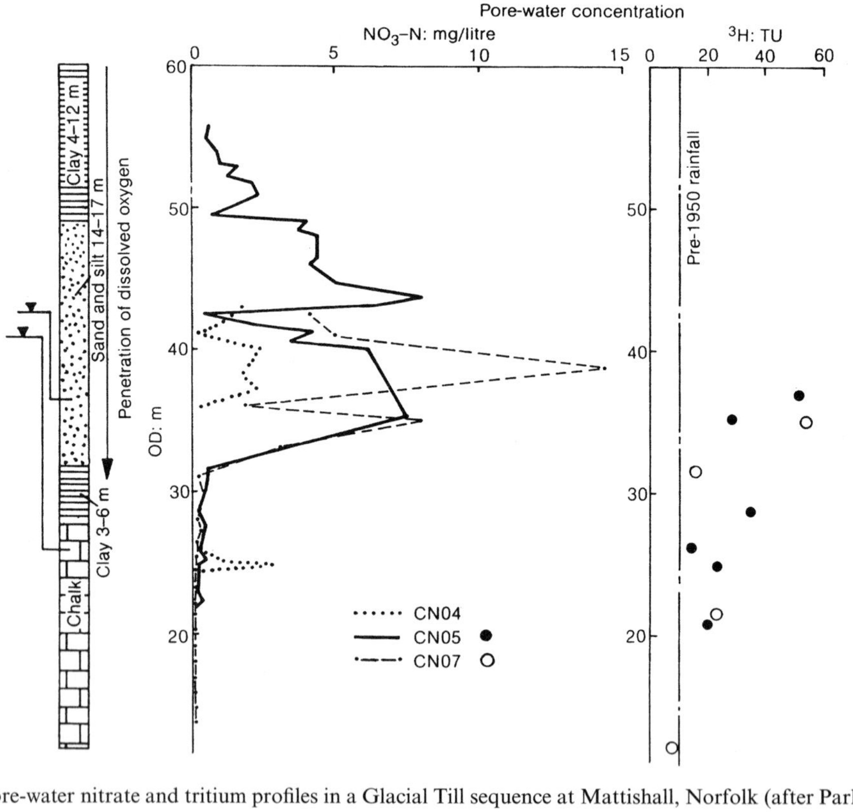

Fig. 8. Pore-water nitrate and tritium profiles in a Glacial Till sequence at Mattishall, Norfolk (after Parker *et al.* 1987).

The hydrogeological behaviour of the superficial Clay-with-Flints deposit, which overlies extensive areas of the Upper Chalk of southern England, has also been subject to recent systematic investigation, including representative sites of detailed field investigation and sampling in the Chilterns of Hertfordshire, on the North and South Downs in East Kent and Hampshire respectively (Klinck *et al.* 1997).

The most widely accepted hypothesis for the formation of this deposit is that it resulted from the cryoturbation of a combination of weathered and locally redistributed Reading Beds and residual Chalk solution deposits, on a sub-Palaeogene surface with signs of both preferential solution and ice wedging. This has resulted in a highly irregular contact with the underlying Chalk. The deposits range up to 10 m in thickness, with rapid lateral variation and thicker accumulation in Chalk pipes. They comprise predominantly stiff red-brown, somewhat silty, clays with unworn flint nodules. The matrix is 45–95% of clay grade (Fig. 6), with kaolinite and smectite (of high cation exchange capacity) as the dominant clay minerals. At the Chalk surface a thin darker clay deposit is commonly present, which is believed to be the product of profile alluviation and is often sheared by slumping into Chalk pipes.

Infiltration capacity tests were undertaken in trial pits to 5 m depth and complemented by laboratory triaxial tests at effective field stress (Fig. 7). Most of the field hydraulic conductivity measurements fell within the range 10^{-6} to 10^{-8} m s^{-1}. The soils on the Clay-with-Flints are not normally subject to agricultural drain installation and some rainfall events (amounting to 20–70% of the total effective precipitation) are likely to exceed infiltration capacity generating interflow or overland flow. The presence of sandy pockets, the irregular Chalk surface and dolines on the Chalk outcrop margin to the Clay- with-Flints cover will mean that this run-off is likely to generate preferential flow in the underlying Chalk vadose zone.

The presence of the Clay-with-Flints will in one sense reduce the vulnerability of the underlying Chalk aquifer as a result of the reaction capacity of clay minerals, and nitrate is also more likely to be reduced in the more clay-rich (reducing) soil environment. However, since the clay layer is often thin and discontinuous, there will also be a tendency for vulnerability to increase where run-off or interflow (polluted with faecal pathogens or pesticides for example) drains at high hydraulic load into the underlying Chalk.

Procedural issues in vulnerability mapping

A number of issues are causing some inconsistency in, and misunderstanding of, British vulnerability mapping. The discussion that follows is designed to clarify these issues.

The substitution of the terms 'major, minor and non-aquifer' for 'highly, moderately and negligibly permeable formations in the vadose zone' in the National Rivers Authority (now Environment Agency) Groundwater Protection Policy for England and Wales has led some people to confuse aquifer vulnerability with aquifer utilization and/or potential yield (and even natural groundwater quality). These are all completely separate and independent issues and should be treated as such, but if this is not recognized, inconsistent vulnerability classifications are likely to result.

Difficulties inevitably arise in situations where important regional aquifers are overlain by thick superficial deposits including local aquifers. A decision has to be made on which aquifer is considered by vulnerability mapping, and this should be done on a consistent basis and stated clearly. It is usually more practicable to map the vulnerability of the regional aquifer, but record where an overlying (usually more vulnerable) local aquifer is present by suitable ornament.

The occurrence of (permanent or intermittent) losing streams presents special problems for vulnerability assessment: first in evaluating the hydrological condition, second in defining the quality of the watercourse, third in appraising stream-bed attenuation capacity and fourth in map representation. It is, however, worth indicating potentially influent sections of streams crossing unconfined aquifers in vulnerability mapping.

Aquifer vulnerability maps are aimed only at giving a first general indication of the potential groundwater pollution risk to allow regulators, planners and developers to make better informed judgements on proposed new developments and on priorities in groundwater quality protection and monitoring. They are based on the best available information at the time of production and will require periodic updating.

In concept and in practice they involve much simplification of naturally complex geological variations and hydrogeological processes. Site-specific questions need to be answered by site-specific investigations, but the same philosophical and methodological approach to the assessment of groundwater pollution risk is normally possible.

Aquifer vulnerability maps are only suitable for assessing the groundwater pollution risk associated with those point-source contaminant discharges which occur at the land surface and in the aqueous phase. Strictly speaking they should not be used for assessing the hazard from DNAPL spillages nor from contaminants injected deeper in the sub-surface (as may be the case for leakage of large underground storage tanks, or for solid-waste landfill leachate).

Key questions for the future

Effect of climate change

The likelihood of accelerated climatic change and its potential effect on Britain have been widely discussed. The balance of opinion suggests that lowland England may experience more frequent and extended warm dry periods in summer and autumn, with higher intensity rainfall episodes in winter.

The possibility arises that this may lead not only to different rates of groundwater recharge but also to changes in recharge mechanisms (as a result of higher rainfall intensity) and in contaminant mobilization/leaching (as a result of dryer, more aerobic and higher temperature soils), and this will need careful consideration.

Improving vadose zone sampling and monitoring

For the understanding of groundwater recharge and pollution vulnerability to progress further, there is a need for innovation and refinement in methods of sampling and monitoring in the vadose zone of aquifers.

Among the requirements that have to be addressed are included:

(a) direct sampling of preferential flow components, with minimum physical disturbance of the media that generates them;
(b) reliable monitoring of changes in the composition of the gaseous phase, at the scale of interest of contaminant degradation processes;
(c) establishing the adaptation of microbial populations to contaminant loads;
(d) sampling/analysis of contaminant metabolites (partial breakdown products);
(e) investigation of physicochemical and biochemical processes on porous fissure walls and in the capillary fringe.

Additionally, more information on the superficial clay cover is required, especially as regards:

(f) the distribution and availability of electron donors (such as reduced iron/sulphur species and organic carbon) to support redox reactions;
(g) the adsorptive properties of their various clay mineral, clay and silt grade fractions.

This paper is published with permission of the Director of the British Geological Survey but the opinions presented are those of the author alone. Grateful recognition is made of numerous colleagues at the British Geological Survey, the former National Rivers Authority and the Institute of Hydrology, who have grappled with the topic of aquifer pollution vulnerability and/or have researched groundwater recharge mechanisms in recent years.

References

Adams, B. & Foster, S. S. D. 1992. Land-surface zoning for groundwater protection. *Journal of the Institute of Water and Environmental Management*, **6**, 312–320.

ADRIS 1995. *Groundwater Protection Strategy for Scotland.* Association of Directors and River Inspectors of Scotland, Perth.

Albinet, M. & Margat, J. 1970. Cartographie de la vulnerabilite a la pollution des nappes d'eau souterraine. *Bull BRGM 2nd Series*, **3**(4) 13–22.

Allen, D. J. & Price, M. 1990. Hydraulic conductivity of the Chalk at shallow depths in north-west Norfolk, United Kingdom. *Proceedings of ICE International Chalk Symposium* ICS-89. Thomas Telford, London, 577–581.

Aller, L., Bennett, T., Lehr, J. H., Petty, R. J. & Hackett G. 1987. *DRASTIC: a standardised system for evaluating groundwater pollution potential using hydrogeologic settings.* US-EPA Report **600/2–87–035**.

Andersen, L. J. & Gosk, J. 1987. Applicability of vulnerability maps. *TNO Committee for Hydrological Research: Proceedings and Information,* **38**, 321–332.

Andrews, R. J., Barker, R. & Meng-Heng, L. 1995. The application of electrical tomography in the study of the unsaturated zone in Chalk at three sites in Cambridgeshire, UK. *Hydrogeology Journal*, **3**, 17–31.

Barker, J. A. & Foster, S. S. D. 1981. A diffusion exchange model for solute movement in fissured porous rock. *Quarterly Journal of Engineering Geology*, **14**, 17–24.

Barraclough, D., Gardiner, C. M. K., Wellings, S. R. & Cooper, J. D. 1994. A tracer investigation into the importance of fissure flow in the unsaturated zone of the British Upper Chalk. *Journal of Hydrology*, **156**, 459–469.

BATES, L. E., BARBER, C., ROSS, J. & VERHOEVEN, J. 1993. Vulnerability of groundwater to pollution: evaluation of the DRASTIC system in the Peel river catchment, northern New South Wales, Australia. *Proceedings of ANU Conference 'Aquifers at Risk'* Canberra, Feb. 1993, 88–93.

BERAN, K. & GERMANN, P. 1982. Macropores and water flow in soils. *Water Resources Research*, **18**, 1311–1325.

CARTER, A. D., PALMER, R. C. & MONKHOUSE, R. A. 1987. Mapping the vulnerability of groundwater to pollution from agricultural practice particularly with respect to nitrate. *TNO Committee for Hydrological Research: Proceedings & Information*, **38**, 382–390.

CHILTON, P. J., STUART, M. E. & GOODY, D. C. 1995. *Occurrence of herbicides in, and the modelling of their transport to, the Chalk aquifer beneath arable land in southern England.* BCPC Monograph **62**, 111–116.

DENI. 1994. *Groundwater vulnerability map of Northern Ireland* (scale 1:250 000). Department of the Environment, Northern Ireland, Belfast.

FOSTER, S. S. D. 1987. Fundamental concepts in aquifer vulnerability, pollution risk and protection strategy. *TNO Committee for Hydrological Research: Proceedings & Information*, **38**, 69–86.

—— 1989. Diffuse pollution of groundwater by agriculture – lessons learnt and future prospects. *Proceedings of AIH Conference 'Recent Advances in Groundwater Hydrology'*, Tampa, 185–194.

—— 1993. The Chalk aquifer – its vulnerability to pollution. *In*: DOWNING, R. A., PRICE, M. & JONES, G. P. (eds) *The Hydrogeology of the Chalk of North-West Europe*. Clarendon, Oxford, 93–112.

—— & ADAMS, B. 1992. Land surface zoning for groundwater protection. *Journal of the Institution of Water and Environmental Management*, **6**, 312–320.

—— & HIRATA, R. A. 1988. *Groundwater pollution risk assessment – a methodology using available data.* WHO-PAHO-CEPIS Report, Lima.

—— & ROBERTSON, A. S. 1977. Evaluation of a semi-confined Chalk aquifer in East Anglia. *Proceedings of the Institiion of Civil Engineers II*, **63**, 803–818.

—— & SKINNER, A. C. 1995. *Groundwater protection: the science and practice of land surface zoning.* IAHS Publication **225**, 471–482.

—— & SMITH-CARINGTON, A. K. 1980. The interpretation of tritium in the Chalk unsaturated zone. *Journal of Hydrology*, **46**, 343–364.

——, CHILTON, P. J. & STUART, M. E. 1991. Mechanisms of groundwater pollution by pesticides. *Journal of the Institute of Water and Environmental Management*, **5**, 186–193.

GARDNER, C. M. K., BELL, J. B., COOPER, J. D., DARLING, W. G. & REEVE, C. E. 1992. Groundwater recharge and water movement in the unsaturated zone. *In*: *Applied Groundwater Hydrology*. Oxford University Press, 54–76.

GEAKE, A. K. & FOSTER, S. S. D. 1989. Sequential isotope and solute profiling in the unsaturated zone of the British Chalk. *Hydrological Sciences Journal*, **34**, 79–95.

HAERTLE, A. 1983. *Method of working and employment of EDP during the preparation of groundwater vulnerability maps.* IAHS Publication **142**, 1073–1085.

HISCOCK, K. M., LOVETT, A. A., BRAINARD, J. S. & PARFITT, J. P. 1995. Groundwater vulnerability assessment: two case studies using GIS methodology. *Quarterly Journal of Engineering Geology*, **28**, 179–194.

HODGSON, J. M., CATT, J. A. & WEIR, A. J. 1967. The origin and development of clay-with- flints and associated soil horizons on the South Downs. *Journal of Soil Science*, **18**, 85–102.

HOLDEN, L. R., GRAHAM, J. A., WHITMORE, R. W., ALEXANDER, W. J., PRATT, R. W., LIDDLE, S. K. & PIPER, L. L. 1992. Results of the national alachlor waterwell survey. *Environmental Science and Technology*, **26**, 935–943.

IAH 1994. *Guidebook on mapping groundwater vulnerability*. IAH Internal Contributions on Hydrogeology **16**.

INESON, J. & DOWNING, R. A. 1967. Some hydrogeological factors in permeable catchment studies. *Journal of the Institute of Water Engineers*, **19**, 59–80.

JACKSON, D. & RUSHTON, K. R. 1987. Assessment of recharge components for a Chalk aquifer unit. *Journal of Hydrology*, **92**, 1–15.

KALINSKI, R. J., KELLY, W. E., BOGARDI, I., EHRMAN, R. L. & YAMAMOTO, P. D. 1994. Correlation between DRASTIC vulnerabilities and incidents of VOC contamination of municipal wells in Nebraska. *Ground Water*, **32**, 31–34.

KLINCK, B. A., BARKER J. A., NOY, D. J. & WEALTHALL, G. P. 1996. *Mechanisms and rates of recharge through glacial till: experimental and modelling studies from a Norfolk site.* BGS Technical Report **WE/96/1**.

——, HOPSON, P. M., LEWIS, M. A., INGLETHORPE, S. D. J., ENTWISLE, D. C. & HARRINGTON, J. F. 1997. *The hydrogeological behaviour of the Clay-with-Flints of southern England.* BGS Technical Report **WE/97/5**.

LE GRAND, H. E. 1983. *A standardised system for evaluating waste disposal sites.* NWWA Report, Worthington, Ohio.

LLOYD, J. W., HARKER, D. & BAXENDALE, R. A. 1981. Recharge mechanisms and groundwater flow in the Chalk and Drift deposits of southern East Anglia. *Quarterly Journal of Engineering Geology*, **14**, 87– 96.

LOAGUE, K. 1994. Regional scale groundwater vulnerability estimates: impact of reducing data uncertainties for assessments in Hawaii. *Ground Water*, **32**, 605–616.

NRA. 1992. *Policy and Practice in the Protection of Groundwater*. National Rivers Authority, Bristol.

NRC. 1993. *Groundwater Vulnerability Assessment: Contamination Potential under Conditions of Uncertainty*. National Academy Press, Washington DC.

PALMER, R. C., HOLMAN, I. P., ROBINS, N. S. & LEWIS, M. A. 1995. *Guide to Groundwater Vulnerability Mapping in England & Wales*. HMSO, London.

PARKER, J. M., BOOTH, S. K. & FOSTER, S. S. D. 1987. Penetration of nitrate from agricultural soils into the groundwater of the Norfolk Chalk. *Proceedings of the Institution of Civil Engineers II*, **83**, 15–32.

ROBINS, N. S., ADAMS, B., FOSTER, S. S. D. & PALMER, R. C. 1995. Groundwater vulnerability mapping: the British perspective. *Hydrogeologie*, **3**, 35–42.

ROSEN, L. 1994. A study of the DRASTIC methodology with emphasis on Swedish conditions. *Ground Water*, **32**, 278–286.

SHAND, P. & BLOOMFIELD, J. 1995. *Mineralisation of shallow fracture surfaces in the Chalk and implications for contaminant attenuation.* BGS Technical Report **WD/95/15**.

SLADEN, J. A. & WRIGLEY, W. 1983. Geotechnical properties of lodgement tills. *In*: EYLES, N. (ed.) *Glacial Geology*. Pergamon, Oxford, 184–212.

THOMPSON, N., BARRIE, I. A. & AYLES, M. 1981. *The Meteorological Office rainfall and evaporation calculation system: MORECS.* Meteorological Office Hydrological Memo **45**.

WARD, R. S., HARRISON, I., LEADER, R. U. & WILLIAMS, A. T. 1997. Fluorescent polystyrene microspheres as tracers of colloidal and particulate materials: examples of their use and developments in analytical technique. *In*: KRANJC, A. (ed.) *Tracer Hydrology*. Balkema, Rotterdam.

WELLINGS, S. R. & COOPER, J. D. 1987. Spatial variability of recharge of the English Chalk. *Agricultural Water Management*, **6**, 243–253.

Identification of the source and fate of nitrate contamination of the Jersey bedrock aquifer using stable nitrogen isotopes

A. R. GREEN*, N. A. FEAST†, K. M. HISCOCK & P. F. DENNIS

School of Environmental Sciences, University of East Anglia, Norwich, NR4 7TJ, UK

Abstract: Measured NO_3^- concentrations in the Jersey bedrock aquifer exceeded 50 mg l^{-1} in 67% of groundwaters sampled in June 1995. The $\delta^{15}N$ values of this dissolved NO_3^- showed a total range in composition from +3.6‰ to +18.4‰ and exhibited a general, island-wide increase from +4‰ in the centre of the island to +9‰ towards coastal districts. The majority of the island is intensively cultivated and most of the sites sampled have $\delta^{15}N$ values of less than +7‰ (mean = +7.7‰, $n = 40$) indicative of an agricultural source for NO_3^-. There is no apparent evidence for point-source contamination from septic tanks. The progressive seaward increase in $\delta^{15}N$ is attributed to an increasing volume of deeper, denitrified groundwater mixing with shallow groundwater. Values of $\delta^{15}N$ greater than +10‰ are restricted to those coastal and valley locations that abstract, due to their low topographic elevation, the greatest proportion of deeper, denitrified groundwater. An alternative explanation is that sewer leakage enhances the observed $\delta^{15}N$ values in southern urban areas, although this is considered unlikely to be occurring at present.

The island of Jersey is the largest of the British Channel Islands with a land area of 117 km^2. The island setting, described previously by Robins & Smedley (1994), comprises a plateau with an elevation of between 60 and 120 m above sea level with a steep topographic rise along the coastline. The plateau, formed largely of Precambrian crystalline rocks, is divided by a series of incised valleys, mostly draining from the higher ground in the north towards the south coast. The temperate maritime climate encourages early flowers and vegetables, with intensive agricultural production sustained by large fertilizer applications. During winter and early spring, applications of nitrogen fertilizer to early-cropping potatoes and horticultural crops may exacerbate the problem of nutrient leaching, with estimates of leaching losses of up to 100 kg nitrogen per hectare expected from Jersey potato crops (Robins *et al.* 1993).

The groundwater resources are affected by both surface pollution and intense exploitation. Mains water supply is principally from surface water storage, but there are large areas which are neither on mains water supply nor mains sewerage, particularly in the rural north of the island. In these areas, well and borehole supplies meet demands for domestic, agricultural and light industrial uses. The main aquifer and isolated perched aquifers occur within a shallow zone of weathering in the bedrock. These aquifers are almost entirely dependent on secondary permeability imparted by dilated fractures, enhanced by east-southeast-trending faults and associated intrusion of dykes. Fracture dilation reduces with overburden pressure. There is very little intergranular permeability. Robins & Smedley (1994) report that boreholes have been drilled up to 84 m below the piezometric surface and penetrate relatively deep transmissive fractures which may permit a modest component of deeper groundwater circulation. Over most of the island, groundwater flow is principally from north to south, and occurs within a shallow zone about 25 m in depth below the piezometric surface.

The fractured bedrock aquifer has a piezometric level which is largely a subdued version of the surface topography. The depth to water is only a few metres beneath much of the island, increasing to 10–30 m beneath the high-elevation coastal areas. Well and borehole yields are modest, typically < 0.5 l s^{-1}, although there are many 'dry' boreholes abandoned on completion. The more fractured and productive parts of the aquifer have transmissivity values ranging from 0.3 to 120 m^2 day^{-1} (Robins & Smedley 1994). Near the coast, boreholes penetrate below sea-level, and some of these have experienced seawater intrusion, notably those north of St Ouen's Bay and in the southeast of the island. However, the effective base of the aquifer is above sea level beneath most of the island and seawater intrusion is not a problem.

The general hydrochemistry of Jersey groundwater is also discussed by Robins & Smedley

*Present address: The Environment Agency, Cobham Road, Ipswich IP3 9JE, UK

† Present address: Shell International Exploration and Production B.V., Volmerlaan 8, P.O. Box 60, 2280 Rijswijk, The Netherlands

GREEN, A. R., FEAST, N. A., HISCOCK, K. M. & DENNIS, P. F. 1998. Identification of the source and fate of nitrate contamination of the Jersey bedrock aquifer using stable nitrogen isotopes. *In*: ROBINS, N. S. (ed.) *Groundwater Pollution, Aquifer Recharge and Vulnerability*. Geological Society, London, Special Publications, **130**, 23–35.

(1994). The chemical composition is controlled by maritime recharge inputs and water–rock interaction, although the effects of anthropogenic pollution, particularly from NO_3^-, are in places severe. The groundwater is principally of a Na–Ca–Cl–HCO_3 type. Groundwaters are mostly acidic and oxidizing in character, and are nearly always undersaturated with respect to calcite, reflecting the small amount of carbonate in the aquifer. A few groundwaters are reducing with low redox potentials, low or zero dissolved oxygen, low NO_3^- and high dissolved iron and manganese concentrations, and occur principally in the southern central and southeast parts of the island, as well as in the valley areas around St Saviour. These reducing waters are thought to have experienced a deeper circulation and longer residence time in which reducing reactions have proceeded, with groundwater discharge along the low-lying southern coastal areas and in valley bottoms.

The combination of intensive agriculture, high livestock densities, shallow groundwater levels and near-surface fissure flow presents a high vulnerability to groundwater contamination by NO_3^-. In some areas, domestic pollution from septic tank discharges is a further potential hazard. In order to identify potential sources of NO_3^-, and the possibility of denitrification, this study employed the technique of nitrogen isotope tracing in the Jersey bedrock aquifer. Groundwater samples were collected in June 1995 from 46 locations and analysed for nitrogen species and stable nitrogen isotope composition. The results of this survey were incorporated with existing hydrochemical data from a previous study by the British Geological Survey (BGS) to enable a fuller interpretation of the nitrogen hydrochemistry.

^{15}N isotope tracing technique

In isotope geoscience, the δ-notation is used to describe the relative abundance of stable isotopes in a sample with reference to the known isotopic ratio in a standard. For example, when comparing stable nitrogen isotopes, $\delta^{15}N$ is defined as $((R_{spl}/R_{std}) - 1)1000$, where $R_{spl} = {}^{15}N/{}^{14}N$ ratio of the sample and $R_{std} = {}^{15}N/{}^{14}N$ ratio of the standard (purified air). Generally, stable isotope data are given in deviations of parts per thousand (‰). Sources of NO_3^- in natural waters may be identified by comparison of the nitrogen isotope ratio of dissolved NO_3^- with that of potential sources (Heaton 1986; Kaplan & Magaritz 1986; Wilson *et al.* 1994). Sources of NO_3^- (and their typical range of $\delta^{15}N$ composition) include: nitrified soil organic nitrogen (+4‰ to +9 ‰); nitrogenous fertilizers (−4‰ to +4‰); and animal and sewage wastes (> +10‰). These ranges incorporate the majority of values associated with each source, although site-specific conditions may cause the ranges to overlap. For example, fertilizer NO_3^- that has exchanged with soil N will have an isotopic value similar to that of nitrified soil organic nitrogen. Thus, a general 'agronomic' source of NO_3^- can be defined with an isotopic composition in the range +4‰ to +9‰. Also, isotopic fractionation by denitrification can alter the original isotope ratios, typically resulting in a depletion of the heavier isotope (^{15}N) in the reaction products, and an enrichment in the residual NO_3^-.

Denitrification is observed to proceed at reduced oxygen levels via a number of microbially mediated steps, the end product of which is normally gaseous nitrogen (N_2) (Korom 1992). The denitrification process has often been treated as a single-step, unidirectional reaction so that changes in the isotopic composition of NO_3^- can be modelled using a Rayleigh fractionation model (Mariotti *et al.* 1988; Fustec *et al.* 1991). Under favourable hydrochemical conditions, the isotopic enrichment of NO_3^- with a decreasing NO_3^- concentration can be predicted by the classic Rayleigh equation, as follows:

$$\delta^{15}N = \delta^{15}N_o + \varepsilon \ln(C/C_o) \quad (1)$$

where $\delta^{15}N$ represents the isotopic composition of reactive NO_3^- at time t, with concentration C; and $\delta^{15}N_o$ represents the isotopic composition of unreacted NO_3^- at time $t = 0$, with concentration C_o. Different values for the isotope enrichment factor, ε, result depending on site-specific conditions such as substrate composition, concentration and availability to microbes of the electron donor, temperature, denitrification rate, type of bacterium involved and growth conditions (Böttcher *et al.* 1990).

Dilution with water with a low NO_3^- concentration may decrease the NO_3^- concentration of a mixed water but will only affect the isotopic composition if the isotopic compositions of the two end-members are different. In this case, if δ and Q are the isotopic composition and the NO_3^- amount, respectively, and a and b refer to two nitrogen sources and m refers to the mixture, then from a consideration of mass and isotopic balances:

$$\delta_m = (Q_a(\delta_a - \delta_b)/Q_m) + \delta_b. \quad (2)$$

Equation 2 shows that progressive mixing between two water types is expressed by a hyperbolic evolution of the isotopic composition of the mixture as a function of the concentration. The two processes of denitrification and dilution can be distinguished from each other in that isotopic fractionation will give a linear relation between $\delta^{15}N$ and $\ln(NO_3^-)$ and a mixing process will give a linear relation between $\delta^{15}N$ and the inverse of the NO_3^- concentration (Mariotti *et al.* 1988).

Field and laboratory methods

The 42 boreholes, two shallow wells and two springs sampled in June 1995 were chosen to provide a good areal coverage of the island (see Fig. 1). Of these sites, 32 had previously been sampled in 1990 by the BGS for which on-site measurements of dissolved oxygen (DO) and specific electrical conductance (SEC_{25}), and laboratory analyses of HCO_3^-, Cl^-, NO_3^-, SO_4^{2-} and total Fe are reported by Robins & Smedley (1991). For the present study, samples were taken either after three borehole volumes had been purged or following 10 mins of pumping and collected in 1 or 2 l polythene containers. A 50 ml sample was also collected and filtered (0.25 μm membrane filter) for analysis of NO_3^-, NO_2^- and NH_4^+. Samples were stored at 4°C prior to analysis. Analysis of NO_3^- was by ion chromatography with an accuracy of ±5%. Dissolved NO_2^- and NH_4^+ were analysed by spectrophotometric techniques with an accuracy of ±10 $\mu g\,l^{-1}$.

Preparation of samples for analysis of $\delta^{15}N$ followed the method of Feast & Dennis (1996). A distillation procedure was used to convert dissolved NO_3^- into ammonium sulphate, followed by combustion to produce N_2. Once purified, the $\delta^{15}N$ of the N_2 produced was measured on a VG Isogas Sira II isotope-ratio mass spectrometer. The method yields a precision better than ±0.2‰, measured and reported relative to purified air according to the accepted isotope terminology.

Results

The results of the 1995 survey carried out for this study are given in Table 1 together with NO_3^- concentration data from the BGS survey of

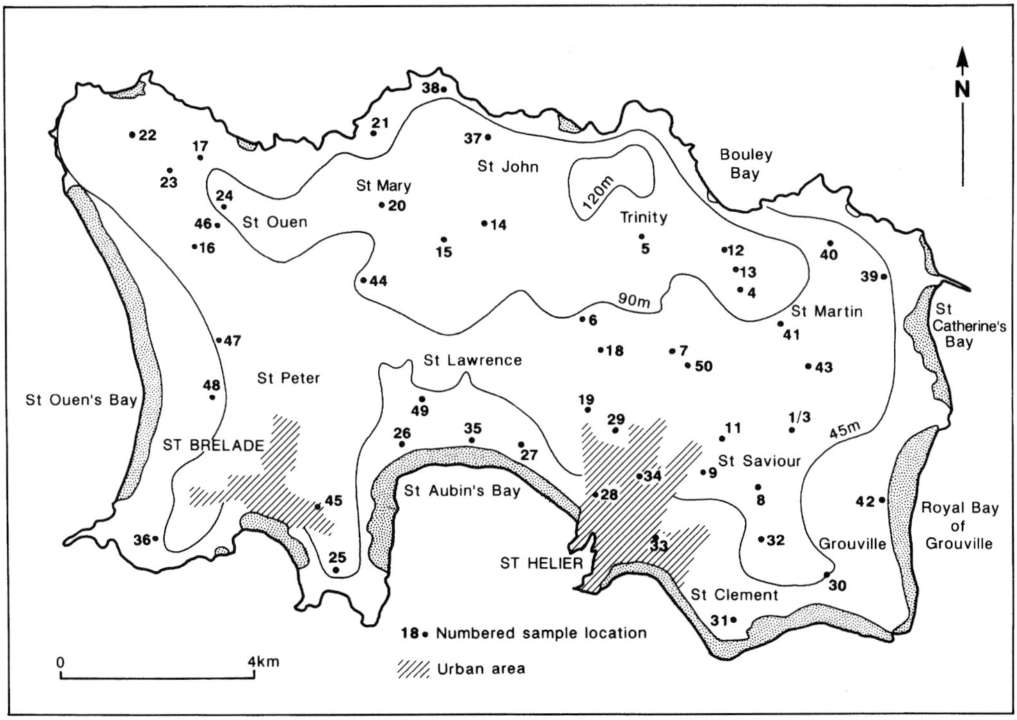

Fig. 1. Location map of Jersey showing surface topography and groundwater sampling locations. Elevation contours are given in metres above sea-level.

Table 1. *Hydrochemical analyses of groundwaters from the Jersey bedrock aquifer*

Sample number	Topographic location	$\delta^{15}N$ (‰)	NO_3^- (1990) ($mg\,l^{-1}$)	NO_3^- (1995) ($mg\,l^{-1}$)	NO_2^- ($\mu g\,l^{-1}$)	NH_4^+ ($\mu g\,l^{-1}$)
1	P		69	62	< 10	56
3	P	6.8		89	< 10	64
4	P	5.6		62		61
5	P	4.8		132	< 10	49
6	P	7.7		62	17	60
7	V	6.0		72	< 10	50
8	P	6.2		133	< 10	56
9	V			64	< 10	56
11	V	11.5	< 1	20		60
12	P	6.4		37	< 10	64
13	P	5.7		75	< 10	45
14	P	3.6		141	52	56
15	P	6.9		108	< 10	41
16	P	9.8		66	< 10	46
17	P	6.0		142	< 10	39
18	P		50	< 1	12	58
19	P	7.4	47	63	< 10	36
20	P	6.0	25	83	< 10	56
21	C	6.5	80	74	< 10	53
22	P	7.3	121	88	< 10	34
23(i)	P	9.0	123	95		68
23(ii)	P	5.9		89	17	170
24	P			11		251
25	P	8.2	49	63	74	86
26	C	15.0	58	19	123	35
27	C	18.4	10	19	10	48
28	C		< 1	< 1	< 10	179
29	V	7.7	52	99	< 10	30
30	C	6.8	217	210	< 10	34
31	C		< 1	2	< 10	
32	V	6.9	110	101	< 10	35
33	C	7.8		61	< 10	34
34	C	8.5	40	39	< 10	13
35	C	11.8	45	26	10	51
36	P	8.7	279	86		34
37	P	6.2	36	106	< 10	29
38	P	8.9	5	16	17	< 10
39	P	6.8	99	76	< 10	26
40	P	7.3	73	69	< 10	31
41	V	6.4	86	90	< 10	38
42	C	9.4		37	13	33
43	V	8.4	124	65	< 10	34
44	V		< 1	< 1	< 10	< 10
45	V	8.0	69	50	< 10	33
46	P	7.5	121	215	< 10	38
47	C		28	22	< 10	33
48	C	5.1	95	144	< 10	15
49	C	5.6	53	66	< 10	23
50	V		2	9	< 10	65

Sample locations are shown in Fig 1. Sample 47 is from the St. Ouen's Bay sand aquifer. All samples are from boreholes except Samples 3 and 12 which are from wells and Samples 32 and 41 which are from springs. $\delta^{15}N$, NO_3^-, NO_2^- and NH_4^+ data were collected in June 1995. Also shown are NO_3^- data collected in summer 1990 by the British Geological Survey (Robins & Smedley 1991). The topographic locations of sampling points are indicated by C, V and P, representing coastal, valley and plateau locations, respectively.

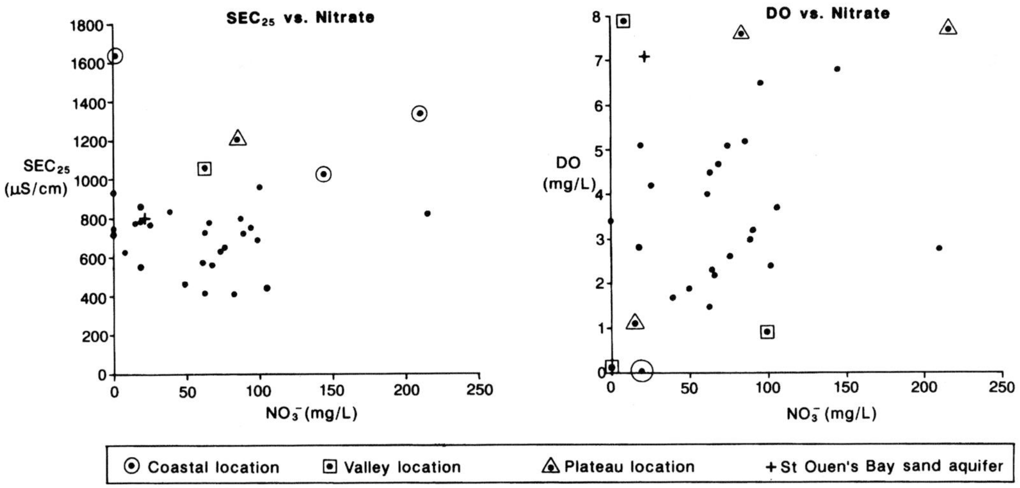

Fig. 2. Cross-plots of specific electrical conductance (SEC_{25}) and dissolved oxygen (DO) versus NO_3^- concentration for Jersey groundwaters. NO_3^- concentration data were collected in June 1995, with the remaining data collected by the British Geological Survey (Robins *et al.* 1993).

1990. Correlating the NO_3^- data from both surveys where sampling locations coincide yields a fairly close relationship ($r^2 = 0.64$, $n = 31$ ignoring Sample 36) and suggests only minor change in groundwater quality between 1990 and 1995. During this five-year period, the mean groundwater NO_3^- concentration for these locations increased slightly by 3.4 $mg\,l^{-1}$. The field and laboratory data are further illustrated in Figs 2 and 3. General observations are that groundwater SEC_{25} and Cl^- increase at coastal locations indicative of a seawater effect, and that DO is highest for plateau locations. Concentrations of HCO_3^-, SO_4^{2-} and total Fe are also elevated in a coastal direction, as previously observed by Robins & Smedley (1994).

The regional distribution of NO_3^- in Jersey groundwaters is shown in Fig. 4. Elevated NO_3^- concentrations occur across much of the island and range from undetected to 215 $mg\,l^{-1}$, with a median value of 66.1 $mg\,l^{-1}$ (mean = 70.6 $mg\,l^{-1}$, $n = 49$). Of the total, 67% of samples exceed the European Commission maximum admissible concentration for drinking water of 50 $mg\,l^{-1}$ as NO_3^- (Fig. 5). The high NO_3^- concentrations appear to decrease in a coastal direction, especially in those samples from the central southern area and in the valley areas around St Saviour. Otherwise, local variations in well and borehole depths and local land-use and agricultural practice, combined with the physical heterogeneity of the aquifer, produce no obvious pattern in the distribution of NO_3^-. The majority of samples analysed for NO_2^- and NH_4^+ were found to be at or below the detection limit, although detectable concentrations were occasionally recorded (Table 1).

The nitrogen isotope data vary between +3.6‰ and +18.4‰ with a median value of +7.1‰ (mean = +7.7‰, $n = 40$), but the greater part (83%) fall within the narrower range of +4‰ to +9‰ (Fig. 6). The majority of groundwaters are therefore within the range commonly attributed to natural mineralization of soil organic nitrogen and often associated with intensively farmed areas with elevated NO_3^- concentrations. The $\delta^{15}N$ values show a distinct spatial trend (Fig. 7). It is evident, from 20 of the sites visited, that the majority of the island is characterized by a $\delta^{15}N$ composition of $<+7$‰, and largely incorporates the area devoted to intensive agriculture. The lowest $\delta^{15}N$ value of +3.6‰ (Sample 14) is located in the central north of the island, coincident with the area of maximum topographic elevation and recharge. Coastal $\delta^{15}N$ values are isotopically heavier, generally in the range +8‰ to +9‰. Although there are fewer points, and despite its proximity to the area of maximum elevation and recharge and low $\delta^{15}N$ groundwater, the north coast also demonstrates this trend.

The most enriched nitrogen isotope ratios are almost exclusively found in the low-lying area around St Aubin's Bay, below the topographic rise and corresponding to the area of lowest NO_3^- concentrations. Sample 11 in the valley area of St Saviour is also isotopically enriched

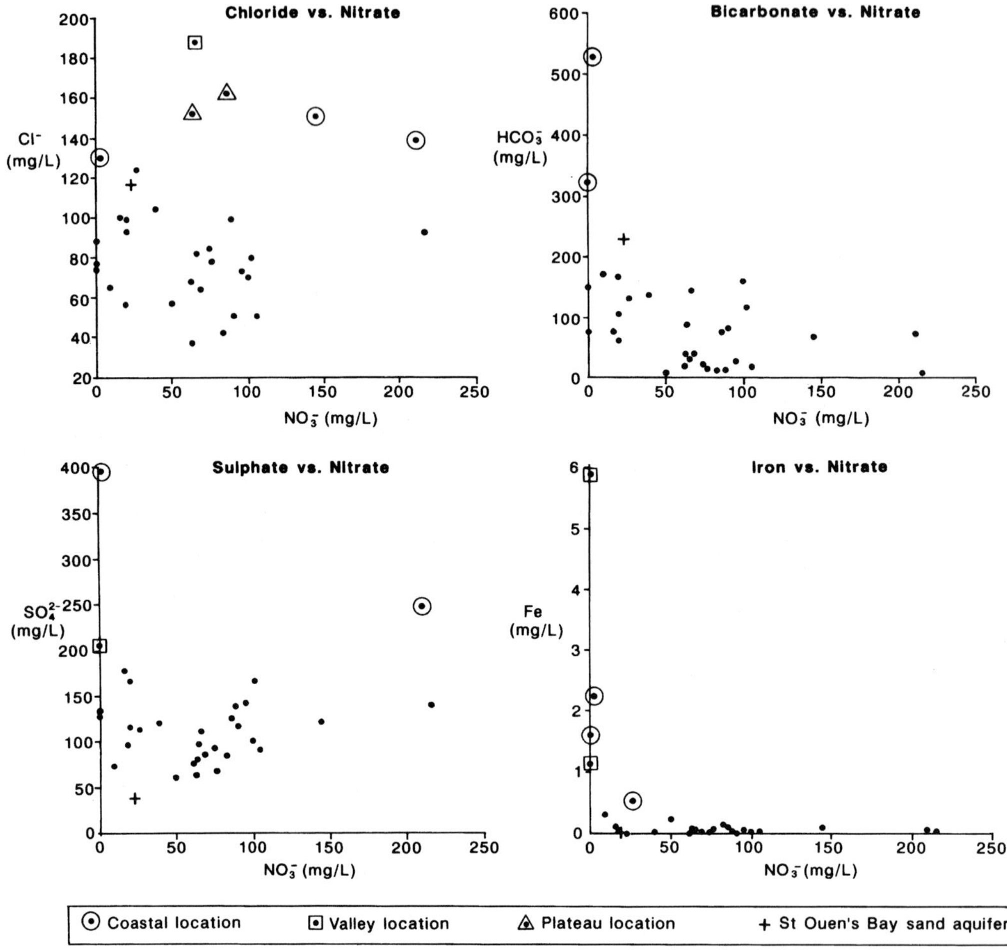

Fig. 3. Cross-plots of Cl^-, HCO_3^-, SO_4^{2-} and total Fe concentrations versus NO_3^- concentration for Jersey groundwaters. NO_3^- concentration data were collected in June 1995, with the other data shown collected by the British Geological Survey (Robins *et al.* 1993).

($\delta^{15}N = +11.5‰$). A cross-plot of $\delta^{15}N$ composition against NO_3^- concentration (Fig. 8) demonstrates that with decreasing NO_3^- concentration the $\delta^{15}N$ composition of the NO_3^- generally increases.

Discussion

The following are considered to explain the observed trends in $\delta^{15}N$: spatial variation in the source of dissolved NO_3^-, denitrification, and groundwater mixing. Taking the first of these, spatial variation in $\delta^{15}N$ values could reflect different land use across the island. Urban areas are unlikely to receive similar nitrogen inputs to agricultural areas. Urbanization may restrict recharge and introduces the possibility of sewage inputs. The decomposition of urea in sewage first leads to high concentrations of NH_4^+ which, under the alkaline conditions of this decomposition, may be lost as gaseous ammonia. At equilibrium, gaseous ammonia should be 34‰ more negative than the aqueous NH_4^+. Kreitler (1979) stated this reaction as the dominant isotopic fractionation effect causing the more positive $\delta^{15}N$ values of NO_3^- derived from animal waste. The partial volatilization of ammonia depleted in ^{15}N during the decomposition of urea in sewage causes the residual NH_4^+ and subsequent NO_3^- to become isotopically enriched.

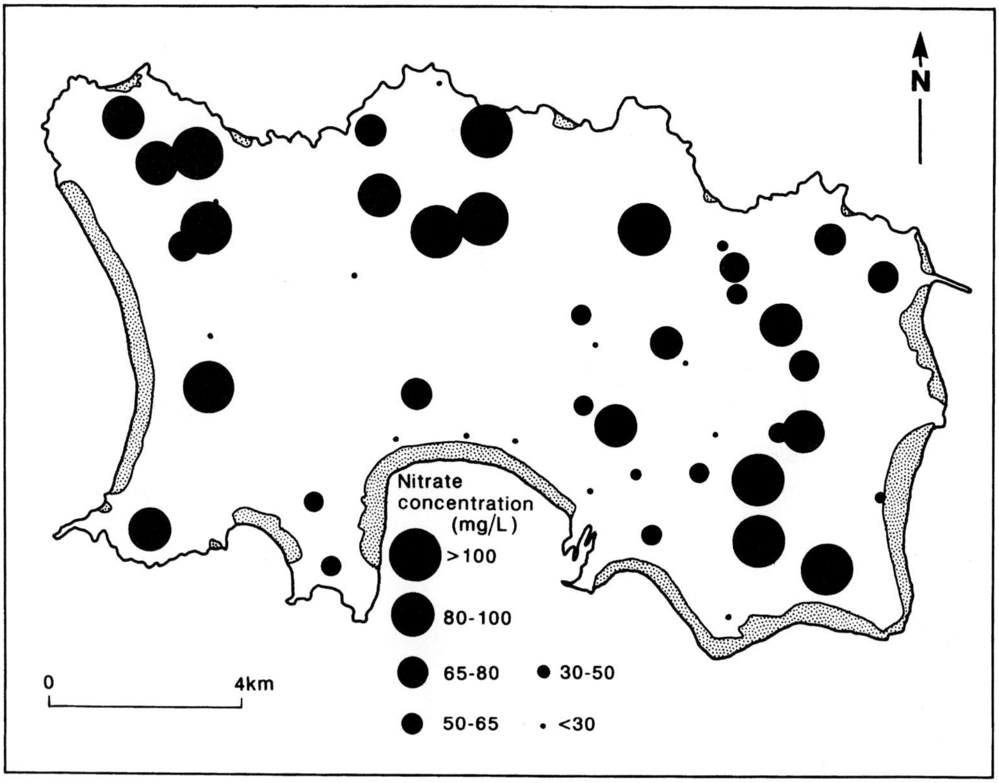

Fig. 4. Regional distribution of NO_3^- concentrations in Jersey groundwaters sampled in June 1995.

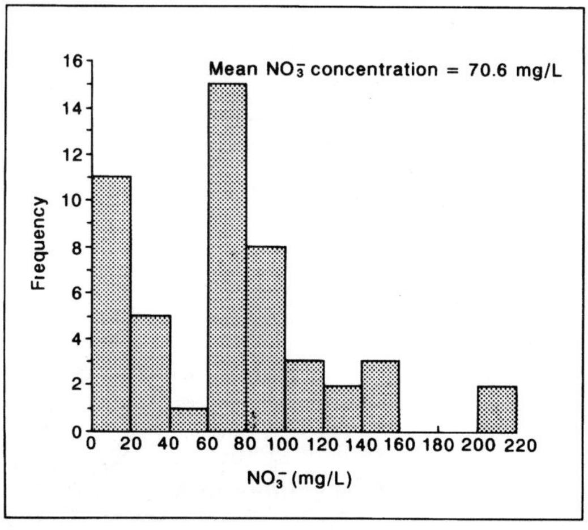

Fig. 5. Histogram of NO_3^- concentrations in Jersey groundwaters sampled in June 1995.

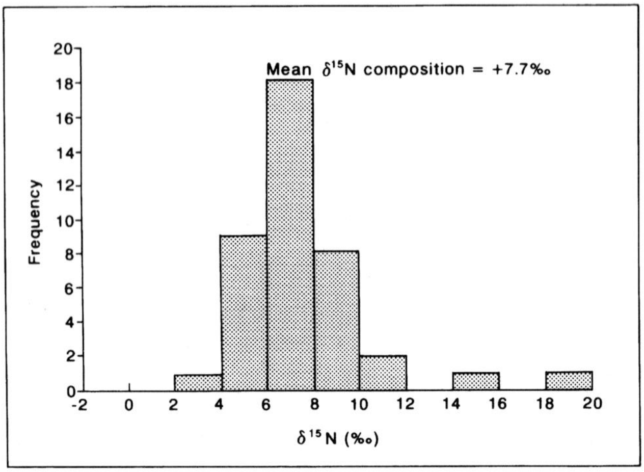

Fig. 6. Histogram of nitrogen isotope composition ($\delta^{15}N$) of Jersey groundwaters sampled in June 1995.

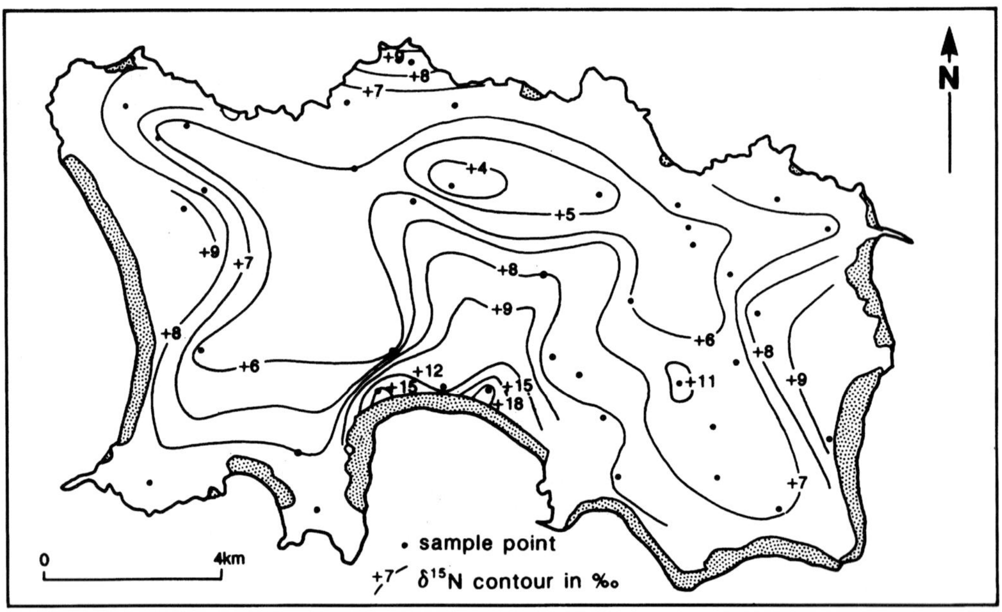

Fig. 7. Regional contours of nitrogen isotope composition ($\delta^{15}N$) of Jersey groundwaters sampled in June 1995.

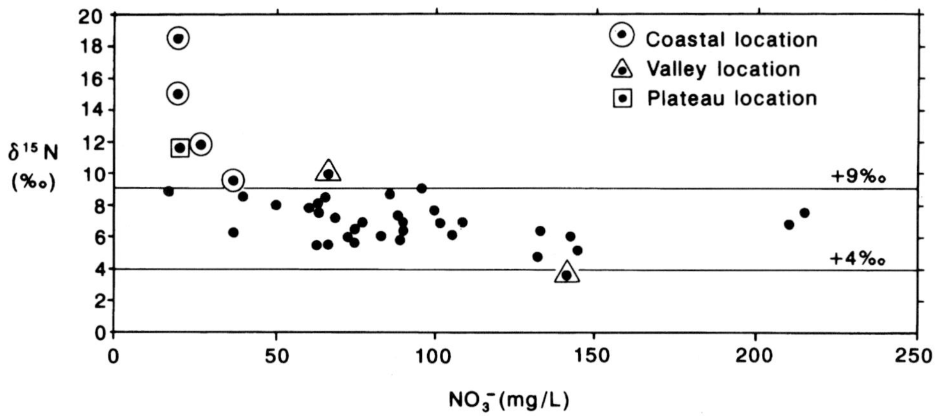

Fig. 8. Cross-plot of nitrogen isotope composition $\delta^{15}N$) against NO_3^- concentration for Jersey groundwaters sampled in June 1995. The band width +4 ‰ to +9 ‰ represents the typical range of $\delta^{15}N$ composition associated with NO_3^- from nitrified soil organic nitrogen.

Extensive leakage from the Jersey sewerage system is unlikely in that the system is relatively new and is maintained under negative pressure, although leakage from older systems is likely. Additionally, those samples with high $\delta^{15}N$ values do not occur in the largest urban areas of St Helier and St Brelades as would be expected if urban contamination were responsible for the observed trend. It is possible that sewage leakage is occurring, but fertilizer application in parks and gardens may counteract the positive $\delta^{15}N$ signature. Septic tank discharges may also give rise to isotopically heavy $\delta^{15}N$ values, but this source of NO_3^- appears unimportant in that septic tanks are dispersed across the island rather than being concentrated in coastal areas.

In a progressively reducing groundwater environment, after the consumption of oxygen, NO_3^- is the next electron acceptor to oxidize organic carbon (heterotrophic denitrification). After NO_3^- concentrations become limited, manganese (IV) and iron (III) and then SO_4^{2-} are reduced. The denitrification process requires a suitable electron donor or donors to complete the dissimilatory reduction of NO_3^- to gaseous N_2, the most likely of which are organic carbon and reduced iron and sulphur species. If NO_3^- is introduced into a manganese/iron or SO_4^{2-} reducing zone, even in the absence of organic carbon, it is thermodynamically unstable (Korom 1992). If appropriate bacteria are present, they may denitrify using reduced manganese, reduced iron and sulphides as electron donors (autotrophic denitrification). The stoichiometry of denitrification reactions can be expressed by the following simple equations (after Korom 1992) which describe a generalized progression of reactions with depth below the water-table.

Heterotrophic denitrification:

$$\begin{aligned} 5C + 4NO_3^- + 2H_2O \\ = 2N_2 + 4HCO_3^- + CO_2 \end{aligned} \tag{3}$$

Here, C represents an arbitrary organic compound with an oxidation state of zero. In this reaction, 1.1 mg of C are required to reduce 1.0 mg of N.

Autotrophic denitrification by reduced iron:

$$\begin{aligned} 2HNO_2 + 6Fe^{2+} + 6H^+ \\ = N_2 + 6Fe^{3+} + 4H_2O \end{aligned} \tag{4}$$

Here, NO_3^- is assumed to be reduced autotrophically to NO_2^- followed by abiotic reduction to N_2 in an environment deficient in organic carbon by the oxidation of Fe^{2+} to Fe^{3+}. In this reaction, 12.0 mg of Fe^{2+} are required to reduce 1.0 mg of N.

Autotrophic denitrification by reduced sulphur:

$$\begin{aligned} 5FeS_2 + 14NO_3^- + 4H^+ \\ = 7N_2 + 10SO_4^{2-} + 5Fe^{2+} + 2H_2O \end{aligned} \tag{5}$$

Here, denitrification is based on the oxidation of pyrite. In this reaction, 1.6 mg of S are required to reduce 1.0 mg of N.

The high concentrations of NO_3^- requiring denitrification would, based on the stoichiometry of Eqs 3–5, demand large quantities of reducing agents within the groundwater environment. Total organic carbon contents of Jersey groundwaters are mostly between 1 and 5 $mg l^{-1}$ (Robins & Smedley 1994) and would, at most, account for a reduction in NO_3^- concentration of 20 $mg l^{-1}$, according to Eq. 3. This oxidation of organic matter would also result in the production of 20 $mg l^{-1}$ of HCO_3^- and the generation of carbon dioxide (CO_2). Further production of HCO_3^-, involving this CO_2, may result from the dissolution of the small amount of carbonate material within the aquifer. Figure 3 shows increased HCO_3^- concentrations at two coastal locations (Samples 28 and 31), suggesting that heterotrophic denitrification is at least in part responsible for this increase.

The concentration of Fe at reducing groundwater sites sampled in this study is < 3 $mg l^{-1}$. This amount of reduced iron would, at most, account for a reduction in NO_3^- concentration of 0.25 $mg l^{-1}$, according to the stoichiometry of Eq. 4. Although quantitatively less important, it is probable, as argued by Robins & Smedley (1994), that additional contributions of reduced iron may be derived from the solid phase of the aquifer involving ferrous iron-bearing minerals such as pyrite, biotite and chlorite. Interestingly, denitrifying bacteria have been identified in groundwater at a number of sites (Robins & Smedley 1994), but were not detected in two of the reducing waters analysed, including Sample 11 in this study. This may be taken as evidence for abiotic reduction of NO_3^- involving the oxidation of Fe^{2+} to Fe^{3+} (Eq. 4), although it is possible that non-identified denitrifiers are responsible.

When compared with groundwater from the north-central area, the reducing groundwaters show an increased concentration of SO_4^{2-}. For example, between Sample locations 20 and 35 there is an increase in SO_4^{2-} concentration of 29 $mg l^{-1}$. Ignoring other sources of SO_4^{2-} such as surface-derived pollution and seawater intrusion, this observed increase would result from a reduction in NO_3^- concentration of 26 $mg l^{-1}$, according to the stoichiometry of Eq. 5. Therefore, it appears that there is a sufficient amount of reducing agents in the aquifer (organic carbon and reduced iron and sulphur species) to account for a reduction in NO_3^- concentration of at least 46 $mg l^{-1}$. For comparison, the actual difference in NO_3^- concentration between Samples 20 and 35 is 57 $mg l^{-1}$.

If denitrification is responsible for the low NO_3^- concentrations observed in reducing groundwaters in southern coastal and valley areas, then the nitrogen isotope data might be expected to conform to the classic Rayleigh equation (Eq. 1). Figure 9a is a plot of $\delta^{15}N$ versus $\ln(NO_3^-)$ for the available data with the line of best fit yielding an isotope enrichment factor of $-3.0‰$ ($r^2 = 0.51$, $n = 40$). There is a clear correlation between the $\delta^{15}N$ value and the $\ln(NO_3^-)$, although there is more scatter at

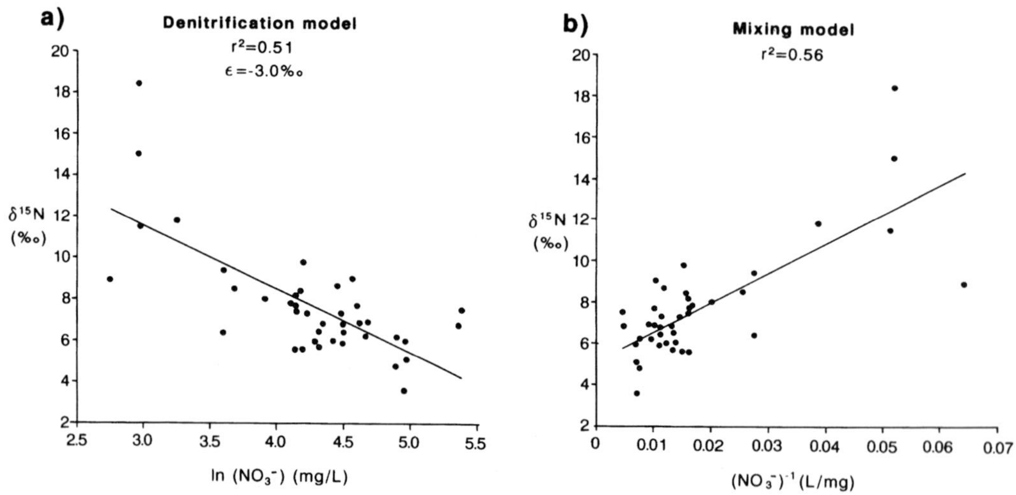

Fig. 9. Cross-plots of nitrogen isotope composition ($\delta^{15}N$) against (**a**) $\ln(NO_3^-)$ and (**b**) $(NO_3^-)^{-1}$ for Jersey groundwaters sampled in June 1995 to illustrate (a) a single-step Rayleigh fractionation denitrification model and b) a mixing model to explain the observed island-wide pattern of increasing $\delta^{15}N$ composition for decreasing NO_3^- concentration.

low NO_3^- concentrations, possibly reflecting greater analytical uncertainty in both parameters. The calculated enrichment factor is smaller than reported for other aquifers, which typically range from -5‰ (the Chalk aquifer of northern France (Mariotti *et al.* 1988)) to −30‰ (the Auob sandstone aquifer in the western Kalahari Desert (Vogel *et al* 1981)) where microbial denitrification is identified. An apparently small enrichment factor could result from a fast denitrification reaction at a high temperature or with increased concentration of electron donors, or more simply could reflect mixing between low- and high-NO_3^- groundwaters with similar isotopic compositions (Feast 1995). A fractionation factor for abiotic reduction of NO_3^- involving the oxidation of Fe^{2+} to Fe^{3+} is not known and so it is not possible to deduce whether this process could explain the small enrichment factor obtained.

The last explanation to be considered for the regional trend in $\delta^{15}N$ values is mixing between end-member groundwaters with different $\delta^{15}N$ and NO_3^- values. Figure 9b is a plot of $\delta^{15}N$ composition versus $(NO_3^-)^{-1}$ for the available data. The line of best fit yields a slightly better coefficient of determination ($r^2 = 0.56$, $n = 40$) compared with the denitrification model. The proportions of high-NO_3^- and denitrified groundwaters abstracted from boreholes situated at the end of the groundwater flowpaths in the coastal and valley areas can be estimated from a consideration of Eq. 2. Taking one end-member groundwater to be shallow recharge water containing 80 mg l^{-1} of NO_3^- with a $\delta^{15}N$ signature of +7‰, and the other end-member to be deeper denitrified water conjectured to contain 5 mg l^{-1} of NO_3^- and a $\delta^{15}N$ signature of +20‰, then, according to Eq. 2, a theoretical mixture of 1:2 would result in a groundwater containing 30 mg l^{-1} of NO_3^- and a $\delta^{15}N$ signature of +8.5‰, similar to coastal groundwater $\delta^{15}N$ values which predominate in the range +8‰ to +9‰. Sample 34 from St Helier measured 39 mg l^{-1} of NO_3^- and a $\delta^{15}N$ signature of +8.5‰, and is therefore close to the modelled groundwater composition.

Further explanation of the NO_3^- and nitrogen isotope data requires a consideration of the physical and chemical heterogeneity of the bedrock aquifer. Evidence of this heterogeneity is demonstrated at the site of Samples 23(i) and (ii) where two boreholes of about equal depth (12–18 m) demonstrate different characteristics. One borehole has a low yield and a tendency to run dry, quickly followed by rapid recharge upon termination of pumping. The other borehole has a greater yield and does not run dry. The low yielding borehole had a higher $\delta^{15}N$ signature (Sample 23(i) = +9.0‰) compared with the better yielding borehole (Sample 23(ii) = +5.9‰) although both had similar NO_3^- concentrations (95 and 89 mg l^{-1}, respectively). This suggests that the low yielding borehole has a contribution of groundwater from a less permeable region of the aquifer where denitrification may be proceeding, while the higher yielding borehole is probably intercepting a fissure and is connected with the more permeable part of the aquifer with a greater degree of groundwater movement. Clearly, any interpretation of the nitrogen isotope data has to take into account the hydrogeological behaviour of the fractured bedrock aquifer.

Generally, and as shown schematically in Fig. 10, the majority of groundwater circulation and abstraction is concentrated in the upper, more weathered horizons of the bedrock aquifer. The aquifer is recharged directly by precipitation and contains groundwater that is mostly oxidizing and chemically immature in nature. This groundwater, which has a short residence time, is typified by a high NO_3^- concentration (50–250 mg l^{-1}) and $\delta^{15}N$ values in the range +4‰ to +9‰ characteristic of an agricultural origin, excluding manure. A deeper component of groundwater exists and is thought to experience progressive denitrification with greater residence time in the aquifer. In the area of lowest elevation, at the foot of the topographic rise, boreholes sample proportionately more of this chemically evolved groundwater. An example of a site exhibiting a greater component of this deeper groundwater is Sample 27 with a low NO_3^- concentration (19 mg l^{-1}) and enriched $\delta^{15}N$ value (+18.4‰). The boundary between the shallow and deeper groundwater types is expected to be gradational and irregular, depending on the development of permeability at depth, as controlled by the geology. The deeper groundwater component may be closer to the surface in areas of poor hydrogeological development.

Conclusions

The bedrock aquifer underlying the island of Jersey is widely contaminated with NO_3^-, with 67% of sites sampled in this study exceeding 50 mg l^{-1} of NO_3^- and maximum concentrations exceeding 200 mg l^{-1}. Stable nitrogen isotope analysis reveals $\delta^{15}N$ values for the groundwater ranging from +3.6‰ to +18.4‰, with the majority (83%) falling in the range +4‰ to +9‰. Over most of the island where NO_3^- concentrations are highest, $\delta^{15}N$ values are between

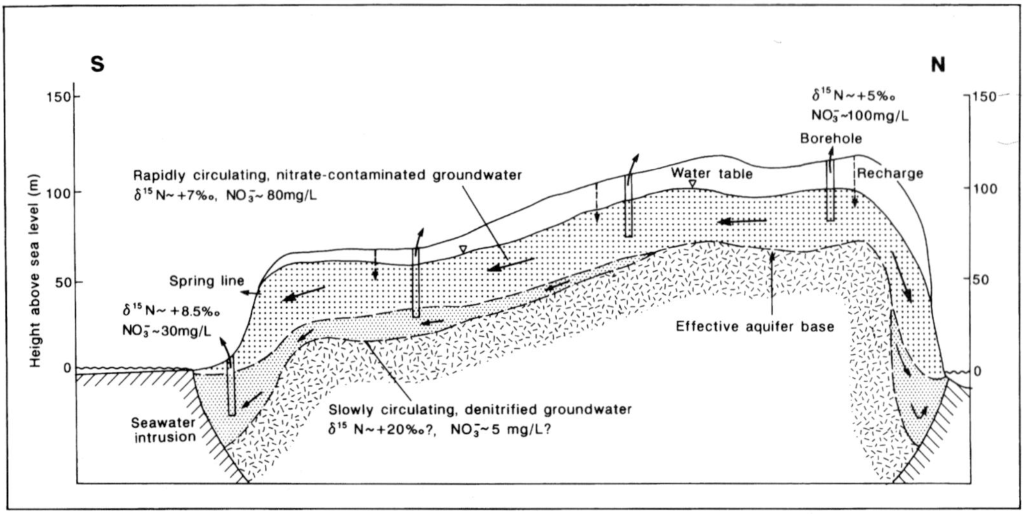

Fig. 10. Conceptual hydrochemical model of the Jersey bedrock aquifer demonstrating the chemical evolution of nitrogen isotope composition ($\delta^{15}N$) and NO_3^- concentration along a hypothetical flowpath from north to south across the island. The vertical scale is greatly exaggerated.

+4‰ and +7‰ and are characteristic of intensively farmed areas. There is very little evidence for direct leaching of fertilizer into groundwater, and apparently little contribution of nitrogen from domestic sanitation systems, including old septic tanks which are common in the north of the island. Thus, it is concluded that the predominantly 'agronomic' isotopic signature obtained for dissolved NO_3^- in groundwater is primarily a result of the intensive agricultural and horticultural practices employed on the island.

Coastal and valley-bottom areas are characterized by reduced NO_3^- concentrations and enriched $\delta^{15}N$ values resulting from the progressive mixing between shallow, rapidly circulating high-NO_3^- groundwater with a slowly circulating, more evolved and denitrified water found deeper in the aquifer. Denitrification is conjectured to proceed as a heterotrophic reaction with the oxidation of organic matter followed by autotrophic denitrification involving reduced iron and sulphur species. The likely sources of the latter are ferrous iron- and sulphide-bearing minerals present in the aquifer material. Compared with the contamination of groundwater by NO_3^- across the island, the process of denitrification is not significant, except for local coastal and valley areas, and is therefore unlikely to provide a long-term natural remediation measure. In southern coastal areas it is conceivable that a component of urban sewage leakage is present, although this nitrogen source is of much less importance and is likely to decrease in significance with technical improvements in sewer design.

We are grateful to Nick Robins and Pauline Smedley of the British Geological Survey for their assistance with this study, both in terms of data provision and fieldwork support. We also thank the two anonymous reviewers for their constructive comments on the manuscript. The Public Services Department, States of Jersey, also gave helpful advice and assistance. This study is published with permission of the Chief Executive, Public Services Department, States of Jersey. The views expressed in this paper are entirely those of the authors.

References

Böttcher, J., Strebel, O., Voerkelius, S. & Schmidt, H.-L. 1990. Using isotope fractionation of nitrate-nitrogen and nitrate-oxygen for evaluation of microbial denitrification in a sandy aquifer. *Journal of Hydrology*, **114**, 413-424.

Feast, N. A. 1995. *Application of Nitrogen and Sulphur Isotope Hydrochemistry in Groundwater Studies.* PhD Thesis, University of East Anglia.

—— & Dennis, P. F. 1996. A comparison of methods for nitrogen isotope analysis of groundwater. *Chemical Geology (Isotope Geoscience Section)*, **129**, 167–171.

FUSTEC, E., MARIOTTI, A., GRILLO, X. & SAJUS, J. 1991. Nitrate removal by denitrification in alluvial ground water: role of a former channel. *Journal of Hydrology*, **123**, 337–354.

HEATON, T. H. E. 1986. Isotopic studies of nitrogen pollution in the hydrosphere and atmosphere: a review. *Chemical Geology (Isotope Geoscience Section)*, **59**, 87–102.

KAPLAN, N. & MAGARITZ, M. 1986. A nitrogen-isotope study of the sources of nitrate contamination in groundwater of the Pleistocene coastal plain aquifer, Israel. *Water Research*, **20**, 131–135.

KOROM, S. F. 1992. Natural denitrification in the saturated zone: a review. *Water Resources Research*, **28**, 1657–1668.

KREITLER, C. W. 1979. Nitrogen-isotope ratio studies of soils and groundwater nitrate from alluvial fan aquifers in Texas. *Journal of Hydrology*, **42**, 147–170.

MARIOTTI, A., LANDREAU, A. & SIMON, B. 1988. ^{15}N isotope biogeochemistry and natural denitrification process in groundwater: application to the chalk aquifer of northern France. *Geochimica Cosmochimica Acta*, **52**, 1869–1878.

ROBINS, N. S. & SMEDLEY, P. L. 1991. *Hydrogeological and hydrogeochemical survey of Jersey*. British Geological Survey, Technical Report **WD/91/15**.

—— 1994. Hydrogeology and hydrogeochemistry of a small, hard-rock island - the heavily stressed aquifer of Jersey. *Journal of Hydrology*, **163**, 249–269.

——, CHILTON, P. J. & STUART, M. E. 1993. *Jersey groundwater – Year 3: further observations and potential sources of pollution*. British Geological Survey, Technical Report **WD/93/28**.

VOGEL, J. C., TALMA, A. S. & HEATON, T. H. E. 1981. Gaseous nitrogen as evidence for denitrification in groundwater. *Journal of Hydrology*, **50**, 191–200.

WILSON, G. B., ANDREWS, J. N. & BATH, A. H. 1994. The nitrogen isotope composition of groundwater nitrates from the East Midlands Triassic sandstone aquifer, England. *Journal of Hydrology*, **157**, 35–46.

The British Nuclear Fuels Drigg low-level waste site characterization programme

ROB SEARS

BNFL, R202, Risley, Warrington, Cheshire, WA12 6AS, UK

Abstract: Site characterization work is being undertaken to underpin the post-closure radiological assessment of the UK's Drigg low-level radioactive waste disposal site. This work is providing high-quality data on the geology, hydrology and hydrogeology of the site, thus ensuring that both the monitoring programme and predictions of the radiological impact are based on a good understanding of the hydrogeology.

The programme has two phases. The first addresses the surface water balance of the site; this has been achieved by constructing a surface water balance model which has been combined with a forest interception model to determine the hydrologically effective rainfall. This model has been calibrated using extensive soil moisture data gathered on-site. The second phase is concerned with improving knowledge of the site geology and hydrogeology. Continuously cored rotary boreholes are being drilled across the site to recover geological data. Additionally, a complementary geophysical programme and trial trench studies have been undertaken. Hydrogeological data are being provided by a series of borehole clusters, drilled using cable percussion rigs, centred around the rotary boreholes. During the drilling of the cable percussion boreholes, extensive hydrogeological and hydrogeochemical data are being acquired. These data, in conjunction with data from weirs on the site streams, will provide the data required to complete the hydrogeological water balance.

The Drigg low-level waste (LLW) facility is located in west Cumbria, between the coast and the Lake District National Park, as illustrated in Fig. 1. The site, which is owned and operated by British Nuclear Fuels Limited (BNFL), receives the vast majority of the radioactive LLW produced in the UK. This includes waste produced by hospitals and general industry as well as that produced by the nuclear industry. The majority of the waste is delivered by rail from the BNFL Sellafield reprocessing plant which is located approximately 7 km north of Drigg.

The site is approximately 1.5 km long by 0.5 km wide, as shown in Fig. 2. It is located 1.0 to 1.5 km inland from the coast and 0.5 km north-west of the River Irt. On the seaward side of the site, the land use comprises dune land and

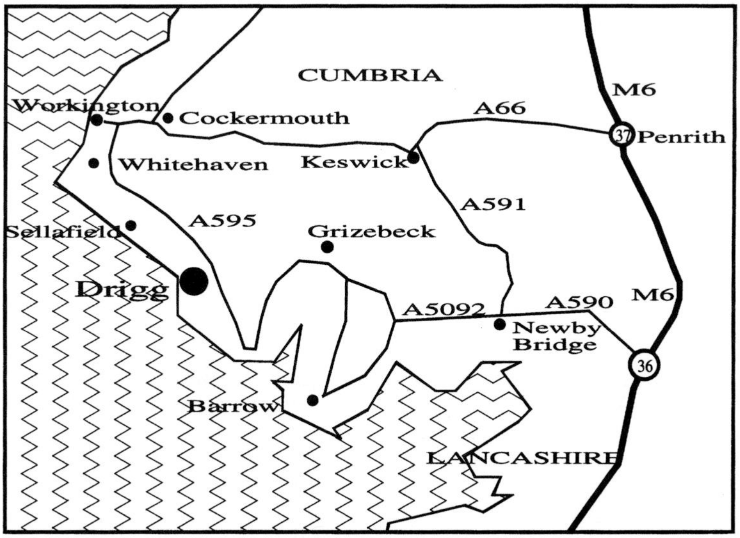

Fig. 1. Location of the Drigg low-level waste facility.

SEARS, R. 1998. The British Nuclear Fuels Drigg low-level waste site characterization programme. *In*: ROBINS, N. S. (ed.) *Groundwater Pollution, Aquifer Recharge and Vulnerability*. Geological Society, London, Special Publications, **130**, 37–46.

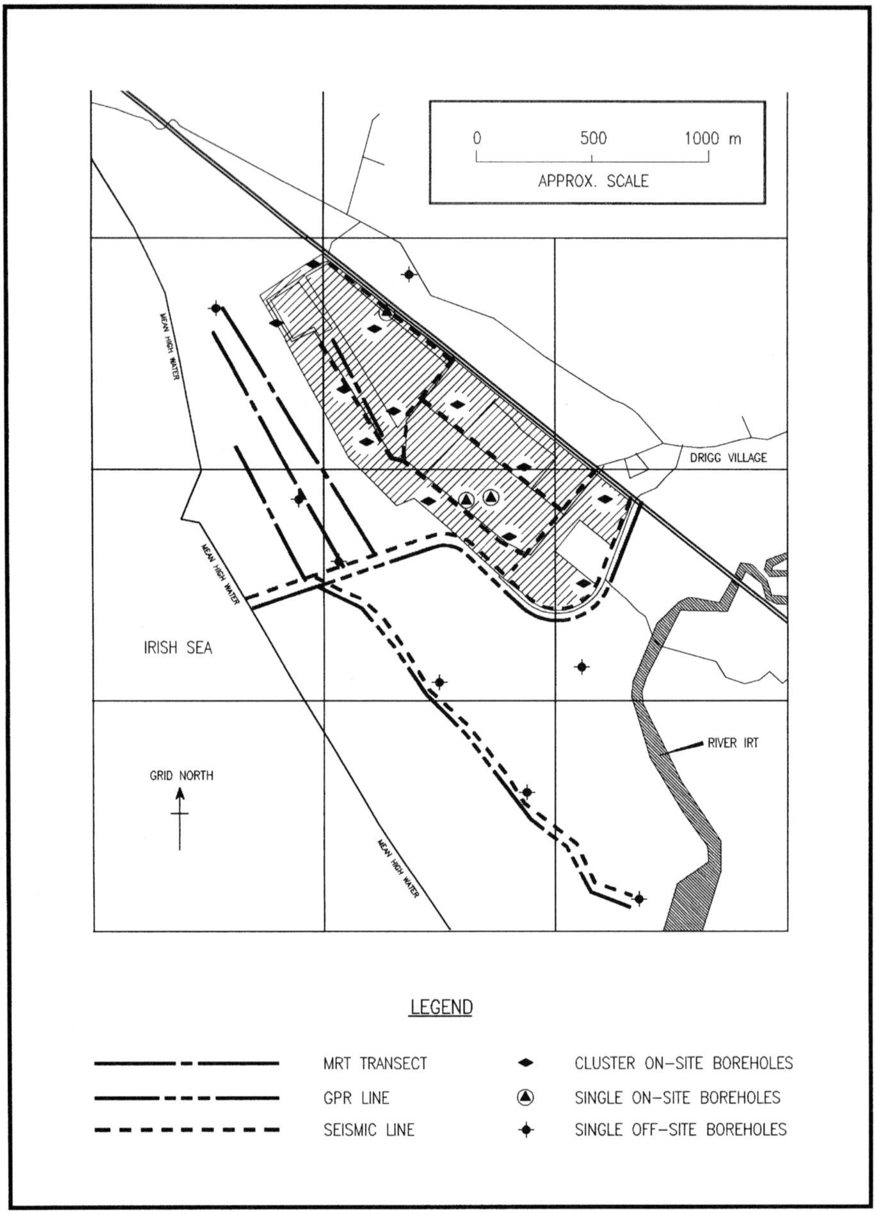

Fig. 2. Map showing borehole locations and position of geophysical survey lines.

rough grazing. On the landward side, the land use comprises pasture. Within the site itself, part is forested with either mature coniferous forest or tree saplings whilst the remaining areas are grassed or hardstanding.

In the northwestern part of the site, trenches have been excavated into which waste was historically tumble-tipped. These trenches have subsequently been covered by an interim cap, incorporating a plastic membrane, to minimize water ingress. Bentonite cut-off walls, keyed into a clay horizon, have been installed around the northwest and northeast sides of the trenches.

Since 1988, waste in steel containers has been placed in a concrete vault. This waste is compacted and surrounded by a grout prior to emplacement in the vault. All rainfall onto the vault floor is collected via drains and discharged to sea together with residual levels of leachate from the trenches.

Existing conceptual model

The existing radiological safety case for the site is based upon a conceptual model of the geology and hydrogeology of Drigg developed by the British Geological Survey (Williams *et al.* 1985). On the basis of an interpretation of 35 borehole logs, the drift geology was divided into eight lithostratigraphic units overlying the Triassic Ormskirk Sandstone. This is illustrated by the simplified cross-section shown in Fig. 3. This sequence also divides the stratigraphy of the drift hydrogeologically into more permeable sandy and gravelly horizons loosely known as aquifers (even-numbered layers G2, G4, G6 and G8) and less permeable, clay-rich horizons (odd-numbered layers G1, G3, G5 and G7).

The bases of the trenches have been excavated into the G5 clay, which was known to support perched water. The existing hydrogeological model assumed that very little leachate could migrate through the G5 clay and gain access to the regional water below. Therefore, attention had concentrated on the overlying G6 perched aquifer and the site streams into which this water could discharge. Less attention had been paid to the deeper units, in particular the G2, which, as Fig. 3 shows, constitutes a significant proportion of the overall drift thickness and probably exerts a hydrogeological influence at the site scale. However, recent borehole and trial pit data have suggested that the G1, G3, G5 and G7 clays may not be continuous. Moreover, work carried out by Eaton & Williams (1993) and NIREX (1997*a*), have suggested that structural controls on the drift stratigraphy may further reduce their ability to resist the downwards flow of groundwater. These structures may have a variety of stratigraphic effects, but can clearly cause breaches in the clay horizons and may cause apparently anomalous lenses of sand or gravel.

Small-scale faulting and structural displacements in the drift have also been observed in the exposed sides of the trenches. Evidence from other investigations in the vicinity of Sellafield indicates that such structural displacements may occur throughout the unconsolidated stratigraphy at Drigg.

A revised conceptual model is now being proposed, whereby water can 'cascade' down between or through the G3, G5 and G7 clays, to the regional aquifer located within the G2 and Triassic Ormskirk Sandstone. A knowledge of the recharge mechanisms to the regional aquifer and groundwater residence times are invaluable in enabling this conceptual model to be incorporated into the safety case.

The Drigg site characterization project was therefore established. The project is designed to run for five years. The first year has been entirely dedicated to data acquisition via a programme of drilling boreholes and conducting geophysical surveys. BNFL is currently at the end of the first year of this programme. The remaining four years will be devoted to collecting high-quality time series datasets of water levels, stream flows, groundwater samples and meteorological variables and to updating the hydrogeological conceptualization of the site.

This paper describes the field techniques employed and discusses some of the early findings.

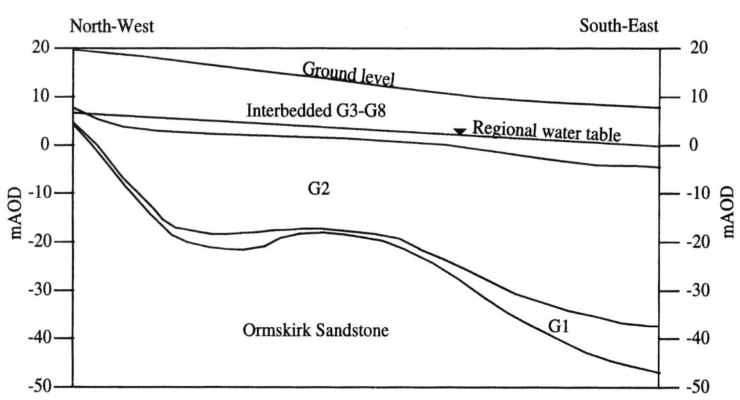

Fig. 3. Simplified cross-section of Drigg geology as proposed by Williams et al. (1985).

Site investigation requirements

The site investigation programme was designed to fulfil the following requirements:

1. In order to further study the geology of the site and its surrounds, high-quality geological data were required. With the benefit of these new data, an interpretation could be carried out based on the regional Quaternary geological framework (NIREX 1997*b*). An event stratigraphical approach was used in order to determine distinctive formations which could be correlated across the site. In order to provide the data, a number of boreholes were drilled. A complementary programme of geophysical surveying and borehole logging was also undertaken and trial trenches were excavated to examine the near-surface structural geology.
2. A revised hydrogeological conceptual model was required, based on the updated geological model. Data for this were provided by drilling boreholes and conducting an extensive programme of soil and water testing.
3. The volume of perched groundwater able to migrate to the regional aquifer will, in part, be controlled by the recharge processes occurring on the site. These can be investigated by studying the vertical gradients within the perched and regional aquifers. In order to do this, boreholes, completed with well screens, were drilled as a series of clusters around the site. Each borehole within a cluster had a well screen completed at a different depth. This allowed all aquifer units to be independently monitored.
4. Perched groundwater that does not permeate down to the regional water-table may flow laterally and could possibly enter the site streams, which are hydraulically connected to the perched aquifers. In order to understand further the processes by which this occurs, a knowledge of the interaction between the site streams and the perched water is essential. Additional boreholes were drilled adjacent to the site streams and data acquired from these boreholes will be used in conjunction with the stream flow data to study this relationship.
5. An understanding of the stream / perched water interaction is also important to establish the component of the hydrologically effective rainfall that discharges to the streams rather than forming recharge to the regional aquifer. Understanding this process will greatly enhance the confidence that can be placed in the site water balance.
6. In order to understand the behaviour of the aquifers underlying the site and the transport of radionuclides within the aquifers, a study of the water-level dynamics will be carried out on the water-level time series dataset. Additionally, chemical analyses of groundwater samples for radionuclides will provide direct evidence of transport within the sediment.
7. Following analysis of the hydrogeological data, certain key boreholes can be added to the network of monitoring boreholes to ensure that the potential migration of radionuclides under the site is adequately monitored.

Previous studies of the site geology had been hampered by inadequate data being collected from the cable percussion drilled boreholes. Owing to the anticipated heterogeneity of the sediments, continuous coring was required to yield sufficient detail to allow an accurate geological model to be created. Therefore, the first borehole drilled in each cluster was drilled using a continuous rotary core technique.

However, the use of a rotary drilling system, with associated drilling polymers, precluded the collection of water samples and the observation of water strikes which were required in order to understand the hydrogeology. Therefore, the second and subsequent boreholes drilled in each cluster were drilled using cable percussion drilling rigs, allowing the collection of hydrogeological data.

Geological and hydrogeological investigation techniques

In view of the above requirements, the site investigation programme was based around the drilling of 12 borehole clusters on the site, at the locations shown in Fig. 2. Each cluster contains between two and four boreholes, comprising the rotary and cable percussion boreholes. Individual boreholes within a cluster have a well screen located at a different depth, allowing a complete horizontal and vertical resolution of the flow system.

Three boreholes were drilled on-site in addition to the clusters; these are also shown in Fig. 2. One borehole was drilled to rockhead to gain further geological information and two boreholes were drilled to approximately 5 m below ground level to investigate the relationship between the site streams and the perched aquifer.

Eight boreholes were drilled off-site at the locations shown in Fig. 2. These were used to gain local geological and hydrogeological information, up-gradient and down-gradient of the site.

In order to study the near-surface geology, three trial trenches were excavated on site. These supplement the numerous trial pits that have been excavated in the past. The trenches have been particularly useful to illustrate the detail of the structural drift features which will not be observed by a point penetration technique such as a borehole.

Extensive geophysical surveying has been carried out to augment the borehole information. Refraction seismic, the micro-resistivity technique (MRT), EM31 and ground-penetrating radar (GPR) have all been utilized, as shown in Fig. 2.

The first and deepest borehole in each cluster was rotary drilled. Coring was carried out using a double-barrelled S size water flush system. The flush provided was Procoat, a degradable polymer. These boreholes were targeted at the Triassic Ormskirk Sandstone which forms the basement rock to the Quaternary drift under the site. The primary aim of these boreholes was to provide geological information to supplement that already known on the site.

Coring through the drift proved to be slow owing to the amount of gravel encountered. Medium-sized gravels spun on the bit face causing extremely low penetration rates. However, overall core recovery was good, except for units with low clay content, where core loss was, not surprisingly, close to 100%. In these zones, it was apparent that the core had been washed by the drilling medium and was not representative. Overall, 853 m of core was cut with a recovery rate of 75%.

The second borehole in each cluster was drilled using a cable percussion drilling rig. The aim of these boreholes was to provide hydrogeological information. These boreholes were drilled without the addition of water, allowing water samples to be taken whilst drilling. The water analyses yielded a hydrogeochemical profile of each borehole. Dry drilling also allowed the accurate identification of water strikes and allowed accurate groundwater heads to be recorded. Dry drilling required the use of specialized shells as shown in Fig. 4. These were equipped with a valvator which retained more sediment within the barrel than experienced with the traditional flange type. The technique proved to be successful even within poorly consolidated sands and gravels. During drilling, soil samples were taken and subsequently analysed in the laboratory. This provided hydrogeological properties data. In addition, rising head tests were conducted to obtain *in situ* permeability data from zones which would not be targeted by a well screen. Standard penetration tests were carried out to obtain *in situ* density data.

The lithological data obtained from these boreholes was used to supplement the core data. It was found that interpreting the data from both drilling techniques in parallel provided a very detailed picture of the geology at each cluster location.

Having drilled two boreholes at each cluster location, a complete identification of aquifer units could be undertaken. This allowed remaining boreholes to be drilled, if required, ensuring that a well screen could be placed in each aquifer unit at each cluster location.

The deepest borehole in each cluster was completed using 75 mm i.d. PVC liners. These were chosen as the smallest liner that would allow geophysical logging of the borehole. Slot sizes for the well screens were selected on the basis of particle size analyses and ranged from 450 μm geotextile wraps to 1mm slotted screens. The remaining boreholes in each cluster were completed using 50 mm i.d. PVC liners. These were selected to minimize the annular volume and, therefore, reduce the volume of purge water that must be removed when sampling. Short screen lengths of 0.5 to 1.0 m were utilized to ensure that selective and representative groundwater samples could be taken. The annular space between the liner and the borehole wall was backfilled using inert, washed and graded sands in the permeable units, bentonite pellets or bentonite slurry in the low-permeability units below the water-table and cement grout in the low-permeability units above the water-table. Furthermore, the top metre of each borehole was backfilled with concrete which was continued to form a base slab for

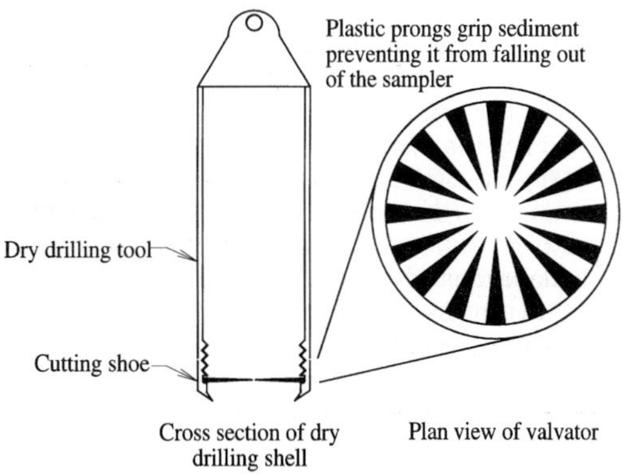

Fig. 4. Diagram of the dry drilling shell used by the project.

the surface brickwork and borehole cover. This ensured that rainwater could not gain access down any voids around the liner. The brickwork was designed to accommodate the data loggers and sample pump ports.

Thirty of the boreholes were fitted with pressure transmitters connected to logging units. These provide a detailed record of the water-level variations in the boreholes. In addition, three of the boreholes were fitted with dedicated sample pumps.

The sample pumps operate using a PVC bladder and are able to pump at extremely low flow rates, thus causing minimal disturbance to the stagnant water that exists within the borehole liner above the well screen. Use of these pumps minimizes the amount of water that must be purged from the borehole whilst ensuring a representative groundwater sample is taken. The volume of water purged is determined by diverting the sample through a flow-through cell and continuing purging until the pH, electrical conductivity, dissolved oxygen and groundwater temperature stabilize.

Water balance

In order to study the water balance, the site was instrumented as shown in Fig. 5. V-notch weirs were installed where streams flow onto and off the site and at all major confluences of streams and drains within the site boundary. Two automatic weather stations (AWS) were installed to measure meteorological variables. A network of neutron probe access tubes was installed in each of the major land-use types, both on-site and off-site, to measure soil moisture conditions. The trench cap was also fitted with tensiometers to measure soil water tension. Finally, three plots were selected in the forested area to carry out a forest precipitation interception experiment, run to determine the interception losses. These were equipped with canopy and ground-level rain gauges as well as stem flow collector devices.

The experimental approach was to determine rainfall at the canopy level, determine the through-fall (the rain falling between the leaves directly to the ground) and the stem flow (the rain collecting in the foliage and then flowing down the trunk to the ground). The latter two processes determine the amount of water which reaches the ground and, therefore, the amount which is available for evapotranspiration and hence for recharge. The experiment was run for three years and concluded that in the forested areas of the site, interception losses amounted to approximately 20% of the rainfall. The experimental data were found to be similar to work carried out by Robins (1974), Leyton *et al.* (1967), Gash & Morton (1978) and Gash *et al.* (1980). The data collected from this experiment were used as the basis for a model of the interception losses, which was subsequently built into a soil moisture balance model.

The soil moisture balance model is a two-soil-layer model based on the Meteorological Office MORECS model (Thompson *et al.* 1981). However, the BNFL model allows for parameter optimization for local conditions. The model accepts rainfall and meteorological variables from the AWS and outputs the hydrologically effective rainfall (HER) for each of the land-use types. The model has been calibrated to three years worth of soil moisture data collected from the neutron probe access tubes. Figures 6 and 7 show examples of the modelled soil moisture deficit against time for the trench cap land-use type and the forest land-use type respectively. From these figures, it can be seen that the fit is generally good although the simulated wetting for the 1991/1992 winter is exaggerated and the simulated deficit for the 1993 summer, for the trench cap grass land use type (Fig. 3), is under-estimated. Work has continued throughout 1995 and 1996 to improve the model.

In order to complete the site water balance, work commenced to link the soil moisture balance model and the land use within the surface catchment area to construct a site water balance that can be validated against the stream flow data.

Hydrostratigraphy

The interpretation of the hydrostatigraphy is presently at an early stage. However, the following account describes the current hydrostratigraphical model.

The basement rock to the Quaternary drift deposits at Drigg is the Triassic Ormskirk Sandstone (Michie 1994). Previous geological studies carried out at Drigg have established that this is an aeolian sandstone comprising interchanging soft and hard bands ranging between 0.5 and 2 m thick. The sandstone has a relatively high transmissivity of 150 to 230 m^2/day, with the highest permeability associated with the soft bands. Generally, the hydraulic conductivity decreases with depth. The top of the sandstone is often completely weathered consisting of 1 to 2 m of fine to medium well-sorted sand.

The drilling programme established that at the southeastern end of the site, where the sandstone

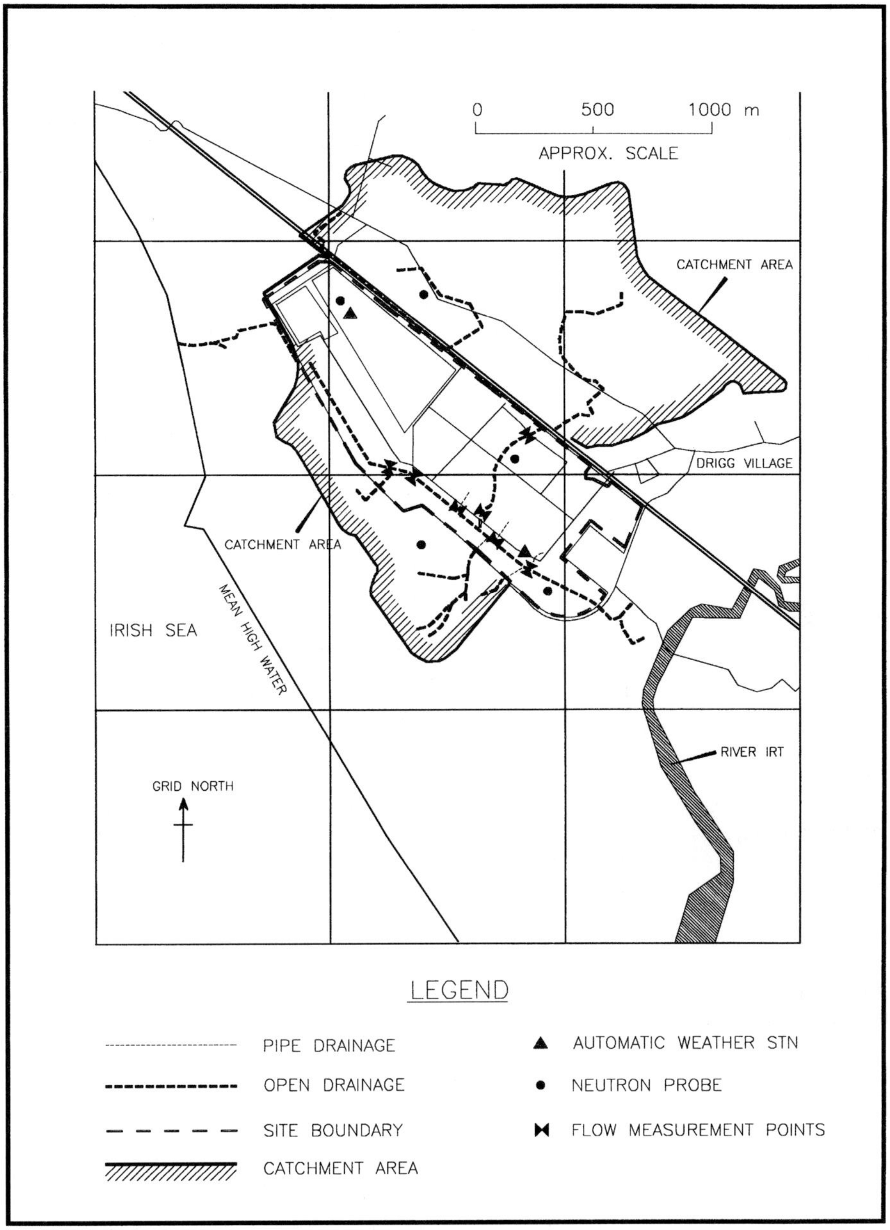

Fig. 5. Map showing hydrometric network and site catchment.

is deepest, there is an over-compacted till deposit. Unconformably overlying this till is a rhythmite/varve sequence which grades up into a grey-green silt. The varve sequence is interpreted to have been deposited in a glaciproximal lake during the early stages of deglaciation. As the water depth increased, so this sequence progressively increased in silt content. Shell fragments were recovered from the top of the silt unit. Preliminary dating of amino acid ratios in these fragments date the sequence at 65 000 years before present. These units predate the Devensian ice advance that was thought to have eroded all sedimentary cover back to the sandstone. However,

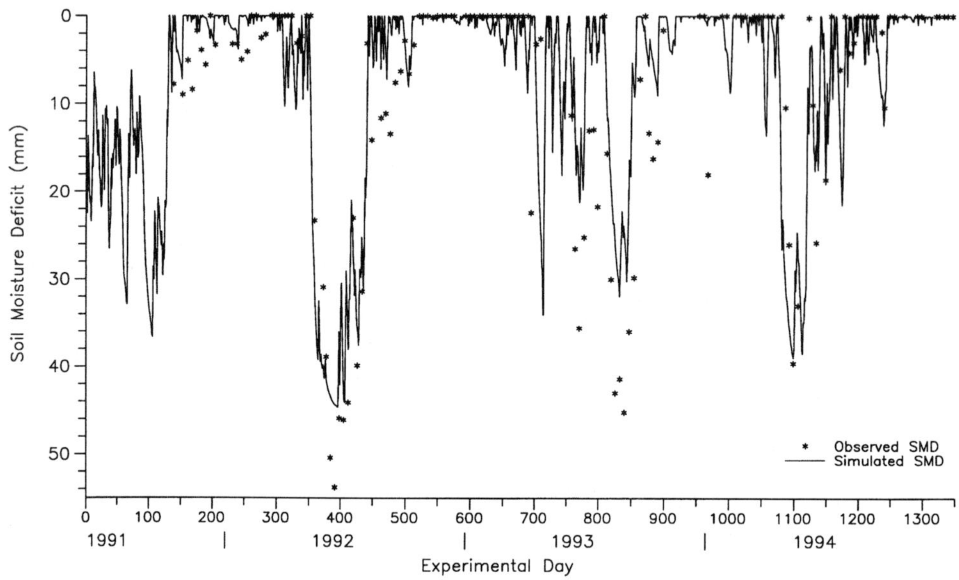

Fig. 6. Soil moisture deficit against time for the trench cap land-use probe, compared with model results.

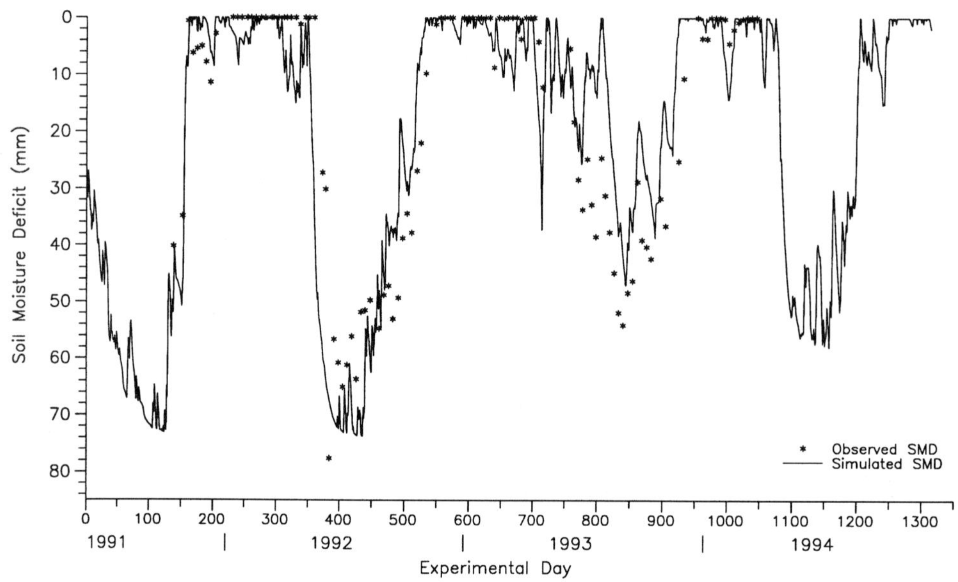

Fig. 7. Soil moisture deficit against time for the forest land-use probe, compared with model results.

these units are only found in the valleys in the sandstone topography, where they would have been protected from the Devensian ice advance, as shown in Fig. 8. In terms of the site hydrogeology, the silt will probably act as an aquitard where it is present, thereby separating the aquifers in the Quaternary drift above and the sandstone below. Elsewhere, the lowermost Quaternary deposits probably form a continuous aquifer unit with the Triassic Ormskirk Sandstone.

Directly overlying the Ormskirk Sandstone across most of the site is a second till which has also undergone over-compaction when compared to the density of the overlying sequences. This till has been associated with the Devensian deglaciation. It is characterised by consisting, mainly, of fine to medium sand bound by a silt/clay matrix and contains a high proportion of poorly sorted gravels. The unit appears to have a highly variable hydraulic conductivity profile, varying from 1 to 100 m/day and probably dominated by preferential flow pathways caused by the reworking or the washing out of fines from the till by sub-glacial water courses.

Lying unconformably over the till, and projecting down into it, is a sequence of braided channel deposits. These are thought to represent glacial outwash deposition from the retreating Wasdale and Eskdale glaciers and probably contain an element of reworked material from the underlying till. This unit also has a highly variable hydraulic conductivity, ranging from 1 to 10 m/day, related to its facies distribution.

Overlying the braided channel deposits is a 20 m sequence of clays, silts, sands and gravels, interpreted to have been deposited on or near a lake or marine margin. There is evidence, in the trial trenches and coastal exposures, that structural events have affected these units. However, there is also significant variability within the units, the clays in particular changing from a stiff clay to sandy clay over relatively short distances. The structure has been tentatively linked to glacial tectonics which has acted to push the near-surface sediments into a series of anticlines accompanied by thrust faulting. Resulting hollows have subsequently provided an environment for the deposition of Holocene sands and peats. Perched aquifers are associated with the hollows and occur as a series of east–west-trending water-bodies. Recharge to the regional aquifer occurs from these upper units, via the perched aquifers. Water is thought to 'cascade' down from the perched water systems until it ultimately reaches the regional aquifer. It is clear that there are many routes by which this can occur, including: overspill off the edges of the discontinuous clay lenses; flow through the clays where they are more permeable; flow through fractures and other preferential pathways within the clays; and flow down man made features.

Conclusions

The Drigg site characterization programme has reached the end of the drilling programme. The interpretation presented is preliminary and attention is now being focused on a detailed interpretation to produce a comprehensive understanding of the site hydrogeology. This will be used to underpin the post-closure radiological assessment programme and provide confidence in the borehole-monitoring network. The project has quantified the various groundwater flow components and can demonstrate evidence for recharge to the regional aquifer.

Figure 8 shows a schematic cross-section running northwest to southeast along the long axis of the site. This illustrates the conceptual hydrostratigraphy described in the paper and shows the potential recharge pathways, passing through

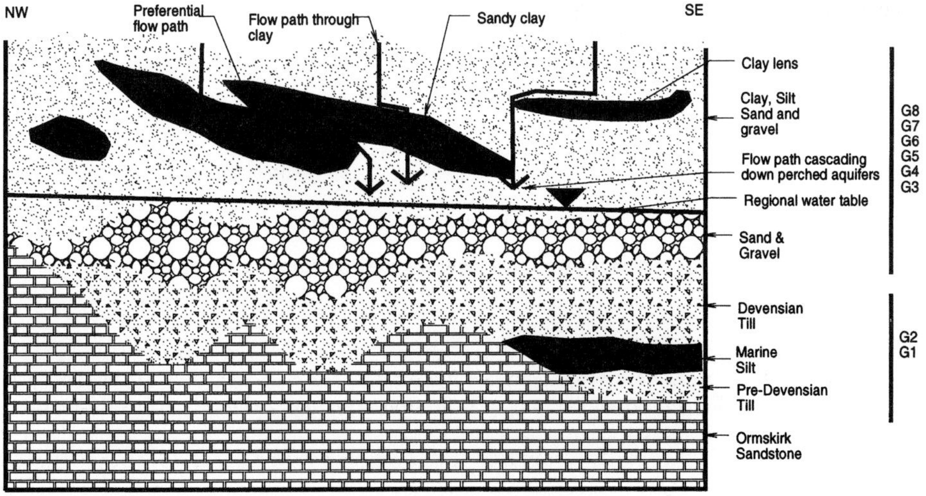

Fig. 8. Schematic cross-section of reinterpreted hydrostratigraphy.

the perched aquifers to the regional aquifer. Once in the regional aquifer, flow occurs within the sand and gravel, till and sandstone. At present, it seems unlikely that the regional aquifer can be divided along stratigraphic horizons. The sand and gravel, till and sandstone all appear to have a similar range in hydraulic conductivity. Therefore, the question becomes that of the transport mechanism within the aquifer and the scale of the preferential pathways within the sand and gravel and the till. Detailed hydrogeological models will now be constructed, utilizing the water level and groundwater quality data acquired over time to understand the flow mechanisms and answer such questions.

References

EATON, G. P. & WILLIAMS, G. D. 1993. *Quaternary geology of the Sellafield Area.* NIREX Report **519**.

GASH, J. H. C. & MORTON, A. J. 1978. An application of the Rutter model to the estimation of the interception loss from Thetford forest. *Journal of Hydrology*, **38**, 49–58.

——, WRIGHT, I. R. & LLOYD, C. R. 1980. Comparative estimates of interception loss from three coniferous forests in Great Britain. *Journal of Hydrology*, **48**, 89–105.

LEYTON, L., REYNOLDS, E. R. C. & THOMPSON, F. B. 1967. Rainfall interception in forest and moorland. *In* SOPPER, W. E. & LULL, H. W. (eds) *Symposium on Forest Hydrology*. Pergamon, Oxford, 163–178.

MICHIE, U. 1994. The geological framework of the Sellafield area and its relationship to hydrogeology. *Quarterly Journal of Engineering Geology*, **29**, S13–S27.

NIREX. 1997*a*. A synthesis of the nature and origin of deformation features within the Quaternary sequences, Sellafield. NIREX Report **SA/97/004**.

—— 1997*b*. *Quaternary evolution of the Sellafield area, Cumbria*. NIREX Report **SA/97/002**.

ROBINS, P. C. 1974. A method of measuring the aerodynamic resistance to the transport of water vapour from forest canopies. *Journal of Applied Ecology*, **11**, 315–325.

THOMPSON, N., BARRIE, I. A. & AYLES, M. 1981. *The Meteorological Office rainfall and evaporation calculation system MORECS*. Meteorological Office, Bracknell, Hydrological Memorandum **45**.

WILLIAMS, G. M., STUART, A. & HOLMES, D. C. 1985. *Investigation of the geology of the low-level radioactive waste burial site at Drigg, Cumbria.* BGS Report Vol. 17, No. 3.

Changes in groundwater chemistry due to rising groundwater levels in the London Basin between 1963 and 1994

I. H. MÜHLHERR, K. M. HISCOCK, P. F. DENNIS & N. A. FEAST*

School of Environmental Sciences, University of East Anglia, Norwich, NR4 7TJ, UK

Abstract: Commencing in the last century, groundwater abstraction has modified groundwater flow and lowered groundwater levels in the Chalk aquifer of the London Basin. Since the 1960s, a decline in pumping has caused a constant rise in groundwater levels and groundwater is now re-entering strata which had formerly been dewatered. In order to assess the effects of the changed hydraulic conditions on the chemical composition of groundwater, Chalk groundwater samples along a transect from Watford to Croydon across the London Basin were analysed for several chemical parameters in 1994, and compared with results from a similar hydrochemical study reported in 1963.

A general deterioration in groundwater quality was identified. Much higher sulphate and alkali and alkaline earth metal concentrations were found in almost the entire study area and are associated with the reflushing of formerly dewatered strata. In the Chalk outcrop areas, high nitrate, sulphate, chloride and sodium concentrations indicate anthropogenic aquifer pollution. Several chemical and physical processes that control the groundwater chemistry down-gradient towards the centre of the Basin were identified. Finally, five groundwater zones established in 1963 on the basis of hydrochemistry were revised to produce three groundwater zones.

The Chalk aquifer of the London Basin has been heavily exploited by private and public water supply boreholes for many decades, thus accelerating natural groundwater circulation and lowering groundwater levels significantly. Since the 1960s, a reduction in abstraction rates in the London area has resulted in a rise in groundwater levels of up to 2.75 $m\,a^{-1}$ (National Rivers Authority (NRA) 1994). Simpson *et al.* (1989) suggest that groundwater levels could return to their original values within the next 30 to 40 years.

The Chalk hydrochemistry of the London Basin was first discussed by Ineson & Downing (1963) who sampled Chalk groundwater along a northwest–southeast line from Watford to Croydon. Ineson & Downing (1963) identified different hydrochemical facies along this line and established groundwater zones based on these facies, as well as proposing hydrochemical processes responsible for the observed changes in groundwater chemistry. The work reported here examines changes in groundwater chemistry resulting from changes in hydraulic conditions within the aquifer over the last three decades and compares these new data with those of Ineson & Downing (1963). Other hydrochemical studies completed in the London Basin since 1963 concentrate on the eastern part of the Basin around the Lee Valley and the area south of it (e.g. Kinniburgh *et al.* 1994) and on the western end of the Basin in Berkshire (e.g. Edmunds *et al.* 1987). The line investigated by Ineson & Downing (1963) offers the unique opportunity to look at hydrochemical changes within an aquifer system on a regional scale due to changed hydraulic conditions over a period of more than three decades.

* Present address: Shell International Exploration and Production B.V., Volmerlaan 8, P.O. Box 60, 2280 Rijswijk, The Netherlands

Geology of the London Basin

The London Basin is a major asymmetrical syncline in Cretaceous and Palaeogene deposits in the southeast of England. Its axis plunges northeastwards into the North Sea. The study area lies in the central part of the Basin as indicated by the line AB in Fig. 1. The line crosses the entire confined zone of the Chalk in the Basin in a northwesterly direction, and terminates in the Chalk outcrops on both the southern and northern rims of the Basin. The dip of the northern limb of the Basin is generally less than 1° and uniform, while the southern limb has a much steeper and less uniform dip with values of up to 55°. A geological cross-section is given in Fig. 2.

The Cretaceous Chalk, which is a fine-grained, fissured limestone, attains a maximum thickness of about 240 m in the Basin, with a general thickness of about 200 m. The Chalk is covered by up to 167 m of Palaeogene deposits with a typical thickness of about 50 m. The main Palaeogene

MÜHLHERR, I. H., HISCOCK, K. M., DENNIS, P. F. & FEAST, N. A. 1998. Changes in groundwater chemistry due to rising groundwater levels in the London Basin between 1963 and 1994. *In*: ROBINS, N. S. (ed.) *Groundwater Pollution, Aquifer Recharge and Vulnerability*. Geological Society, London, Special Publications, **130**, 47–62.

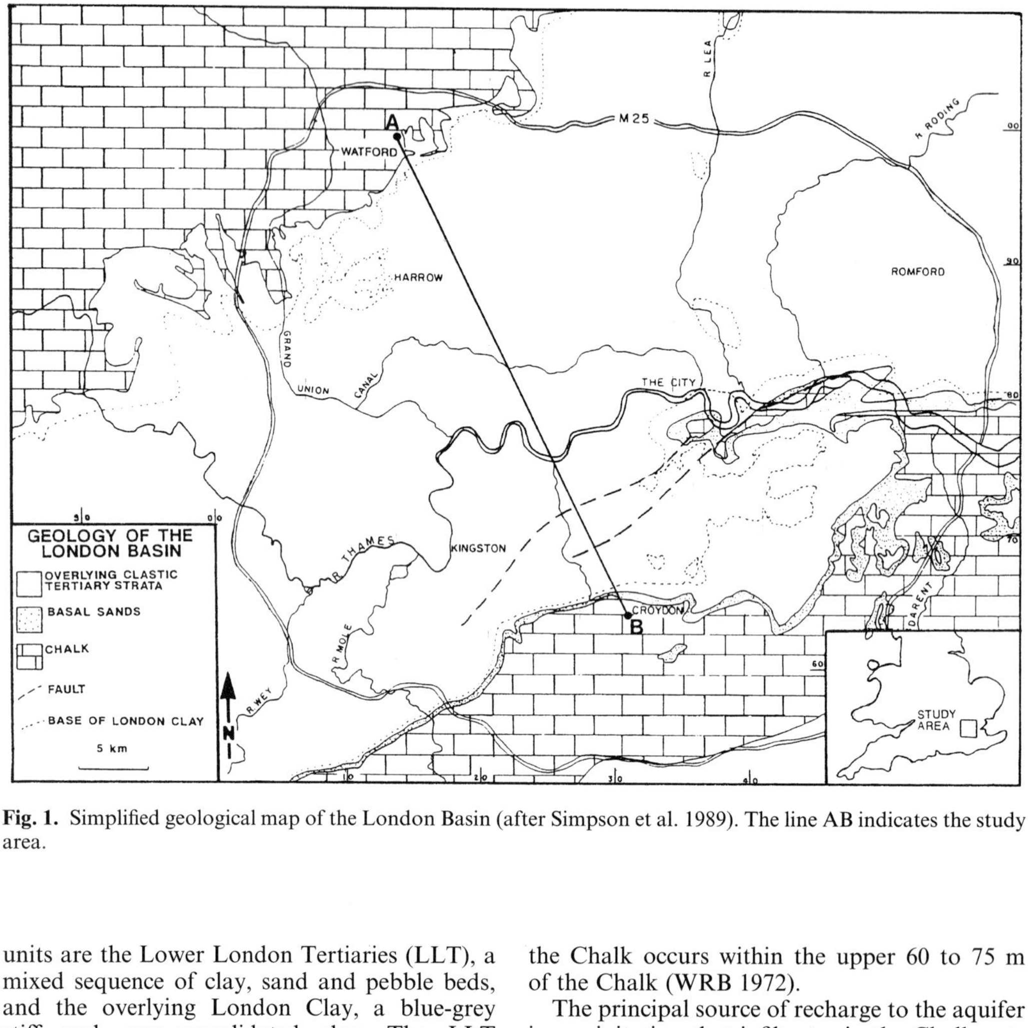

Fig. 1. Simplified geological map of the London Basin (after Simpson et al. 1989). The line AB indicates the study area.

units are the Lower London Tertiaries (LLT), a mixed sequence of clay, sand and pebble beds, and the overlying London Clay, a blue-grey stiff and over-consolidated clay. The LLT include the Basal Sands which comprise sand or coarser grade material and lie between the top of the Chalk and the first clay horizon exceeding 2 m in thickness (Water Resources Board (WRB) 1972). The Basal Sands have an average thickness of approximately 7 m in the study area.

Hydrogeology of the London Basin

The main aquifer within the London Basin comprises the Chalk which is in hydraulic continuity with the overlying sandy strata of the Basal Sands, thus forming a single hydrogeological unit. This unit is underlain by the Gault Clay and overlain by the London Clay, with both considered to be low-permeability strata or aquitards. Most of the groundwater flow through the Chalk occurs within the upper 60 to 75 m of the Chalk (WRB 1972).

The principal source of recharge to the aquifer is precipitation that infiltrates in the Chalk outcrop areas to the north and south of the Basin, and leakage of surface waters through more permeable zones of the London Clay (Morgan 1987). Estimates of the average annual recharge to the aquifer range from 90 to 140 $mm\,a^{-1}$ (WRB 1972). Under pre-abstraction conditions, the River Thames and springs along its bank were the main discharge points from the aquifer. Today, groundwater discharge is entirely accounted for by abstraction boreholes within the Basin.

Total groundwater abstraction in the confined area of the Basin has been steadily declining since its peak of $227 \times 10^6\ m^3\,a^{-1}$ in 1940 to $190 \times 10^6\ m^3\,a^{-1}$ in 1965 and around $121 \times 10^6\ m^3\,a^{-1}$ at present (Wilkinson & Brassington 1991). Groundwater abstraction has altered natural groundwater flow directions within the aquifer

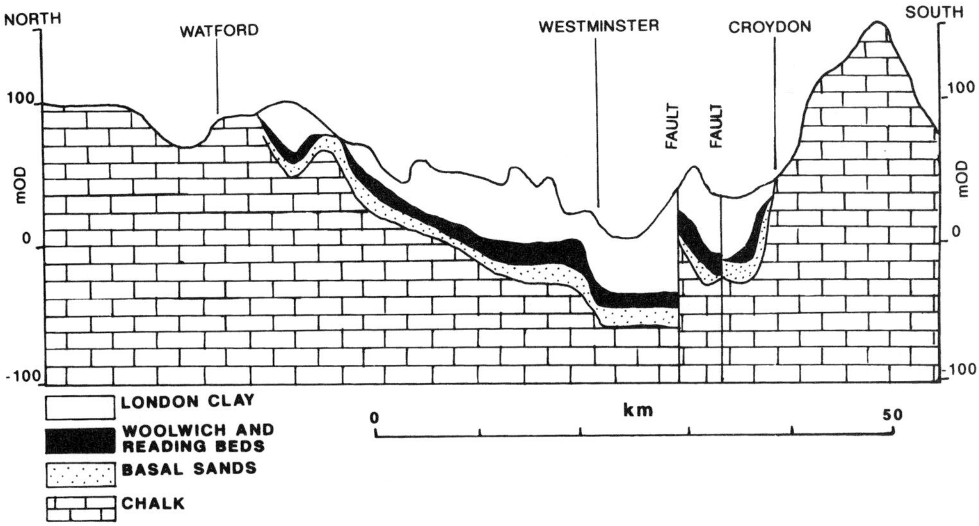

Fig. 2. Generalized geological cross-section through the London Basin (after Wilkinson & Brassington 1991).

system significantly. A maximum decline of 95 m in groundwater levels due to abstraction was recorded at Trafalgar Square in central London in 1967 (Simpson *et al.* 1989). Since the early 1970s, groundwater levels have been rising at a rate of about 1 m a^{-1} (Simpson *et al.* 1989). The NRA recorded a maximum rate of rise of 2.75 m a^{-1} in central London in 1994.

Measured transmissivity values range from up to 1100 m^2 day^{-1} along river valleys to 75 m^2 day^{-1} beneath interfluves in the outcrop area and from 370 m^2 day^{-1} to as low as 15 m^2 day^{-1} for the confined region (WRB 1972). The transmissivity of the Chalk generally decreases towards the centre of the Basin. Groundwater flow within the Chalk consists of two components which are in continuous exchange with each other. There is quite rapid flow through the enlarged secondary fissures and very slow flow through the aquifer matrix and fine primary fissure network (Price 1987). Storage coefficients for the Chalk vary locally in the Basin. In the outcrop areas values range from 10^{-2} to 5×10^{-2} and in the confined area values vary between 5×10^{-5} and 10^{-3} (WRB 1972).

Little is known about the regional variations in hydraulic conductivity of the Basal Sands. Local variations in lithology, especially the degree of consolidation and cementation, influence the hydraulic properties of the Basal Sands and the degree of hydraulic connectivity with the Chalk aquifer. Furthermore, dewatering of the Basal Sands and the subsequent consolidation may have changed its properties (WRB 1972). Values for hydraulic conductivity might be as high as 200 m day^{-1} for gravel layers and as low as 2 m day^{-1} for layers of fine sand, with an average hydraulic conductivity for the sand matrix of around 10 m day^{-1}. The average storage coefficient for the Basal Sands varies between 2×10^{-2} and 10^{-1} (Wilkinson & Brassington 1991).

Hydrochemistry of the London Basin

Regional trends

In their early study, Ineson & Downing (1963) established a sequence of five Chalk groundwater zones based on hydrochemical changes observed in the direction of groundwater flow. Variations in hydrochemistry are a function of residence time, flow conditions and aquifer properties. Leakage inputs along the flow path also affect the Chalk hydrochemistry. Generally, Ineson & Downing (1963) describe the Chalk groundwater as evolving down-gradient from a calcium bicarbonate-type into a sodium bicarbonate-type with an increase in chloride and variable sulphate content (see summary below, with reference to Table 1 and Table 2, which also incorporates the interpretation of later studies).

Zone 1. This zone comprises Chalk outcrop water characterized by increasing calcium and bicarbonate concentrations due to calcite solution under high free carbon dioxide contents, and relatively high nitrate concentrations.

Table 1. *Analytical results of Ineson & Downing (1963)*

Sample number	Site name	Na^+ ($mg\,l^{-1}$)	Ca^{2+} ($mg\,l^{-1}$)	Mg^{2+} ($mg\,l^{-1}$)	HCO_3^- ($mg\,l^{-1}$)	Cl^- ($mg\,l^{-1}$)	SO_4^{2-} ($mg\,l^{-1}$)	Total alkalinity ($mg\,l^{-1}$ as $CaCO_3$)	Carbonate hardness ($mg\,l^{-1}$ as $CaCO_3$	Non-carbonate hardness ($mg\,l^{-1}$ as $CaCO_3$	Groundwater zone after Ineson & Downing (1963)
a	Berry Grove	15	115	5	299	20	39	245	245	64	1
b	Northolt	–	–	–	335	22	–	275	275	45	2
c	Glaxo Research	105	66	27	396	71	78	325	277	0	2
d	Stonebridge Park	241	34	26	342	136	215	280	189	0	3
e	Willesden	274	27	12	361	173	241	294	116	0	5
f	Brentford	182	10	2	275	96	62	225	33	0	4
g	Lotts Road	110	46	13	327	41	74	270	168	0	4
h	Merton	21	66	15	262	14	24	215	215	12	3
i	Mitcham	12	100	3	274	15	24	225	225	35	2
j	Croydon	7	94	3	250	15	24	205	205	40	2
k	Surrey Street	13	77	4	221	13	10	181	181	18	1

Sampling locations are shown in Fig. 3

Table 2. *Summary of the five groundwater zones established by Ineson & Downing (1963), and incorporating the interpretation of later studies*

	Zone 1 Outcrop zone	Zone 2 Start of confined zone	Zone 3 Ion exchange zone	Zone 4 Reduction zone	Zone 5 Central zone
Water type	Ca-HCO_3	Ca-HCO_3	Na, Ca-HCO_3	Na-HCO_3	Na-HCO_3, Cl
Processes	Solution	Solution. Leakage from overlying strata	Ion exchange in Chalk. Leakage from overlying strata	Chemical reduction. Leakage from overlying strata	Pore-water diffusion
Cations	Increase in all species	Increase in all species, especially in Mg, K	Increase in Na and Mg, Ca low	Na increase, Ca and Mg low	Increase in all species, especially Na
Anions	Increase in all species	Increase in all species, notably in SO_4	SO_4 increase, slight HCO_3 decrease	Decrease in SO_4, increase in HCO_3	Increase in all species, especially Cl

Zone 2: This zone begins where the Chalk aquifer becomes confined. Calcite dissolution continues, with acidic run-off from Tertiary deposits supporting dissolution at some locations (Smith *et al.* 1976). Alkalinity reaches its maximum value, and all major ions show an increase in concentration. Leakage from overlying strata causes a slight rise in sulphate, magnesium, potassium and strontium.

Zone 3. A sharp decrease in the calcium content of groundwater marks the start of this zone. Magnesium increases initially, but thereafter decreases, whereas sodium concentrations continue to rise. Ineson & Downing (1963) attributed these changes in major cations to ion exchange. However, Morgan (1987) argues that there are not enough ion exchange sites within the Chalk to accomplish the observed changes in groundwater chemistry, and suggests that the increase in magnesium and decrease in calcium might be due to recrystallization of the Chalk, with the release of impurities such as magnesium. This view is supported by Edmunds *et al.* (1987) who note a similar increase of strontium along a Chalk groundwater flowpath (strontium is even more strongly rejected during recrystallization).

Sources of sodium, other than from ion exchange in the Chalk, include equilibration between pore and fissure water, and ion exchange in the Basal Sands, which are known to have a higher ion exchange capacity than the Chalk (WRB 1972).

Zone 4. This zone has the lowest measured calcium and magnesium concentrations, while sodium and bicarbonate are increasing. This change is ascribed to leakage of calcium and magnesium sulphates from the London Clay, with the formation of sodium sulphate solutions by ion exchange in the LLT and the upper part of the Chalk. The additional sulphate is subsequently reduced, with bicarbonate increasing as a reaction product.

Zone 5. Located in the centre of the Basin, this zone shows an increase in all ionic species, and is particularly enriched in chloride, presumably released from the Chalk matrix. Ineson & Downing (1963) suggested that this is a zone of mixing with original, near-marine composition saline groundwater.

Further work by Smith *et al.* (1976) and Downing *et al.* (1979) based on carbon-14 and $\delta^{18}O$ and δ^2H measurements, showed that the Chalk groundwater is of meteoric origin with an average mixing age of 10 000 years in the centre of the Basin.

Sampling and anayltical methods

Field methods and procedures

Sampling took place in June 1994 at the sampling locations shown in Fig.3. Where access to boreholes sampled by Ineson & Downing (1963) was no longer possible, the closest accessible borehole was sampled instead. Twenty boreholes were sampled, with samples 1 and 20 corresponding to points A and B, respectively, on the line AB shown in Fig.1. Sample 1 was also used as the starting point of the hydrochemical cross-sections shown in Figs 6–9.

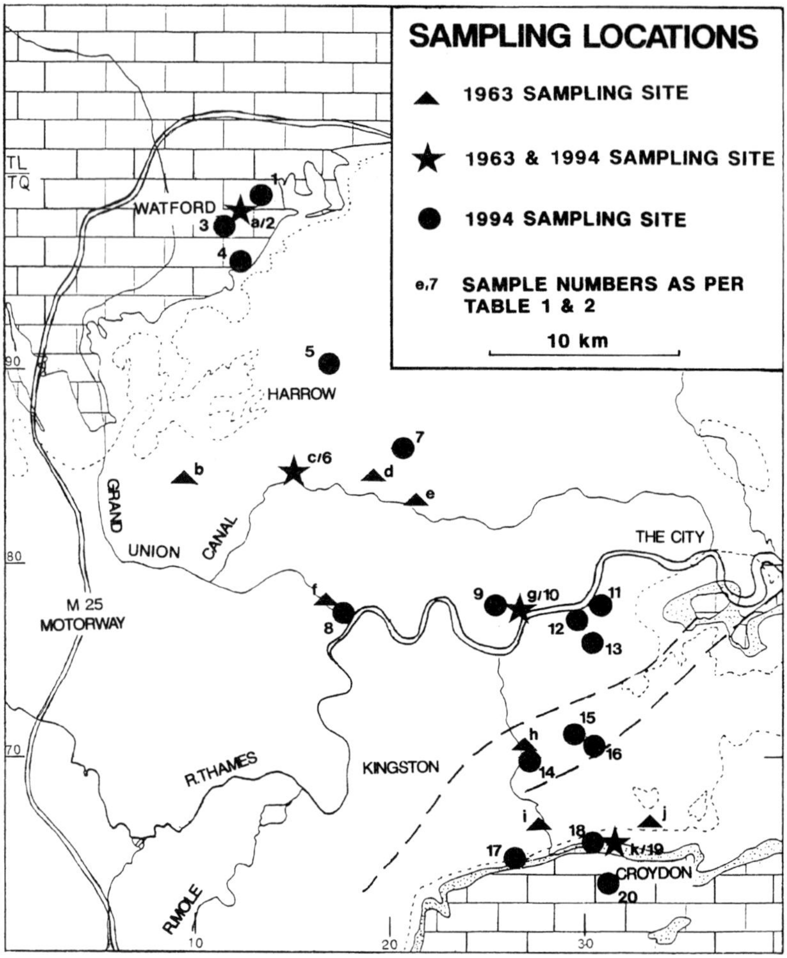

Fig. 3. Sampling locations along section AB (see Fig. 1) in 1963 (Ineson & Downing study) and 1994 (this study).

The selection of sampled boreholes included public and private water supplies, as well as some observation boreholes. Where possible in the field, a flow-through cell was attached to the borehole sampling tap to measure well-head parameters (redox potential, dissolved oxygen, temperature, pH and electrical conductivity). An alkalinity titration was carried out at each sampling site. Groundwater samples for major ion analysis were filtered using an on-line syringe system with Sartorius 0.2 μm cellulose acetate filter paper and collected in 50 ml plastic bottles. Winchester glass bottles of 150 ml volume were filled completely to the top with unfiltered groundwater for δ^2H, δ^{13}C and δ^{18}O analysis, with samples for δ^{13}C analysis poisoned with mercuric chloride solution. Samples for δ^{15}N analysis were collected in 500 ml plastic bottles.

Major and minor ion analysis

Samples for cation analysis were acidified with nitric acid prior to analysis. A Varian Spectr AA-10 atomic absorption spectrometer was used to determine cation concentrations. Boron concentrations were measured with a Thermo Jarrell Ash PolyScan 61E ICP-spectrometer. Ammonium was determined spectrophotometrically. Chloride, nitrate and sulphate were determined with a Dionex Autoion System 12 Analyser ion chromatograph. Accuracy for cation measurements was within ±10%, while anions were measured with an accuracy of ±5%. Bicarbonate concentrations were calculated from the results of the alkalinity field titration.

Stable isotope analysis

Samples for δ^2H, $\delta^{13}C$, $\delta^{15}N$ and $\delta^{18}O$ analysis were prepared in accordance with, respectively, Coleman *et al.* (1982), Bishop (1990), Feast & Dennis (1996), and Epstein & Mayeda (1953). $\delta^{13}C$, $\delta^{15}N$ and $\delta^{18}O$ isotopic ratios were measured with a VG Isogas SIRA Series II isotope ratio mass spectrometer, and δ^2H using a VG Micromass 602D isotope ratio mass spectrometer.

Isotopic ratios are reported in the standard δ-notation, where $\delta = (R_{sam}/R_{std} - 1)1000$, and R is either $^2H/^1H$, $^{13}C/^{12}C$, $^{15}N/^{14}N$ or $^{18}O/^{16}O$ of the sample (sam) or the standard (std). δ^2H and $\delta^{18}O$ values are referenced to Vienna Standard Mean Ocean Water (VSMOW), $\delta^{13}C$ ratios to the Vienna Pee Dee Belemnite (VPDB) and $\delta^{15}N$ compositions to air nitrogen. Analytical precision is ±0.7‰ for δ^2H, ±0.05‰ for $\delta^{13}C$, ±0.2‰ for $\delta^{15}N$ and ±0.03‰ for $\delta^{18}O$.

Results

A summary of the borehole locations and the well-head data is given in Table 3. All analytical results and some calculated parameters are shown in Table 4. Data for borehole 8 clearly shows signs of surface contamination.

Major cations

Calcium is the dominating cation along most of the section AB, while sodium dominates in the central part of the Basin. Magnesium is usually the third most abundant cation with its concentration approaching that of calcium at some locations. Potassium is the fourth most abundant cation with concentrations up to 17.9 mg l^{-1}.

Table 3. *Well-head parameters measured in 1994.*

Sample number	Site name	pH	Temperature (°C)	Electrical conductivity (μS cm^{-1})	Alkalinity (mg l^{-1} as $CaCO_3$)	Dissolved oxygen (mg l^{-1})	Redox potential (mV)
1	Wall Hall	7.24	11.2	815	268	–	–
2	Berry Grove	6.72	12.3	938	300	2.8	315
3	Bushey Hall	7.00	12.0	925	305	–	–
4	Bushey No.4	7.05	11.0	852	288	–	–
5	Brazie Diary	7.19	12.8	764	290	–	–
6	Glaxo Research	7.23	14.0	1180	330	0	< 125
7	Neasden	7.30	–	1500	278	–	–
8	Syon Park	6.72	11.9	1313	205	4.6	306
9	Earls Court	8.60	–	1206	271	–	–
10	Lotts Road	7.60	14.0	737	224	–	–
11	Unigate Diaries	7.35	13.1	1230	266	0.2	169
12	Battersea	7.20	–	1142	242	–	–
13	Savoy Hotel	7.00	–	1225	290	–	–
14	Merton Abbey	7.90	–	463	226	–	–
15	Streatham	7.10	–	540	233	–	–
16	Modeluxe Linen	7.30	12.6	530	225	0.6	81
17	Sutton No.2	7.08	12.5	758	240	5.0	400
18	Philips No.2	7.14	13.0	651	245	5.4	258
19	Surrey Street	7.06	12.0	741	265	3.5	239
20	Waddon	6.98	12.0	777	240	3.5	240

Sampling locations are shown in Fig. 3

Table 4. *Chemical and isotopic analysis results and calculated parameters for samples collected in 1994*

Sample number	Na^+ (mg l^{-1})	K^+ (mg l^{-1})	Ca^{2+} (mg l^{-1})	Mg^{2+} (mg l^{-1})	Sr^{2+} (mg l^{-1})	Fe_{tot} (mg l^{-1})	NH_4^+ (µg l^{-1})	B^+ (µg l^{-1})	HCO_3^- (mg l^{-1})	Cl^- (mg l^{-1})	NO_3^- (mg l^{-1})	SO_4^{2-} (mg l^{-1})	δ^2H VSMOW	$\delta^{18}O$ VSMOW	$\delta^{13}C$ VPDB	$\delta^{15}N$ Air	$SI_{calcite}$	$SI_{dolomite}$	SI_{gypsum}
1	31.0	4.3	122.6	5.9	1.0	<0.1	0	162	220	58.0	33.4	52.8	−49.21	−7.43	−14.31	10.1	0.04	−1.09	−1.65
2	38.3	4.7	135.0	8.9	1.1	<0.1	0	204	246	69.2	35.8	60.0	−49.04	−7.56	−15.32	10.7	−0.39	−1.80	−1.58
3	32.6	4.3	140.9	8.2	1.1	<0.1	0	173	250	58.4	41.1	64.8	−50.04	−7.52	−15.54	10.0	−0.09	−1.26	−1.53
4	35.0	5.6	164.0	–	–	<0.1	30	–	236	56.0	38.0	101.0	–	–	–	9.6	−0.03	–	−1.29
5	52.2	10.6	57.5	51.7	7.4	<0.1	341	205	238	35.0	0	73.0	−47.47	−7.60	−2.04	–	−0.30	−0.47	−1.86
6	95.8	15.8	78.0	77.2	8.3	<0.1	212	332	270	82.4	0	186.0	−43.21	−7.49	−1.97	–	−0.12	−0.06	−1.43
7	238.0	17.9	96.0	44.3	5.0	0.2	323	502	228	119.0	0	269.0	–	–	–	–	−0.08	−0.32	−1.20
8	48.4	13.9	168.1	28.1	1.0	<0.1	0	237	168	101.9	130.1	186.0	−49.13	−7.18	−18.00	–	−0.52	−1.67	−1.09
9	214.0	6.1	70.0	7.0	1.1	1.1	643	513	222	216.0	0	63.1	–	–	–	–	1.11	1.38	−1.85
10	100.4	6.6	39.0	11.1	1.6	0.4	742	229	184	50.0	0	87.0	–	–	–	–	−0.12	−0.62	−1.77
11	119.0	11.1	20.0	33.9	3.7	<0.1	–	227	218	142.0	0	177.0	−47.00	−7.20	−12.80	–	−0.49	−0.75	−1.23
12	91.0	8.7	117.0	26.7	3.5	<0.1	446	151	198	136.0	0	179.0	–	–	–	–	−0.10	−0.68	−1.16
13	63.4	6.4	165.0	27.5	3.1	1.1	225	90	238	125.0	1.0	161.0	–	–	–	–	−0.08	−0.77	−1.87
14	11.5	5.0	70.0	11.8	2.8	0.2	220	43	185	16.0	0.5	39.0	–	–		–	0.45	0.28	−1.94
15	10.9	4.1	78.0	11.2	2.6	<0.1	116	80	191	17.0	0	49.0	–	–	–	–	−0.29	−1.27	−1.81
16	28.0	7.2	56.0	17.0	2.6	0.1	–	60	184	18.0	0	31.0	−51.00	−7.40	−3.80	–	−0.25	−0.85	−2.13
17	15.2	2.5	118.6	2.8	0.9	<0.1	0	52	198	35.5	46.7	34.3	−50.89	−7.28	−14.38	6.8	−0.14	−1.75	−1.82
18	10.2	2.0	105.9	2.6	0.7	<0.1	0	29	201	27.6	24.1	22.1	−48.79	−7.38	−14.27	6.5	−0.10	−1.65	−2.04
19	12.6	3.3	121.6	4.0	0.6	<0.1	0	41	217	30.9	32.1	31.1	−49.23	−7.48	−15.05	5.9	−0.12	−1.56	−1.86
20	16.0	1.9	116.8	2.6	0.7	<0.1	0	46	197	40.6	30.8	34.0	−48.73	−7.36	–	6.1	−0.25	−2.01	−1.83

Stable isotope data are given in ‰. Saturation indices (SI) for calcite, dolomite and gypsum were calculated using the computer program WATEQ (Truesdell & Jones 1973). Sampling locations are shown in Fig. 3

Minor cations

Strontium concentrations varied between 0.6 and 8.3 mg l^{-1}. Fe_{tot} values were below 0.1 mg l^{-1} for most of the boreholes sampled, with higher concentrations of up to 1.1 mg l^{-1} measured in the very reducing central part of the Basin. Boron displayed an almost identical concentration pattern to that of sodium. Ammonium was only detected in the reducing, central part of the Basin where concentrations reach 742 μg l^{-1}.

Major anions

Bicarbonate is the dominating anion for almost the entire section AB. Sulphate concentrations of up to 269 mg l^{-1} exceed bicarbonate contents at some locations. High chloride concentrations are present in the centre of the Basin where a value of 216 mg l^{-1} was recorded. Nitrate concentrations decrease rapidly after entering the confined part of the aquifer. The very high nitrate concentration of 130.1 mg l^{-1} of borehole 8 is related to surface water contamination, as is probably the low chloride concentration of 50 mg l^{-1} at borehole 10.

Carbonate chemistry

Carbonate hardness and alkalinity are fairly similar for most of the boreholes. Ion-rich leakage from overlying strata and, to a limited extent, solution of non-carbonate minerals such as gypsum, lead to an increase in non-carbonate hardness down-gradient. Ion exchange in the LLT, which act as a preferential flowpath as the transmissivity of the Chalk decreases, may cause the sharp decline in non-carbonate hardness observed in the centre of the Basin. Saturation indices of calcite are generally slightly negative. Saturation indices for dolomite rise towards the centre of the Basin indicating incongruent solution of the Chalk matrix, although ion-rich leakage into the Chalk also contributes to this trend. The slight under-saturation with respect to calcite might be due to the prevailing fissure flow in the Chalk which does not allow extensive equilibration time between the aquifer matrix and mobile groundwater. Positive indices may then reflect reduced fissure flow or ion-rich leakage.

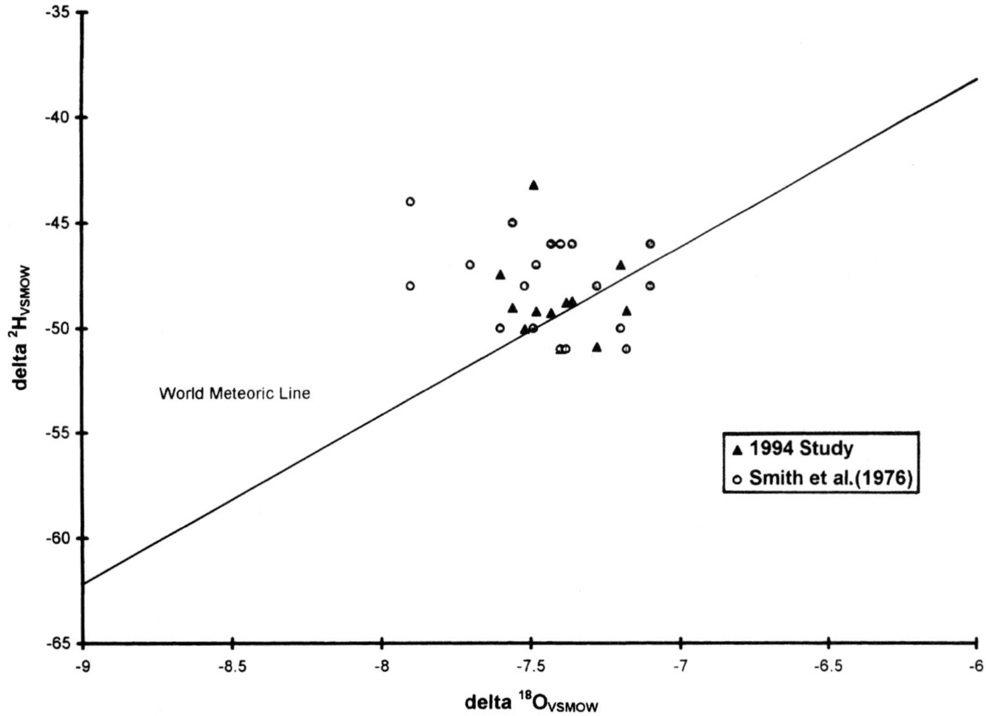

Fig. 4. δ^2H versus $\delta^{18}O$ cross-plot of stable isotope analysis results given in Table 4. The data of Smith *et al.* (1976) are shown for comparison.

Stable isotopes

The stable isotope data compare well with the data of Smith *et al.* (1976) and Downing *et al.* (1979). δ^2H and $\delta^{18}O$ ratios display only small variations and plot very close to the World Meteoric Line ($\delta^2H = 8\delta^{18}O + 10$, see Fig. 4). The close clustering of all samples suggests that recharge temperatures during the Holocene have been fairly constant.

$\delta^{13}C$ values reflect the increasing maturity of the groundwater. In the direction of groundwater flow the water equilibrates with the aquifer matrix which has a $\delta^{13}C$ value of +2.35‰ (Smith *et al.* 1976). The maximum value of −1.97‰ recorded for sample 6 suggests a groundwater residence time of several thousand years. The isotopically light value of −12.8‰ at boreholes 8 and 11 might be due to replenishment of the aquifer by river water, with this view supported by the isotopically heavy δ^2H and $\delta^{18}O$ ratios measured for these samples.

$\delta^{15}N$ values measured for samples containing nitrate in the southern outcrop area vary between +5.9 and +6.8‰ and samples from the northern outcrop area range in composition from +9.6 to +10.7‰. These values are in the range commonly associated with a soil organic nitrogen source (Heaton 1986). The slight isotopic enrichment of the samples from the northern outcrop may be partly influenced by the input of isotopically heavy nitrogen from a sewage source.

Major changes since 1963

The sharp increase in groundwater sulphate concentration is the most striking change in hydrochemistry that has occurred since 1963. Oxidation products from sulphide minerals are added to the abstracted groundwater as groundwater re-enters formerly dewatered strata (Kinniburgh *et al.* 1994). Overall there is an increase in almost all ions analysed. The rise in sodium, chloride and sulphate levels and the fairly high nitrate concentrations in the outcrop area can be linked to anthropogenic aquifer pollution.

Hydrochemical facies

No overall change has occurred in the general sequence of hydrochemical facies along the groundwater flowpath. Figure 5 shows the 1994 major ion data projected on to an expanded Durov diagram (Lloyd & Heathcote 1985) and illustrates the change from a calcium bicarbonate groundwater type in the outcrop area towards a sodium bicarbonate type with high chloride concentrations in the centre of the Basin. The sequence of groundwater composition is not the same for the southern and northern parts of section AB. Local variations in aquifer conditions seem to influence the sequence significantly. Flavin & Joseph (1983) point out the influence of varying thickness and composition of the Basal Sands on abstracted groundwater composition. Enhanced surface leakage may also alter the composition locally. Ca–Mg–Na bicarbonate groundwater with variable sulphate content, Ca bicarbonate sulphate water and Na sulphate groundwater types can be found along the transitional section of the sampled line.

Discussion

Several processes that influence the groundwater chemistry of the Chalk aquifer operate simultaneously as the groundwater flows down-gradient. The effects of these processes are generally clearer on the less steeply dipping northern limb of the Basin.

Reflushing of formerly dewatered strata

The rise in Chalk groundwater levels of up to 30 m in the study area since 1963 has now reached the Basal Sands. The Basal Sands have a high potential for sulphide oxidation when dewatered. Kimblin (1991) argues that redox conditions within the Chalk were not significantly affected by the dewatering to produce the amount of sulphate which is now apparent in abstracted groundwater. Sulphide oxidation in the Chalk aquifer would have been limited to fissure surfaces, which make up only a small fraction of the total surface area of the Chalk.

The rise in non-carbonate hardness also suggests that the Basal Sands are the source of the sulphate. Kinniburgh *et al.* (1994) state that the acidic conditions produced by the oxidation of sulphide minerals in the Basal Sands lead to the disintegration of other minerals, with groundwater entering strata that have experienced oxidation becoming enriched in non-carbonate hardness and monovalent ions derived from this mineral destruction. The comparison between potassium, sodium and sulphate concentrations in 1994 is rather close (Fig. 6), and further supports the suggestion that the Basal Sands are the main source of oxidation products.

Enhanced ion-rich leakage from the overlying London Clay as proposed by Morgan (1987) is also a potential sulphate source. An increase in

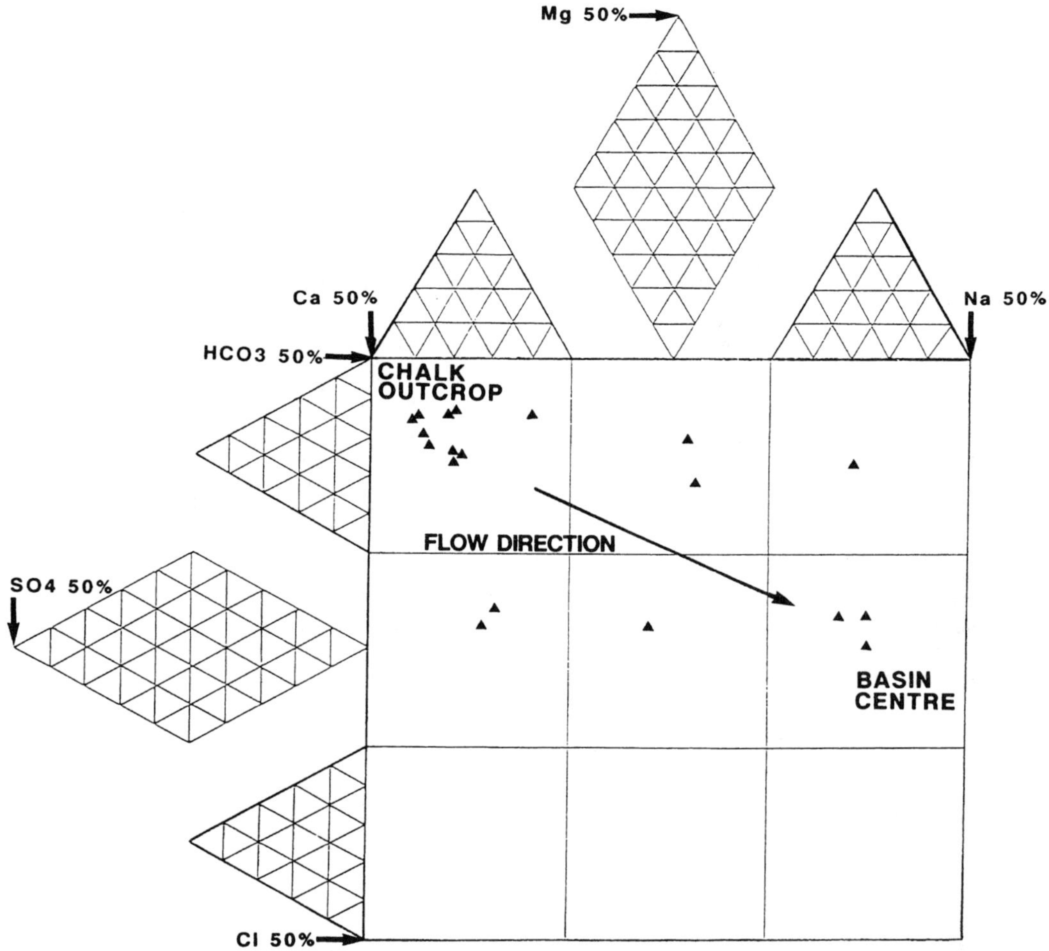

Fig. 5. Expanded Durov diagram of major ion analysis results given in Table 4. Groundwater evolves from a Ca-HCO_3 water type in the outcrop area towards a Na-HCO_3, Cl water type in the centre of the Basin. Various intermediate sub-types are found in the transitional zone.

sulphate due to solution of gypsum as described by Edmunds *et al.* (1987) can be ruled out on the basis of more or less constant saturation indices for gypsum along the section AB.

Ion exchange

The sequence of local cation maxima after Chalk groundwater has entered the confined part of the Basin (see Fig. 7) suggests that cation exchange is modifying the groundwater composition. The local peaks of magnesium, sodium and potassium are accompanied by a steady rise in sulphate. This suggests that ion exchange is taking place within the Basal Sands and not within the Chalk, as proposed by Ineson & Downing (1963), which has only a very limited amount of exchange sites (Edmunds *et al.* 1987; Morgan 1987),

Equilibration with pore water

As the residence time of groundwater in the aquifer increases, older pore-water of marine origin starts to equilibrate with the fissure water. This process adds sodium, boron and chloride to the groundwater (Edmunds *et al.* 1987, 1992). Figure 8 shows the relationship between these three elements. Sodium and boron concentrations compare very well with each other. The

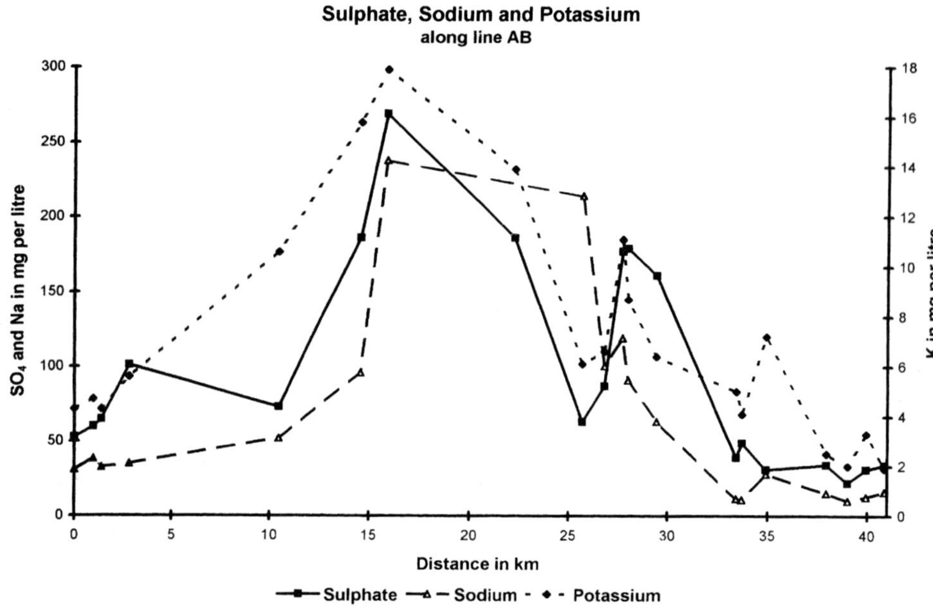

Fig. 6. Comparison of sulphate, sodium and potassium concentrations along the section AB (see Fig.1) for data given in Table 4. The Basal Sands are considered to be the main source of oxidation and mineral destruction products explaining the increase in potassium, sodium and sulphate contents of abstracted Chalk groundwater.

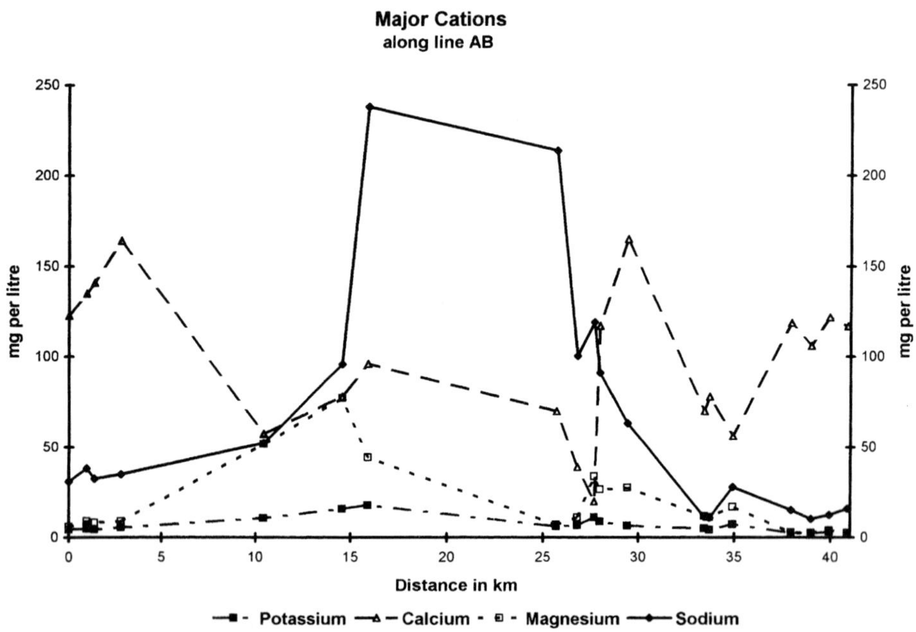

Fig. 7. Sequential peaks of calcium, magnesium, potassium and sodium (data given in Table 4) indicating ion exchange along the section AB (see Fig. 1).

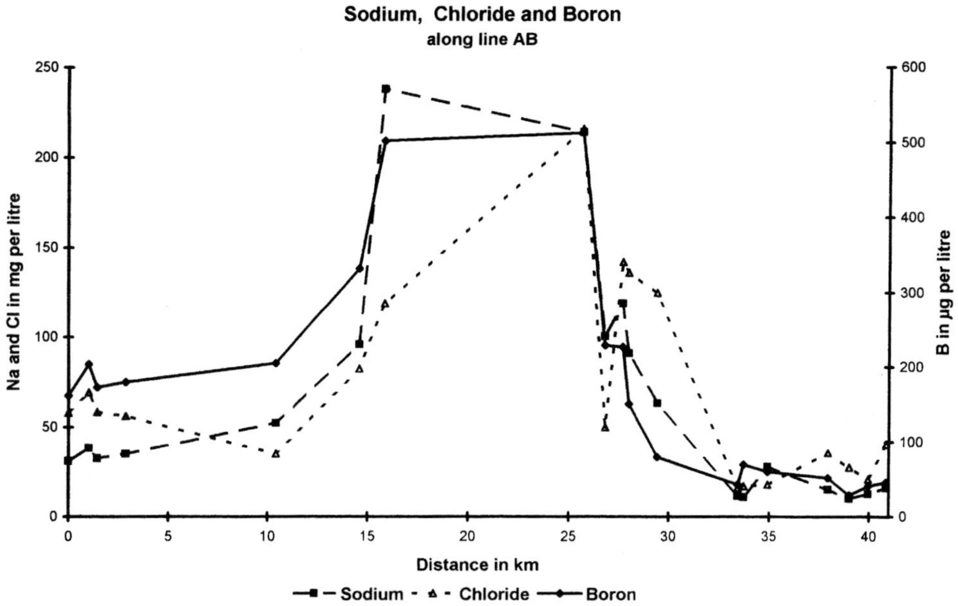

Fig. 8. Influence of pore-water on Chalk hydrochemistry along the section AB (see Fig. 1) from a consideration of the spatial pattern of sodium, chloride and boron concentrations given in Table 4.

relationship between these two elements and chloride is very close in the centre of the Basin, but is less close along the transitional sections between the outcrop areas and the centre, presumably as a result of ion exchange. The ancient pore-water in the centre of the Basin is characterized by low sulphate and high ammonium contents reflecting reducing conditions within the Chalk matrix. Borehole 9 seems to have the strongest pore-water influence. Borehole 7, however, displays a high pore-water input combined with a high sulphate content leached from the Basal Sands.

Incongruent solution of carbonates

Calcium, bicarbonate, magnesium and strontium concentrations are affected by this process. The Chalk tends to recrystallize in the presence of groundwater with a high calcium concentration, and in doing so expels impurities such as magnesium and strontium (Edmunds *et al.* 1987). This process is shown in Fig. 9 by a rise in the magnesium and strontium concentrations in the direction of groundwater flow, and reflects the continuous dissolution and precipitation of carbonate material. In the centre of the Basin sodium becomes the dominant cation and the recrystallization process decreases significantly. Magnesium and strontium reach very low concentrations as the incongruent solution of the Chalk matrix nearly ceases. Incongruent solution is also indicated by the increase of dolomite saturation indices and the gradual enrichment in $\delta^{13}C$ values.

Groundwater zonation

Three groundwater zones can be established based on the analytical results collected in 1994 (borehole 8, which has been contaminated by surface water, is not considered). The zones reflect the major changes in hydrochemistry along a groundwater flowpath, with further zonation thought inappropriate based on the processes identified. A summary of the three zones, including the corresponding sampling points, is given in Table 5.

Zone 1. This zone is located in the outcrop areas and at the very start of the confined zone. It is charaterized by solution processes which causes an increase in all ions and a saturation index for calcite close to zero. High nitrate values are present in the outcrop area.

Zone 2. This transitional zone starts shortly after the groundwater has entered the confined zone. Hydrochemical processes are diverse and

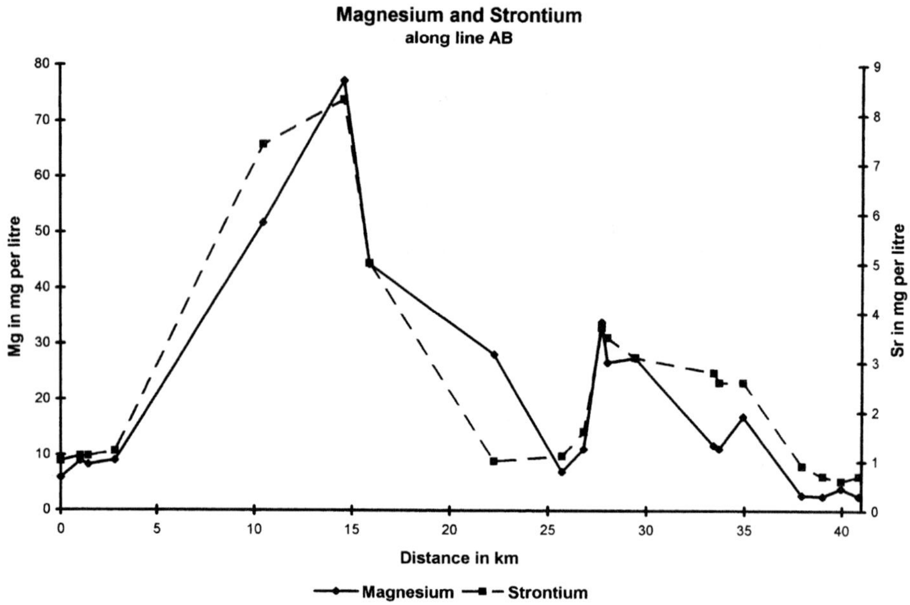

Fig. 9. Pattern of magnesium and strontium concentrations (data given in Table 4) showing the effect of incongruent solution on Chalk groundwater chemistry along the section AB (see Fig. 1).

Table 5. *Summary of the three groundwater zones identified from the data collected in 1994*

	Zone 1 Outcrop zone	Zone 2 Transitional zone	Zone 3 Central zone
Water type	Ca-HCO_3	Ca-HCO_3 Ca, Mg, Na-HCO_3 Ca-HCO_3, SO_4 Na-SO_4 Na-HCO_3	Na-HCO_3 with high Cl
Processes	Solution Aquifer contamination	Cation exchange in Basal Sands Incongruent solution of carbonates Pore-water diffusion Surface leakage	Pore water diffusion Reduction
Cations	Increase in all species	Sequential peaks of Ca, Mg, K and Na Increase in Mg and Sr Increase in Na and B	Increase in Na and NH_4 Slight increase in Ca Decrease in Mg and Sr
Anions	Increase in all species	Increase in HCO_3 and SO_4	Increase in Cl Decrease in SO_4
Effect of rising water levels	Only minor, masked by anthropogenic aquifer pollution	Large increase in SO_4, Na and K	Only minor, slight increase in SO_4
Corresponding 1994 sampling points	1, 2, 3, 4, 17, 18, 19, 20	5, 6, 12, 13, 14, 15, 16	7, 9, (10), 11

operate simultaneously within this zone. The groundwater enters strata that had formerly been dewatered. Sulphate and many other ions are leached, especially from the Basal Sands, and added to the groundwater. Cation exchange within the Basal Sands subsequently modifies the ion-enriched groundwater. Calcium is removed from solution, followed by the removal of magnesium and potassium. The final result of cation exchange is a sodium-enriched groundwater. Sodium, chloride and boron begin to enter the mobile fissure water via diffusion from old pore-water. Incongruent solution of the Chalk matrix takes place at the same time. Less pure Chalk containing magnesium and strontium is dissolved and pure calcium carbonate is precipitated. Surface leakage leads to local modifications in groundwater chemistry.

Zone 3: This zone occurs at the centre of the Basin. Chemical reduction leads to a decrease in sulphate and high ammonium contents. Diffusion of solutes from pore-water within this zone further enriches the groundwater with sodium, boron and chloride. Reverse ion exchange leading to a rise in calcium may occur within this zone. No further recrystallization of the Chalk takes place.

Conclusions

Groundwater abstraction from the Chalk aquifer has modified groundwater flow in the London Basin significantly since it commenced in the middle of the last century. A constant rise in groundwater levels has been observed within the study area since the late 1960s, with groundwater re-entering strata that had formerly been dewatered by over-abstraction. Hydrochemical parameters have been used in this study to identify different processes within the aquifer, with significant changes in the composition of groundwater being related to the changed hydraulic conditions of the last three decades.

A general deterioration of groundwater quality was noticed. Sulphate concentrations within the confined, transitional part of the aquifer have risen significantly, and alkali and alkaline earth metals have shown an associated increase. Anthropogenic aquifer pollution has produced relatively high concentrations of nitrate, sodium, chloride and sulphate in the outcrop areas. The nitrogen isotope composition of the measured nitrate suggests a slight sewage influence in the northern outcrop area.

Calcite dissolution, solution of oxidation products, ion exchange, surface leakage, incongruent solution of carbonates, redox reactions and equilibration with pore-water were identified as the processes controlling the chemical composition of the Chalk groundwater. Sulphate is mainly added to the groundwater component flowing through the Basal Sands, with only a minor contribution from the component flowing through the Chalk. Most of the ion exchange appears to occur in the Basal Sands and not within the Chalk as proposed by Ineson & Downing (1963).

The five groundwater zones identified by Ineson & Downing (1963) have been revised, and three modified groundwater zones established. A hydrochemical zone with strong connate saline water influence in the centre of the Basin as put forward by Ineson & Downing (1963) was not identified. However, a certain degree of pore-water influence, causing a maximum chloride concentration of 216 $mg\,l^{-1}$, was found.

The groundwater quality within the study area will probably be further modified as groundwater levels continue to rise, although exact predictions cannot be made without more information regarding the Basal Sands. Measurement of stable sulphur isotopes may be useful in identifying the sources of sulphate now being added to the abstracted groundwater.

The authors would like to thank Thames Water Utilities, Three Valleys Water Company and the then NRA (Thames Region) for their co-operation while carrying out this study. The views expressed in this paper are entirely those of the authors.

References

BAXTER, K. M. 1989. *Report on the 1988 baseline quality surveys, London Basin.* Thames Water Report UC **487/1**.

BISHOP, P. K. 1990. Precipitation of dissolved carbonate species from natural waters for ^{13}C analysis - A critical appraisal. *Chemical Geology (Isotope Geoscience Section)*, **80**, 251–259.

COLEMAN, M. L., SHEPHERD, T. J., DURHAM, J. J., ROUSE, J. E. & MOORE, G. R. 1982. Reduction of water with zinc for hydrogen isotope analysis. *Analytical Chemistry*, **54**(6), 993–995.

DOWNING, R. A., PEARSON, F. J. & SMITH, D. B. 1979. The flow mechanisms in the Chalk based on radio-isotope analyses of groundwater in the London Basin. *Journal of Hydrology*, **40**, 67–83.

EDMUNDS, W. M., COOK, J. M., DARLING, W. G., KINNIBURGH, D. G. & MILES, D. L. 1987. Baseline geochemical conditions in the Chalk aquifer, Berkshire, U.K.: a basis for groundwater quality management. *Applied Geochemistry*, **2**, 251–74.

——, Darling, W. G., Kinniburgh, D. G., Dever, L. & Vachier, P. 1992. *Chalk groundwater in England & France: hydrochemistry and water quality*. British Geological Survey Research Report SD/92/2.

Epstein, S. & Mayeda, T. 1953. Variation of ^{18}O content of waters from natural sources. *Geochimica et Cosmochimica Acta*, **4**, 213–224.

Feast, N. A. & Dennis, P. F. 1996. A comparison of methods for nitrogen isotope analysis of groundwater. *Chemical Geology (Isotope Geoscience Section)*, **129**, 167–171.

Flavin, R. J. & Joseph, J. B. 1983. The hydrogeology of the Lee Valley and some effects of artifical recharge. *Quarterly Journal of Engineering Geology*, **16**, 65–82.

Heaton, T. H. E. 1986. Isotopic studies of nitrogen pollution in the hydrosphere and atmosphere: a review. *Chemical Geology (Isotope Geoscience Section)*, **59**, 87–102.

Ineson, J. & Downing, R. A. 1963. Changes in the chemistry of groundwaters of the Chalk passing beneath argillaceous strata. *Bulletin of the Geological Survey Great Britain*, **20**, 176–192.

Kimblin, R. T. 1991. *Groundwater – sediment reactions in the north London aquifer system with specific reference to artificial groundwater recharge*. PhD thesis, University of Reading.

Kinniburgh, D. G., Gale, I. N., Smedley, P. L. *et al.* 1994. The effects of historic abstraction of groundwater from the London Basin aquifers on groundwater quality. *Applied Geochemistry*, **9**, 175–195.

Lloyd, J. W. & Heathcote, J. A. 1985. *Natural Inorganic Hydrochemistry in Relation to Groundwater: an Introduction*. Clarendon, Oxford.

Morgan, D. J. 1987. *Projects B1 and B2 – Final Baseline Survey Report for the London Basin*. WRc Internal Report **CO 1597–M/EV 8525**.

National Rivers Authority. 1994. *Rising Groundwater Levels in the Chalk-Basal Sands Aquifer of the Central London Basin*. National Rivers Authority, Reading, Progress Report, March 1994.

Price, M. 1987. Fluid flow in the Chalk of England. *In*: Goff, J. C. & Williams, B. P. J. (eds) *Fluid Flow in Sedimentary Basins and Aquifers*. Geological Society, London, Special Publication, **34**, 141–156.

Simpson, B., Blower, T., Craig, R. N. & Wilkinson, W. B. 1989. *The Engineering Implications of Rising Groundwater Levels in the Deep Aquifer Beneath London*. Construction Industry Research and Information Association, London, Special Publication **69**.

Smith, D. B., Downing, R. A., Monkhouse, R. A., Otlet, R. L. & Pearson, F. J. 1976. The age of groundwater in the Chalk of the London Basin. *Water Resources Research*, **12**, 392–404.

Truesdell, A. H. & Jones, B. R. 1973. *WATEQ, a computer program for calculating chemical equilibria of natural waters*. US Geological Survey, National Technical Information Service, **PB-220 464**.

Water Resources Board. 1972. *The Hydrogeology of the London Basin*. Water Resources Board, Reading.

Wilkinson, W. B. & Brassington, F. C. 1991. Rising groundwater levels – an international problem. *In*: Downing, R. A. & Wilkinson, W. B. (eds) *Applied Groundwater Hydrology*. Oxford University Press, 35–53.

Groundwater resource development and protection considerations for the Ogallala Formation in Ogallala, Nebraska

WILLIAM J. GABRIEL[1], JOHN M. MASON[1] & PAUL F. GOTTLER[2]

[1]*O'Brien & Gere Engineers, Inc., 5000 Brittonfield Parkway, Syracuse, New York 13221, USA*

[2]*O'Brien & Gere Engineers, Inc., 12600 West Colfax Avenue, Suite A-260, Denver, Colorado 80215, USA*

Abstract: The Ogallala Formation is the primary stratigraphic unit of the High Plains Aquifer which comprises the largest fresh-water aquifer in the United States. The Ogallala Formation covers approximately 450 500 km^2 extending from South Dakota to west Texas and eastern New Mexico. This study focuses on the hydrogeological characteristics and resource management of the Ogallala Formation in the type locality of Ogallala, Nebraska. For this study, the geology of the Ogallala Formation was evaluated using outcrops and sub-surface data. The hydrogeological and geochemical properties of the aquifer were also characterized to assess groundwater flow and analyte distributions. The study revealed two distinct hydrostratigraphic units separated by an aquitard. Groundwater analyses indicated that the upper unit contained high concentrations of sulphate which would preclude its use as a potable water source, but that it could be used for industrial or agricultural purposes. In contrast, water in the lower unit was of high quality, and it is currently used for drinking water supply. However, water supply well construction allowed the hydraulic connection of these two units. The study concluded that adequate aquifer characterization prior to developing this groundwater resource is essential for proper development. Specifically, the proper design of pumping wells to limit the potential for mixing of non-potable water from the upper zone with high quality water within the lower unit should be considered.

As the general population increases, so does the demand for natural resources such as fresh-water. In 1990 fresh-water use in the United States was estimated at 1285 Mm^3 day^{-1}. Of this total, 300 Mm^3 day^{-1} was obtained from groundwater aquifers. The largest contiguous source of groundwater in the continental United States is the High Plains Aquifer. The Ogallala Formation, which is the primary stratigraphic unit within the High Plains Aquifer, covers approximately 450 500 km^2, extending from South Dakota to west Texas and eastern New Mexico (Weeks *et al.* 1984).

To meet the increasing demand for fresh-water supplies, the development and protection of potable groundwater resources becomes increasingly important. An adequate understanding of the aquifer geology and geochemistry are key to this development and protection. This paper presents an evaluation of the geology and geochemistry of the Ogallala Formation in Ogallala, Nebraska, and emphasizes the importance of aquifer and land-use characterization for the development and protection of groundwater resources.

Background

The city of Ogallala, which the formation is named after, is located in Keith County, Nebraska, at latitude 41 07′ 36″ N and longitude 101 44′ 16″ W (NDEC 1992). The area surrounding the city is mainly used for agricultural purposes with cattle ranching and farming. Within the city, there are some light manufacturing facilities and commercial industries. A landfill and stockyard are located in the western portion of the city.

According to a 1990 census, the population of Ogallala is 5095. Potable water for the city is provided from nine municipal groundwater supply wells which are installed in the Ogallala Formation. Other uses of fresh-water in the area include industrial, irrigation and livestock. The Ogallala Formation aquifer constitutes the sole source of potable groundwater in the area.

Area physiography and geology

The city of Ogallala is situated along the banks of the South Platte river at an elevation of approxi-

GABRIEL, W. J., MASON, J. M. & GOTTLER, P. F. 1998. Groundwater resource development and protection considerations for the Ogallala Formation in Ogallala, Nebraska. *In*: ROBINS, N. S. (ed.) *Groundwater Pollution, Aquifer Recharge and Vulnerability*. Geological Society, London, Special Publications, **130**, 63–70.

mately 981 m (a.m.s.l.). The topography of the area is marked by a relatively flat-lying plateau dissected by the South Platte River Valley. The valley is approximately 4.3 km in width with an approximate change in elevation of 36 m. The area has a clear, dry climate with low humidity and a high evaporation rate, which is typical of moderately high plateaus.

The geology of the area consists of Tertiary (Palaeogene) and younger sediments and sedimentary rocks. In the study area, the Brule Clay, which is the upper member of the White River Group (Oligocene), occurs below the Ogallala Formation. The Brule Clay is composed of grey to pink, massive, silty clay with thin sandy zones. This formation has been interpreted as a water-laid loess (Wenzel & Waite 1941).

The Ogallala Formation, which is late Miocene and Pliocene in age, unconformably overlies the Brule Clay. Based on outcrop and borehole logging information, the Ogallala Formation consists of light to medium grey, well-sorted, unconsolidated, very fine to fine sand with traces of silt and medium sand. Sediments of the Ogallala Formation were encountered to depths of approximately 61 m during drilling in the study area. Partial calcareous cementation of the Ogallala sands occurs at various depths across the area, but these layers do not appear to be continuous across the site. A layer of red-brown clayey silt occurs at a depth of approximately 30 m within the city of Ogallala. This clay unit is approximately 3 to 6 m thick and can be correlated across the city.

Unconformably overlying the Ogallala Formation are unconsolidated sediments of Pleistocene and Recent age. These sediments are composed of silts, sands and gravels which represent terrace and flood-plain deposits. The thickness of these sediments varies from a few metres to approximately 46 m (Wenzel & Waite 1941). A schematic geologic cross-section of the study area is shown in Fig. 1.

Aquifer evaluation methods

To evaluate the Ogallala Formation aquifer in the study area, three nested groundwater monitoring wells were installed in the vicinity of, and hydraulically up-gradient of, a municipal water supply well. These wells are designated MW-1S/D, MW-2S/D and MW-3S/D. The municipal water supply well is screened at a depth of approximately 55 m to 61 m below the ground surface and is actively operated periodically during the year. City records indicate that the well can pump at rates of 0.05 $m^3 s^{-1}$ for sustained periods. Typically, this municipal water supply well operates during the summer months when the groundwater demand is greater. The shallow monitoring wells are screened at approximately 6 m to 12 m below the ground surface to assess the shallow water-bearing zone. The deep monitoring wells are screened at approximately 49 m to 55 m below the ground surface in the Ogallala Formation aquifer to assess the deep water-bearing zone. The deep monitoring wells penetrate a 3 to 6 m thick clay unit, which is encountered at a depth of approximately 30 m. During completion of the deep monitoring wells, bentonite clay was placed around the well casing at the depths of the clay unit to seal the borehole.

Hydrogeological evaluation

Groundwater elevation measurements were collected quarterly from the three nested well locations during a two-year period.

The data indicates that there is a slight upward flow potential from the deep water-bearing zone to the shallow, suggesting that the clay unit separating the shallow and deep zones acts as a confining layer and that the shallow and deep groundwater zones are not hydraulically connected. However, a downward flow potential was noted during the June and September 1994 sampling events in nested monitoring wells MW-2S/D and MW-3S/D, which are located close to the municipal water supply well. These downward flow potentials correspond to times of prolonged pumping of the municipal water supply well.

Groundwater flow maps prepared using the March 1994 and June 1994 groundwater elevation data illustrate periods of inactive and active use of the municipal water supply well on the shallow and deep water-bearing zones. The March groundwater flow data indicate that, when the municipal water supply well is not pumping, the groundwater flow in the deep and shallow water-bearing zones is to the east under a hydraulic gradient of approximately 0.002 (Figs 2 and 3, respectively). When the municipal water supply well is pumping, as represented by the June 1994 data (Fig. 4), deep groundwater flows to the northeast and the hydraulic gradient increases to approximately 0.008. Similarly, during pumping, shallow groundwater flow changes direction from the east to the east-northeast under a hydraulic gradient of approximately 0.003 (Fig. 5). The data suggest that hydraulic connection exists between

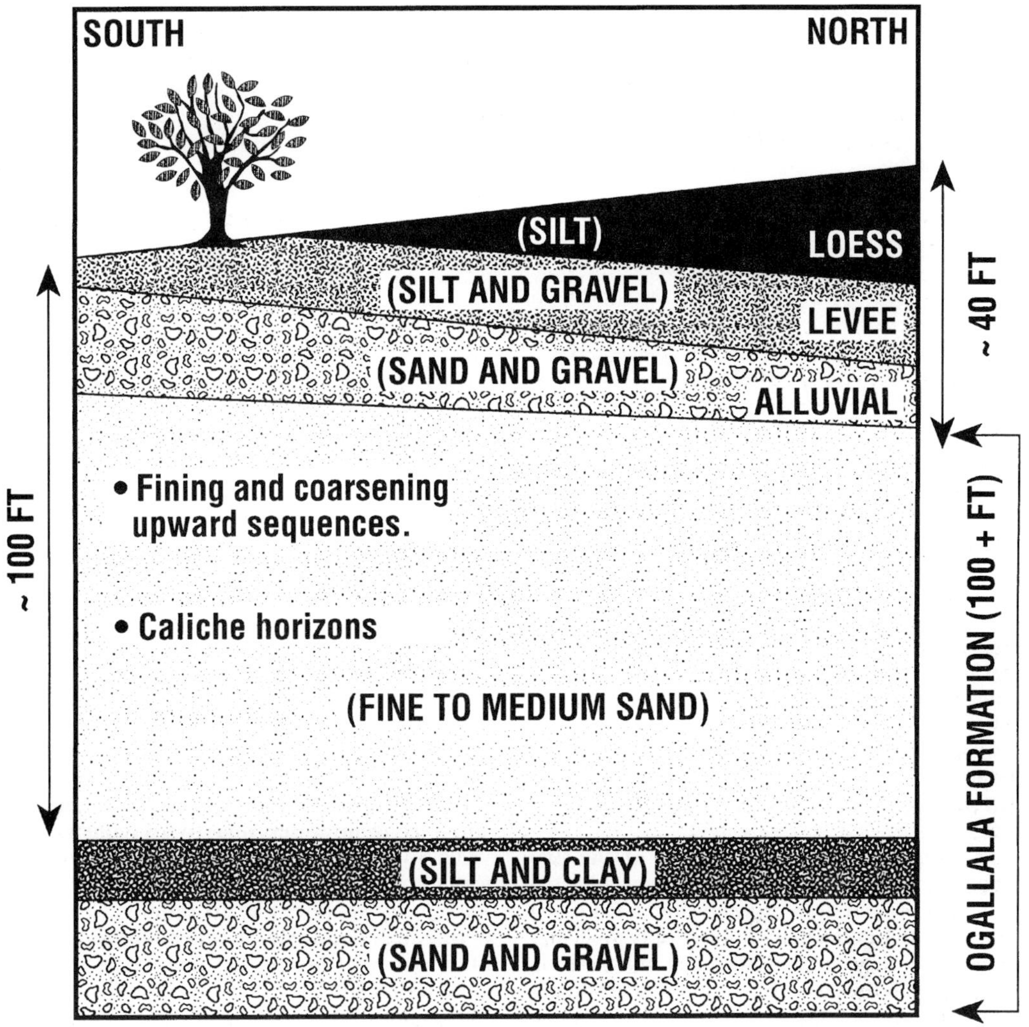

Fig. 1. Schematic geological cross section of the study area.

the shallow and deep groundwater zones in the vicinity of the municipal water supply well.

A review of historical records indicated that during construction of the municipal wells, the annular space between the well casing and the borehole, which penetrates the clay confining unit, was backfilled with gravel to the ground surface. The records do not indicate that any type of seal was emplaced at the clay confining layer to minimize mixing between the shallow and deep water-bearing zones. This type of construction hydraulically connects the shallow and deep water-bearing zones during active operation of the municipal water supply well, as shown in Fig. 6. The connection between the two zones explains the similarity in shallow and deep groundwater flow patterns exhibited during pumping of the municipal well.

Groundwater chemistry

To provide representative information on the conditions of the shallow and deep water-bearing zones, basic field parameters, such as pH, temperature and specific conductance, were recorded from the three nested monitoring well locations. The data collected quarterly from March 1994 to December 1995 were averaged and the results are summarized in Table 1.

Table 1. *Field parameter data summary*

	Groundwater parameter		
Well number	Average temperature (°C)	Average pH	Average specific conductance (μmhos/cm)
MW-1S	12.7	7.4	831
MW-1D	13.8	7.8	289
MW-2S	13.1	7.3	994
MW-2D	14.1	7.7	250
MW-3S	12.1	7.3	902
MW-3D	13.5	7.9	263

A review of the data indicates that the shallow water-bearing zone has a lower temperature, lower pH, and a higher specific conductance than the deep water-bearing zone. These data support the assumption that the clay confining unit separates the shallow and deep groundwater zones.

To further assess the conditions of the two zones, groundwater samples were collected from shallow and deep groundwater monitoring wells and the municipal water supply well and analysed for major cations (calcium, magnesium, potassium and sodium) and anions (carbonate, bicarbonate, chloride and sulphate). The analytical results for the shallow and deep water-bearing zones were averaged and the results are summarized in Table 2.

The parameter-specific data reveal that sulphate concentrations measured in the shallow zone exceed the federal drinking water standard of 250 $mg\,l^{-1}$ established by the United States Environmental Protection Agency, rendering it unsuitable as a drinking water source. The shallow water-bearing zone could, however, be used for irrigation, industrial, or other non-potable applications. The water quality in the deep water-bearing zone is considered potable by these standards and is appropriate as a drinking water source.

The data summarized in Table 2 indicate that the shallow and deep water-bearing zones are chemically different and support the conclusion that the clay unit separates the two zones. It was anticipated that the municipal well would exhibit a mixed proportion of the anion and cation concentrations due to its construction which hydraulically connects the shallow and deep groundwater zones. The data, however, do not directly support this assumption. This may be due to the volume of water pumped by the municipal well from the deep water-bearing zone, which would be significantly greater than that pumped from the shallow zone through the borehole.

Though not supported by the anion and cation data, the hydraulic connection between the two water-bearing zones is evident due to the recent detection of volatile organic compounds in the municipal wells (NDEC 1992). In the study area, these compounds were detected in the shallow water-bearing zone monitoring wells, but not in the deep water-bearing zone monitoring wells which are located adjacent to, and screened at the same depth as, the municipal water supply well intake. The detection of these compounds in the municipal water supply well suggests that the compounds are migrating through the unsealed borehole to the municipal well pump intake.

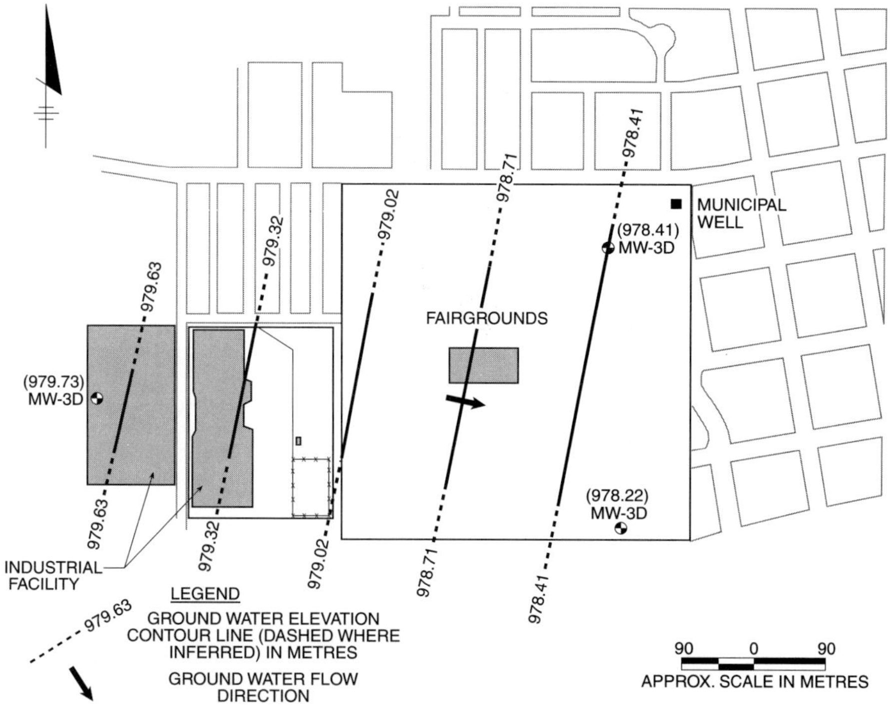

Fig. 2. Deep groundwater flow map (March 1994).

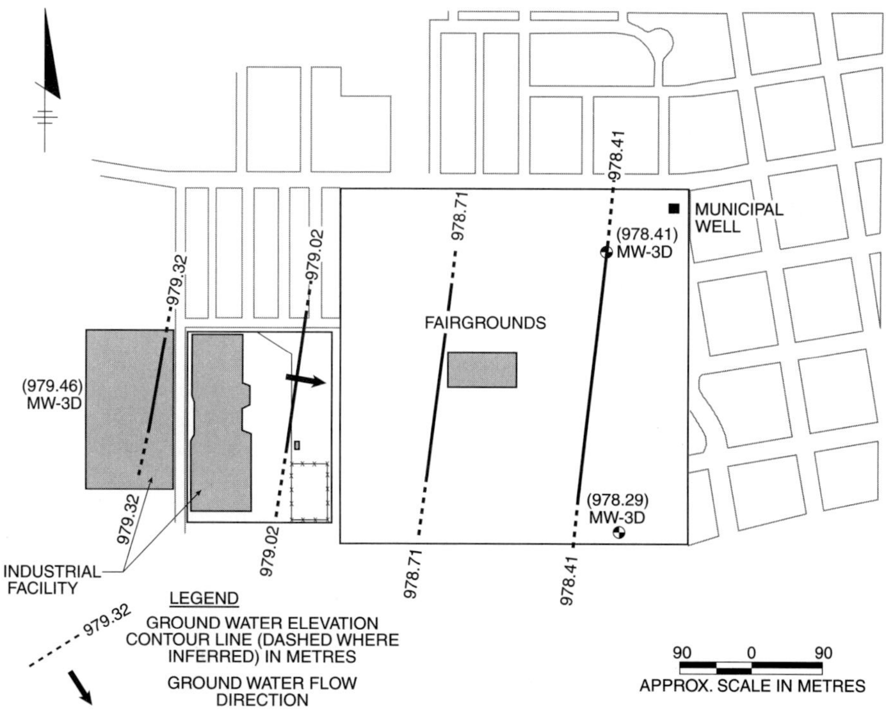

Fig. 3. Shallow groundwater flow map (March 1994).

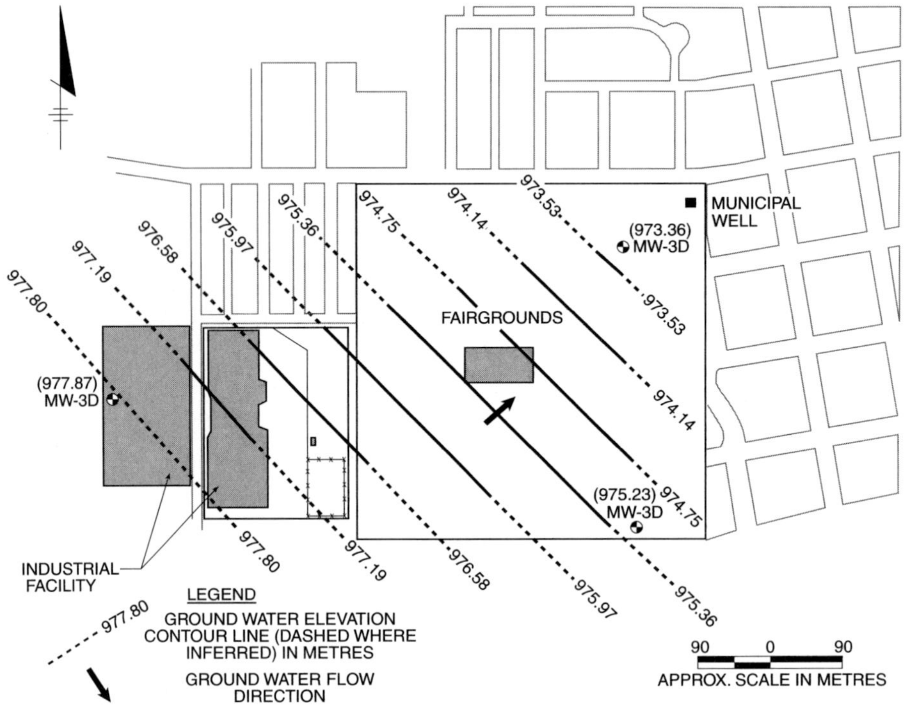

Fig. 4. Deep groundwater flow map (June 1994).

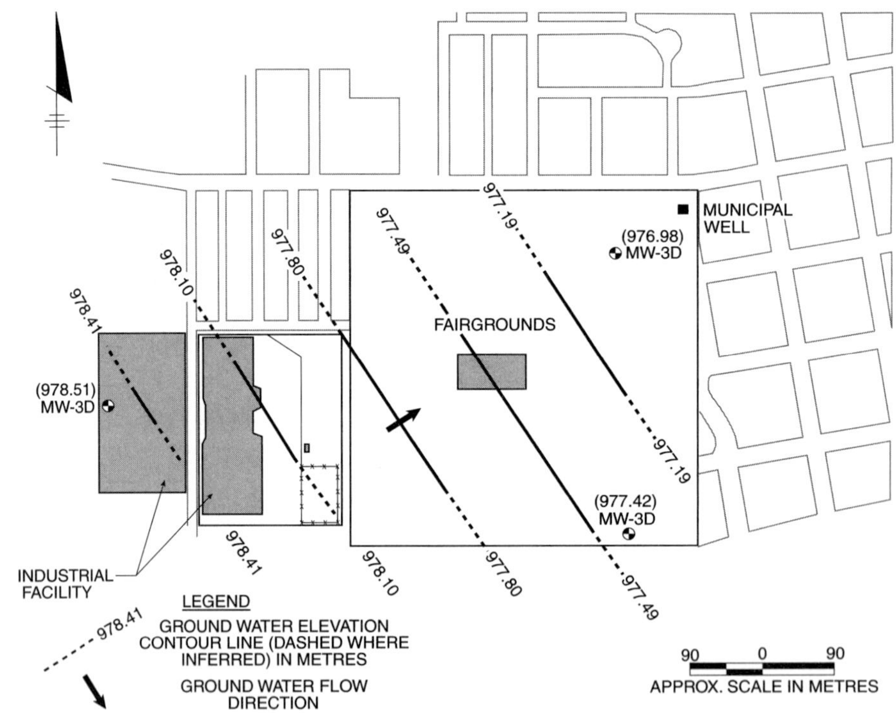

Fig. 5. Shallow groundwater flow map (June 1994).

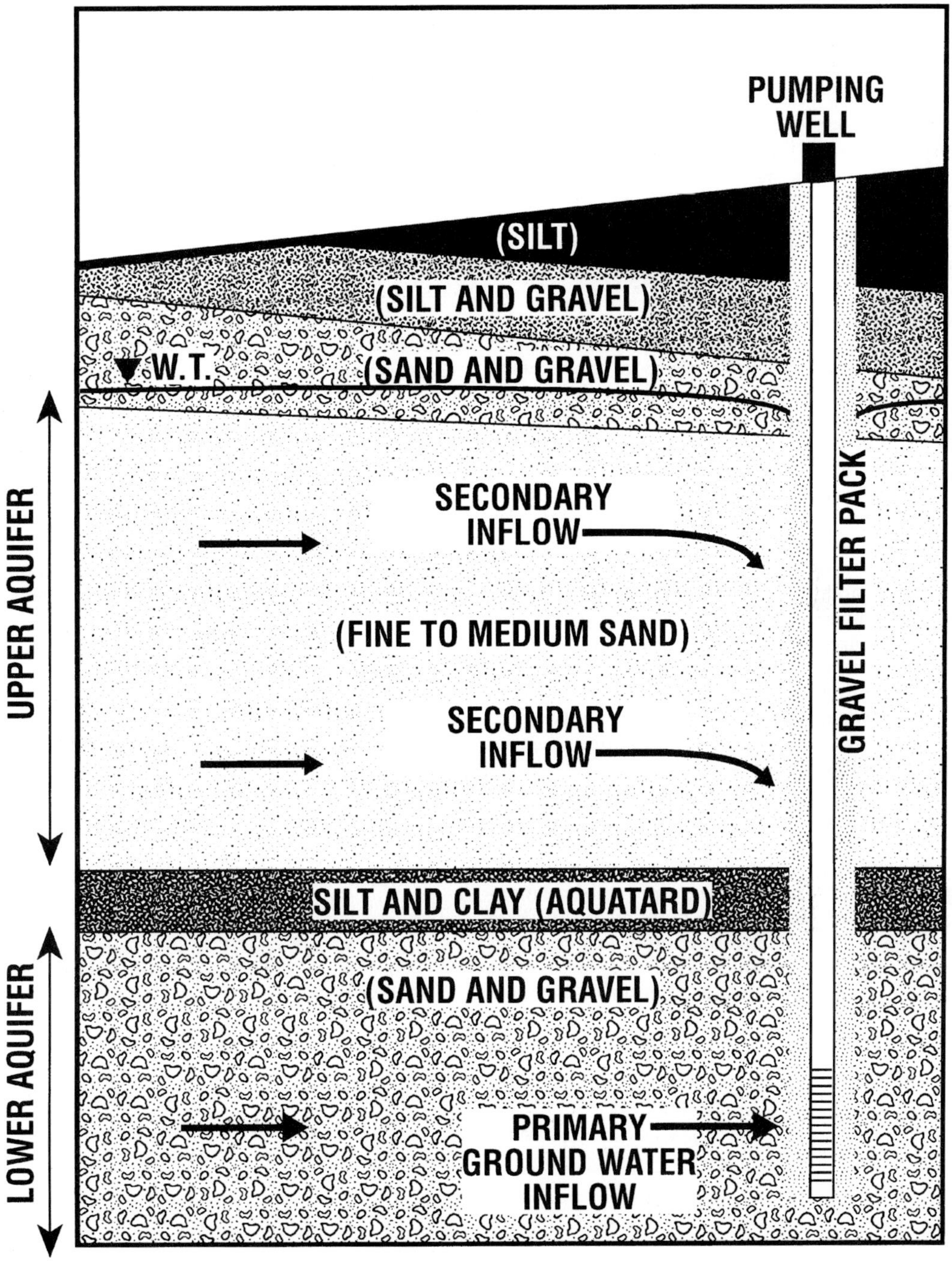

Fig. 6. Primary and secondary groundwater flow to the municipal supply well (WT, water table)

Table 2. *Inorganic analyte results*

	Cation concentration (in mg/l)				Anion concentration (in mg/l)			
	Calcium	Magnesium	Potassium	Sodium	Carbonate	Bicarbonate	Chloride	Sulphate
Shallow wells	123	22	17	138	< 1	330	39	270
Deep wells	28	8	8	23	< 1	140	6	23
Municipal well	41	8	8	18	< 1	140	4	24

Conclusions and implications

The results of this study underscore the necessity for proper aquifer evaluation prior to developing a groundwater resource. In this case study, the potential for impacting the municipal wells could have been minimized by characterizing the geology and groundwater chemistry of the aquifer prior to designing and installing the municipal water supply well system. This characterization would have provided the data necessary to modify the basic construction of the municipal wells to protect the deep, potable water source from the upper, non-potable water-bearing zone. In this manner, the deep water-bearing zone would provide a protected potable water source to the community while the shallow zone could be used for industrial and agricultural purposes.

The situation and implications presented in this study are not uncommon in developing urban areas throughout world. In areas where groundwater is relied upon as a sole source of potable water, the need for proper resource management and protection is more acute. In these sensitive areas, the introduction of any contaminants to the groundwater system may necessitate pre-treatment prior to distribution or in many cases may result in well abandonment. If the aquifer is of limited extent and capacity or additional water supply sources are not readily available, the ability of the community to expand will be restricted. For communities with limited financial resources, impacts such as these can be devastating. In most cases, groundwater impacts can be prevented or minimized through proper land-use planning and resource management. This can only be accomplished, however, if the value of the resource is recognized and protected during the initial stages of development.

References

NEBRASKA DEPARTMENT OF ENVIRONMENTAL CONTROL. 1992. *Site Investigation Report – Ogallala Water Supply*. Ogallala, Nebraska.

WEEKS, J. B., GUTENTAG, E. D., HEIMES, F. J. & LUCKEY, R. R. 1984. *Summary of the High Plains regional aquifer-system in parts of Colorado, Kansas, Nebraska, New Mexico, Oklahoma, South Dakota, Texas, and Wyoming*. U.S. Geological Survey Professional Paper 1400-A.

WENZEL, L. K. & WAITE, H. A. 1941. *Ground Water in Keith County, Nebraska*. United States Geological Survey Water Supply Paper 848, 22–23.

Aquifer susceptibility to side-effects of groundwater exploitation

B. ADAMS & A. M. MACDONALD

Hydrogeology Group, British Geological Survey, Wallingford, Oxon, OX10 8BB, UK

Abstract: Aquifer vulnerability has become a 'cornerstone' of groundwater protection policies, taking into account, as it does, aspects of pollution resulting from applied contaminant load. However, this is not the whole story and groundwater resource degradation can also result from other impacts of development such as piezometric level decline, saline intrusion and subsidence. It is generally the case that such impacts are normally diagnosed after they have occurred. Nevertheless, it is evident that certain aquifers are more susceptible to these impacts (often referred to as 'over-exploitation') than others.

As a first step towards a method to determine aquifer susceptibility to such impacts, a diagnostic method is introduced as a means to provide managers and planners with an additional method of evaluating potential aquifer degradation as a result of development.

The method, as presented, is at an early stage of development and cannot yet be used for inter- regional comparisons. Used critically, however, the authors believe it provides a useful tool for giving a first estimate of aquifer susceptibility. Planned future developments of the methodology are also presented.

'Aquifer and groundwater overexploitation is a relatively new, ambiguous and controversial concept in Hydrogeology' (Custodio 1992), and its effects are being increasingly reported in the literature (e.g. Simmers *et al.* 1992, Pulido Bosch *et al.* 1989). Given that 'overexploitation' and, hence, its various impacts are widespread, some measure of the susceptibility of aquifers to this effect would be useful. If the potential impacts of excessive exploitation can be identified before they occur, or at least before they become significant, then management measures can be taken to avoid or mitigate their worst effects.

In themselves, the component parts of the diagnostic method discussed in this paper present nothing new to the practising hydrogeologist. However, the authors believe that the integration of these parts has several benefits, mainly:

(1) it facilitates the development of a conceptual model of a region or system using concepts already available to hydrogeologists, and helps identify the hydrogeological factors important in determining resource sustainability;
(2) it provides a tool to planners responsible for regional development to enable them to assess (in broad terms) the long-term inter-relationship between such development and the regional groundwater resources.

The purpose of this paper is to present an initial step to a widely applicable methodology and to initiate discussion of the potential of such a 'tool'. For example, the concept of susceptibility could be developed in the same manner, and be as useful as that of vulnerability to pollution. The term 'susceptibility' is used to avoid confusion with the term 'vulnerability', which is now generally used specifically in the context of pollution (Adams & Foster 1992; Vrba & Zaporozec 1994). An indication of further work required in the development of this approach is also given.

Concept of over-exploitation

The concept of aquifer over-exploitation is one that is poorly defined, is possibly incapable of definition (Foster 1992) and, because a full hydrogeological assessment is generally only possible after an aquifer has been stressed, is normally only diagnosed after it has occurred. A number of terms have been used in the literature over the years to convey the concept (Adams & MacDonald 1995) but close examination shows that there is possibly no one suitable generic term. In general, definitions of over-exploitation tend to depend on a concept of 'undesirable results'; however, this will be perceived differently by the exploiter, an affected third party, the licensing authority, and environmentalists. Thus over-exploitation is a relative concept dependent upon the criteria used to define it: qualitative, economic, social, ecological etc.

Participants of a recent International Association of Hydrogeologists' Congress on the topic (Simmers *et al.* 1992) and an associated workshop concluded that while the concept of over-exploitation was useful in public relations terms, it was incapable of precise definition. In

Adams, B. & MacDonald, A. M. 1998. Aquifer susceptibility to side-effects of groundwater exploitation. *In*: Robins, N. S. (ed.) *Groundwater Pollution, Aquifer Recharge and Vulnerability*. Geological Society, London, Special Publications, **130**, 71–76.

these circumstances care needs to be taken in the use of the word, hence its use in this paper has generally been avoided.

Evidently, where excessive or irrational aquifer exploitation results in a range of differing impacts, some consideration of the collective 'result' should be made. Thus, a consistent means of evaluating the advantages and disadvantages of any particular groundwater development plan is required which is capable of equating the benefits provided by the use of the exploited water with any 'undesirable' impacts such exploitation may have. Potential for such an approach lies within economic analysis (e.g Young 1992).

Indicators of exploitation side-effects

As noted previously, over-exploitation is generally only diagnosed *a posteriori*, and discussion of many examples from various parts of the world can be found in the literature (e.g. Pulido Bosch *et al.* 1989; Simmers *et al.* 1992). From a general review of the literature it is evident that exploitation can result in a variety of side-effects depending on the degree or scale at which it has occurred. The main effects which can result from excessive or irrational development include: reduction in piezometric levels, land subsidence and deterioration of water quality.

Groundwater-level decline

It is common for over-exploitation to be defined as the condition which exists when total groundwater abstraction exceeds the recharge, giving rise to significant groundwater-level decline. However, all aquifers, whether excessively exploited or not, experience some degree of water-level decline as a response to any degree of groundwater abstraction. When recharge is greater than abstraction a transition state (during which water levels will decline) will always occur between the onset of pumping and a subsequent steady state (when water levels stabilize). The length of this transition period depends upon a number of factors including the magnitude of the difference between recharge and abstraction, and intrinsic parameters of the aquifer (Custodio 1992). It is therefore important to consider at what stage these falling water levels become serious and what level of drawdown is acceptable; this is a decision which ideally should be based on consideration of both the physical and socio-economic consequences of water-level decline. Evidently, in many situations 'unacceptable' drawdown can occur, even when recharge exceeds the total groundwater abstraction; environmental impacts, such as dewatering of wetlands and reduction of river flows can be a particular manifestation of declining groundwater levels in certain situations.

Land subsidence

As the thickness of sedimentary deposits builds up during the process of deposition, the increasing weight of the overburden compresses the underlying strata. At each stage, the system tends towards equilibrium where the weight of the overburden is balanced by the effective intergranular stress in the skeleton of the formation in combination with the pore-water pressure. Groundwater abstraction has the effect of decreasing the pore-water pressure thus increasing the effective stress from the overlying strata on the aquifer's matrix. When the increase in effective stress is greater than a critical value (the preconsolidation stress), the resulting compaction of the sediments is mainly inelastic and therefore not recoverable.

Dewatering of any sedimentary strata leads to an increase in effective stress upon the matrix. However, coarse-grained sandy aquifer strata form a rigid matrix skeleton which generally resists compaction, whereas finer-grained clayey strata are more compressible and hence more prone to compaction. Where relatively coarse-grained aquifers are bounded by fine-grained aquitards, groundwater abstraction from the coarse layers can induce leakage from the aquitards; the resulting delayed dewatering of the aquitards can result in greater compaction than occurs in the aquifer. Thus in a multilayered system consisting of coarse-grained aquifers separated by clayey aquitards, cumulative compaction of the aquitard layers can result in significant subsidence at the ground surface.

During the compaction process a state can be reached where the porosity of the original material is reduced to a point beyond which no significant future compaction will occur. The term 'stable depth' has been used to define the point at which the weight of the overburden corresponds to the loading required to achieve this state (Prokopovich 1976). Thus, stable and unstable fields can be defined as occurring below and above the stable depth, respectively. In areas where erosion processes remove portions of the uppermost 'unstable' strata, the potential for subsidence is reduced. Conversely subsidence is more probable in areas of deposition in which the entire thickness of the more 'unstable' section is preserved. This consideration has not been explicitly included in the diagnostic approach presented in this paper, but is worthy of future consideration.

Groundwater quality

The abstraction of significant amounts of groundwater can lead to the deterioration of water quality within an aquifer by such means as induction of contaminant flow as a result of a new hydraulic head distribution, saline intrusion, and geochemical evolution of groundwater.

The aspect of induced pollution is directly related to consideration of aquifer vulnerability to pollution. As evaluation of aquifer vulnerability has been widely discussed elsewhere (Adams & Foster 1992; Vrba & Zaporozec 1994), it is not included in this discussion. Indeed, the authors believe that the concept of aquifer susceptibility to exploitation side-effects will become a 'tool' similar to that of aquifer vulnerability to pollution.

Saline intrusion is an important consideration for aquifers adjacent to the sea or other saline water-bodies. The mobility of such saline water depends upon the local hydraulic gradients (enhanced by groundwater abstraction), the permeability of the aquifer and the presence or otherwise of hydraulic barriers. As with the discussion of declining water levels, a consideration of the time period involved in movement of a saline front is important; movement of a few years would normally be a matter of concern, indicating a high susceptibility to the side-effects of exploitation, but movement over hundreds or thousands of years could well be acceptable in the context of long-term management strategies.

Upconing of deeper saline water can also be a problem; groundwater abstraction can disrupt an existing hydraulic regime beyond a critical point, resulting in upconing of saline water into the exploited aquifer (Schmorak & Mercado 1969; Miroshnikov 1973).

Another way in which the intrusion of water with dissimilar hydrochemistry can affect an aquifer is to alter the physical properties of the aquifer. These physical changes can manifest themselves in changes in porosity and permeability and result from diagenetic processes brought about by water–rock interaction (Goldenburg *et al.* 1986). Such processes can be non-reversible and irrevocably damage the fabric and hydraulic properties of the aquifer.

Changes induced in the groundwater hydrochemistry due to water–rock reaction may also have detrimental health implications when the aquifer concerned is used for potable supply. One extreme example is that of arsenic and its deposition from or solution in groundwater in certain environments, depending on the local redox conditions (e.g. Korte 1991). Under natural conditions, all aquifer systems undergo redox changes as groundwater moves along flow lines. Timescales of redox change vary widely, the main control being the nature of the aquifer (amounts of reducing constituents: organic carbon, reduced iron, sulphide). Thus redox processes may be more rapid in marine argillaceous sediments than in red-bed sandstones, for example (Edmunds *et al.* 1984; Edmunds 1994). However, whilst groundwater exploitation greatly enhances flow rates, study has shown that redox zones (as well as other hydrochemical zones) are comparatively stable during quite drastic changes in the groundwater regime (Edmunds *et al.* 1984). Thus it is felt that development of diagnostics of susceptibility which involve a consideration of changes in quality as a result of movement of redox boundaries is not likely to be helpful.

A first step in the development of diagnostics

In the preceding sections various physico-chemical parameters which affect the susceptibility of an aquifer to exploitation side-effects have been discussed. This section combines some of these contributing factors to formulate a set of diagnostics to determine aquifer susceptibility to excessive exploitation. Necessarily only a rough guideline can be given since the many idiosyncrasies unique to particular locations cannot be accounted for; used critically, however, a general guideline can be a useful tool for giving a first estimate of such susceptibility.

As the side-effects resulting from aquifer exploitation can involve a variety of detrimental impacts, it was thought expedient to establish three scales of susceptibility which relate to groundwater-level decline, saline intrusion and subsidence, respectively. This approach is considered to be more pragmatic than having a single measure of susceptibility which would necessarily involve a large number of interactive variables. Groundwater pollution in its general sense is not included as this is dealt with elsewhere under a consideration of aquifer vulnerability to pollution.

The technique is based on assigning numerical values to each of up to four aquifer features relevant to groundwater decline, saline intrusion or subsidence, and then summing to give an overall grade of susceptibility to the particular impact of exploitation under consideration. For each factor, three grades are possible: high, moderate or low (Tables 1, 2 and 3). Once the grades have been determined, Tables 1, 2 and 3 can be used to attribute numbers for each parameter. As only relative values are used in the separate grading tables, the diagnostics as presented cannot be used for inter-regional comparisons; future work will need to establish ranges and weights for the separate parameters (see Conclusions). Equally, the method cannot be used for absolute assessment of the effects in detail. Users on a regional scale, however, will develop their own perception of the grading as their experience in application of the method increases. As groundwater-level decline must also precede saline intrusion and/or subsidence, it is recommended that for any aquifer the susceptibility to water-level decline is quoted first and is qualified by the other two.

Table 1. *A method for estimating the susceptibility of an aquifer to groundwater decline using four aquifer attributes*

	Index of susceptibility to groundwater decline for various aquifer attributes			
	Hydraulic diffusivity	Heterogeneity	Annual recharge	Volume of aquifer
High	3	3	1	1
Moderate	2	2	3	2
Low	1	1	6	3

Three different score bands for the sum of the indices are suggested: <8, 8–11, and >11, denoting, respectively, slight, significant and grave susceptibility to water-level decline.

Table 2. *A method for determining the susceptibility of an aquifer to saline intrusion using three aquifer attributes*

	Index of susceptibility to saline intrusion for various aquifer attributes		
	Hydraulic gradient towards the coast	Permeability	Effectiveness of hydraulic barriers
High	1	3	1
Moderate	2	2	2
Low	3	1	3

Three different score bands for the sum of the indices are suggested: <5, 5–7, and >7, denoting, respectively, slight, significant and grave susceptibility to saline intrusion. The technique assumes there to be a source of saline water near to the exploited aquifer.

Table 3. *A method for estimating the susceptibility of an unconsolidated aquifer to subsidence*

	Index of susceptibility to subsidence for an unconsolidated aquifer		
	Stratification of aquifer	Combined thickness of aquitards	Compressibility of aquitards
High	3	3	3
Moderate	2	2	2
Low	1	1	1

Three different score bands for the sum of the indices are suggested: <5, 5–7, >7, again denoting slight, significant and grave susceptibility to subsidence. NB If the aquitard compressibility cannot be estimated then aquifer stratification and aquitard thickness only should be estimated and score bands of <3, 3–5, <5 be used.

In Tables 1–3, weighting of the separate aquifer attributes is implicit in the scoring system used; see particularly annual recharge in Table 1. However, future work will need to identify explicit weighting and also specific limits for the high, moderate and low ranges of the separate attributes to facilitate widespread application of the methodology leading to inter-regional comparison of aquifer susceptibility (see Conclusions).

As an example of the use of the tables, consider Table 1. An aquifer with high hydraulic diffusivity (3), low heterogeneity (1), low annual recharge (6), and a small volume (3) will have an overall index of susceptibility to groundwater decline of 13, and thus be deemed to have grave susceptibility to groundwater decline.

Chronology

It would be extremely convenient if each excessively exploited aquifer followed the same pattern of degradation. Unfortunately this is not the case. Some aquifers are more susceptible to subsidence than saline intrusion, others will experience saline intrusion after only a small decline in water levels, and still others can experience severe dewatering without any decrease in quality or subsidence. The chronology of the impacts of exploitation depends critically on the unique conditions of individual aquifers.

Once the susceptibility of an aquifer to the three different impacts of exploitation has been determined, a more coherent chronology can be proposed. For example, if an aquifer is gravely susceptible to subsidence, but only slightly susceptible to saline intrusion, then it is reasonable to expect that subsidence will occur first, i.e. detrimental effects having the higher rating are likely to happen first.

Generally, however, groundwater-level decline is required to occur before any other detrimental effects can develop. Only after water levels have declined sufficiently will the hydraulic gradients be reversed, allowing saline water to contaminate the aquifer. Likewise, water levels have to decrease to allow leakage from the aquitards and therefore consolidation and subsidence. As a general rule, therefore, water-level decline will occur first followed by either saline intrusion or subsidence, depending on the susceptibility of the aquifer to both.

The effect of water-level decline can often manifest itself in different ways. In the first instance, there can be a general decrease in water levels, rendering pumping from abstraction boreholes less efficient. This may be followed by reduction of surface discharges, including springs or rivers drying up or the drainage of wetlands. After all these surface discharges have been eliminated, water levels can continue to decline, leading to other detrimental effects. It follows that falling water levels in the aquifer (not in pumped boreholes) are a warning to developers that further, more serious degradation might follow.

Conclusions

A first step in the development of a diagnostic approach to the determination of aquifer susceptibility to exploitation side-effects (aquifer susceptibility) has been presented. In its current form the methodology should be seen as a first step in the development of an approach with potential for wide application as a planning and management tool. As such, it should only be used with great caution for inter-regional comparisons of aquifer susceptibility.

In order for the methodology to become widely applicable, a significant effort must be made to develop specific weights for the individual parameters used in Tables 1–3 and specific ranges for the current relative designations of high, moderate and low. One way in which this could be achieved is by application of scaling techniques and the Delphi procedure (Dee *et al.* 1973); this approach is currently being considered by the authors in a research project funded by the Overseas Development Administration.

Further potential for development of the method is provided by current research being carried out by the British Geological Survey and the Natural Resources Institute (UK). This study is considering the interdependency of hydrogeology and socio-economic factors in groundwater resources degradation and the mitigation of the impacts of such degradation. In the future it may be possible to add socio-economic considerations to the diagnostic approach.

The separate elements of the approach described in this paper contain nothing new for the practising hydrogeologist. However, the combination of these elements provides a diagnostic methodology which is accessible to the non-hydrogeologist, and as such can provide an important element in the general planning process. The authors envisage that, in certain circumstances, the methodology can be as useful to planners as the concept of aquifer vulnerability to pollution, and future developments could include the production of aquifer susceptibility maps. It will probably find most application in regions having rapidly developing economies which are highly dependent upon groundwater.

The concepts discussed in this paper were developed during a project financed by the British Overseas Development Administration as part of their Technology Development and Research Programme (Adams & MacDonald 1995). This paper is published with the permission of the Director, British Geological Survey, National Environment Research Council.

References

ADAMS, B. & FOSTER, S. S. D. 1992. Land-surface zoning for groundwater protection. *Journal of the Institution of Water and Environmental Management*, **6**, 312–320.

—— & MACDONALD, A. M. 1995. *Over exploitation of aquifers – Final Report*. British Geological Survey Technical Report **WC/95/3**.

CUSTODIO, E. 1992. Characterisation of aquifer overexploitation: comments on hydrogeological and hydrochemical aspects: the situation in Spain. *In*: SIMMERS, I., VILLARROYA, F. & REBOLLO, L. F. (eds) *Selected Papers on Aquifer Overexploitation*. International Association of Hydrogeologists, Heise **3**, 3–28.

DEE, N., BAKER, J., DOBNY, N. & DUKE, K. 1973. An environmental evaluation system for water resource planning. *Water Resources Research*, **9**, No. 3, 523–535.

EDMUNDS, W. M. 1994. Indicators in the groundwater of rapid environmental change. *In*: BERGER, A. R. (ed.) *Geoindicators*. Balkema, Rotterdam.

——, W. M., MILES, D. L. & COOK, J. M. 1984. A comparative study of sequential redox processes in three British aquifers. *In*: ERIKSSON, E. (ed.) *Hydrochemical Balances of Freshwater Systems*. International Association of Hydrological Sciences, Publication **150**, Wallingford, 55–70.

FOSTER, S. S. D. 1992. Unsustainable development and irrational exploitation of groundwater resources in developing nations – an overview. *In*: SIMMERS, I., VILLAROYA, F. & REBOLLO, L. F. (eds) *Selected Papers on Aquifer Overexploitation*. International Association of Hydrogeologists, Heise **3**, 321–336.

GOLDENBURG, L. C., MANDEL, S. & MARGARITZ, M. 1986. Water-rock interaction at the mixing zone between chemically different bodies of groundwater-implications for management of sandy aquifers. *Hydrological Sciences Journal*, **31**(3), 413–425.

KORTE, N. E. 1991. Naturally occurring arsenic in groundwater of the midwestern United States. *Environmental Geology and Water Sciences*, **18**(2), 137–141.

MIROSHNIKOV, Y. 1973. *Costa limite del agua dulce en los pazos Imperfectos del litoral*. Voluntad hidraulica, ano XX, No 27, La Habana, 26–34

PROKOPOVICH, N. P. 1976. Some geologic factors determining land subsidence. *Bulletin of International Association of Engineering Geology*, Kefield, West Germany, 75–81.

PULIDO BOSCH, A., CASTILLO MARTIN, A. & PADILLA BERITEZ, A. 1989. La sobreexploitación de acuiferos. *National Congress in Almeria*, 11–14 December 1989.

SCHMORAK, S. & MERCADO, A. 1969. Upconing of fresh water–sea water interface below pumping wells, field study. *Water Resources Research*, **5**(6), 1290–1311.

SIMMERS, I., VILLARROYA, F. & REBOLLO, L. F. (eds) 1992. *Selected Papers on Aquifer Overexploitation*, Puerto de la Cruz, Tenerife (Spain), 15–19 April 1991. International Association of Hydrogeologists, Heise.

VRBA, J. & ZAPOROZEC, A. (eds) 1994. *Guidebook on Mapping Groundwater Vulnerability*. International Association of Hydrogeologists, Heise. International Contributions to Hydrogeology. No. **16**.

YOUNG, R. A. 1992. Managing aquifer over-exploitation: Economics and policies. *In*: SIMMERS, I., VILLARROYA, F. & REBOLLO, L. F. (eds) *Selected Papers on Overexploitation*. International Association of Hydrogeologists, Heise **3**, 199–222.

A geophysical investigation of saline intrusion and geological structure beneath areas of tidal coastal wetland at Langstone Harbour, Hampshire, UK

ROBERT J. McDONALD[1], NICHOLAS R.W. RUSSILL[1], MARIOS MILIORIZOS[2] & JONATHAN W. THOMAS[1]

[1]*TerraDat (UK) Ltd, PO Box 319, Cardiff, CF1 3UJ, UK*
[2]*Department of Earth Sciences, University of Wales, College of Cardiff, PO Box 914, Cardiff, CF1 3YE, UK*

Abstract: Experiments are described that test how tide and geological structure affect saline intrusion beneath an area of coastal wetland in Hampshire, southern England. Resistivity tomography and time-dependent ground conductivity surveys were carried out at two closely located survey sites. Resultant maps and sections show clear geophysical anomalies that can be attributed to tidal saline intrusion and to the position of geological structures. Results suggest that migration of saline groundwater is, in places, out of phase with the tidal cycle, and is concentrated along sub-surface tidal channels, gravel aquifers and depressions in Chalk bedrock topography. We conclude that the seawater–fresh-water boundary in the studied wetland environment is complicated by shallow-level saline flow along geologically definable routes within the Quaternary deposits above the less permeable Chalk bedrock.

In 1996, TerraDat (UK) Ltd carried out a field experiment at two locations in the Langstone Harbour area (Fig. 1) to assess the applicability of modern geophysical methods to the investigation of tidally affected wetlands. The experiment was designed to identify key features that can be targeted by reconnaissance geophysical methods for an environmental study or planned monitoring programme.

Our aims were three-fold: (i) to map the distribution and geometry of shallow saline intrusion and locate concealed geological structure; (ii) to investigate the geophysical properties of a coastal wetland environment; (iii) to investigate the relationships between the location and geometry of saline intrusion and the tide and geological structure beneath wetland.

With reference to previous investigations of saline groundwater by McNeill (1988) and Kauahikaua (1989) our experiments relied on the well-constrained relationships between groundwater and geophysical character and the noted changes in geophysical character with time of a small groundwater system (Barker 1990, 1997).

The Coastal wetland environment

Saltmarsh is a geographic definition of one type of coastal wetland which is found throughout extra-tropical regions wherever there is a combination of low wave energy and a surplus of fine sediment (Carter 1988). Wetland usually occurs landwards of the low energy, high tidal flat environment and is fed by tidal channels thus exposing the environment to fresh- and salt-water groundwater flow.

Sedimentological examples of sequences including marsh deposits have been described by Evans (1965), Evans & Collins (1975), and Hayes (1980). Wright & Platt (1995) discuss the definition of wetland by Tarnocai (1979) and state that wetland is characterized by a water-table close to the surface and occurs in both tidal and non-tidal areas. A definable groundwater boundary was thus anticipated at shallow depth beneath the survey areas. The tidal control is known from work by Allen (1992) who described sedimentation in tidally influenced marshes in the Severn Estuary, and Pye (1992) who described saltmarshes on the barrier coastline of north Norfolk. Deltaic marshes are described by Aqrawi & Evans (1994).

Ancient marsh deposits as part of a tidal flat sequence have been described by Greer (1975) and Rust (1977) and a notable example is found in the Marine Bracklesham Group and Barton Beds of the Upper Eocene to Lower Oligocene strata on the Isle of Wight (Anderton *et al.* 1979; Curry 1965; Murray & Wright 1974). Another account of similar strata is given by Richards (1994) who described the transgression of an ancient, estuarine channel and tidal flat complex.

The tidal control on saltmarsh sedimentation is investigated further by Allen (1994) who

McDonald, R. J., Russill, N. R. W., Miliorizos, M. & Thomas, J. W. 1998. A geophysical investigation of saline intrusion and geological structure beneath areas of tidal coastal wetland at Langstone Harbour, Hampshire, UK. *In*: Robins, N. S. (ed.) *Groundwater Pollution, Aquifer Recharge and Vulnerability*. Geological Society, London, Special Publications, **130**, 77–94.

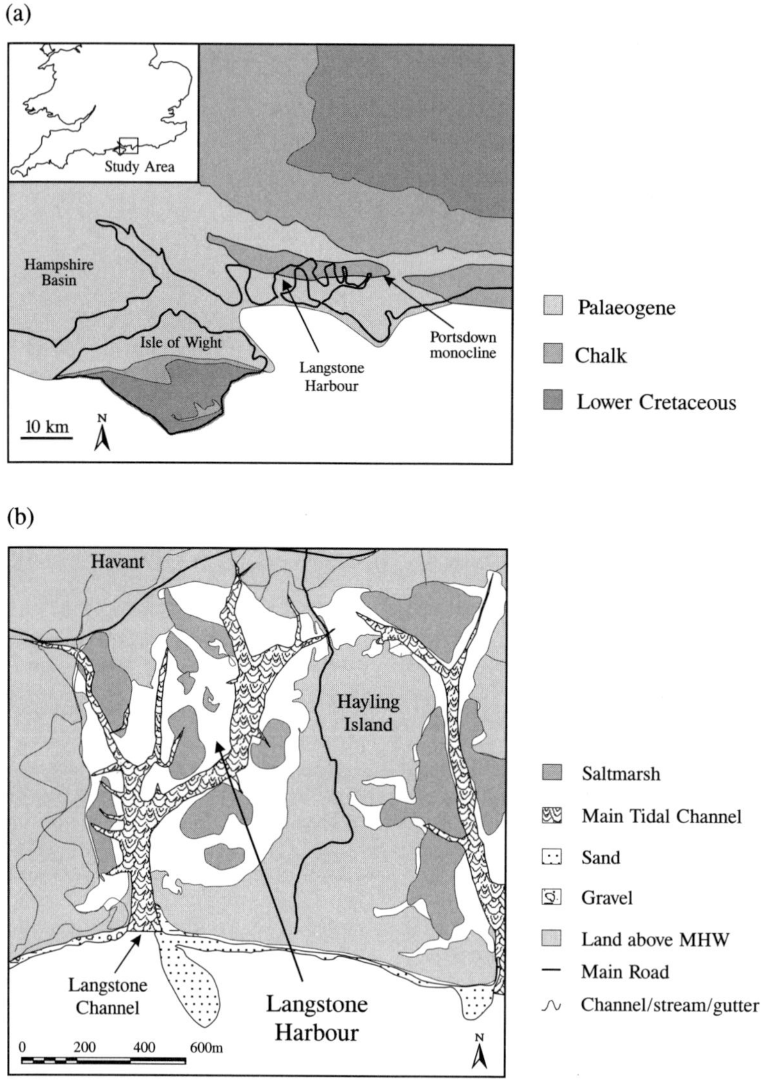

Fig. 1. (**a**) Location and geological setting of the study area. The specific locations of the two survey sites are not shown due to local sensitivities. (**b**) Geomorphological map of the Langstone Harbour area.

describes an exploratory model for the surface hydraulic and sedimentary regimes of tidal saltmarshes and suggests that tidal range as well as sediment type control the character of channel networks. Our paper sub-divides the tidal and channel controls on groundwater by analysis to reconstruct a reliable geophysical interpretation.

In addition to the geological data above, examples of allied work include an estuarine study by Reed (1989) on patterns of sediment deposition in subsiding coastal saltmarshes in relation to winter storms, and the study of Stoddart *et al.* (1989) on tidal control on marsh sedimentation and accretion. Our experiment was carried out during favourable weather conditions; however, a more extensive study should include work to assess the effects of storm recharge.

The study area and timing are characterized by flood tide seawater influx via channels, low wave energy, location by a modern harbour, dry vegetated ground conditions and some nearby human developments. In addition, vegetation on site shows fresh-water to seawater zonation.

Groundwater saline intrusion

Much work has been done in marine chemistry, on the salinity of the coastal seawater column and the chemical and physical parameters controlling salinity which are beyond the scope of our paper. A simple definition of saline water is, however, given by Owen *et al.* (1991) who stated that saline water in intrusions is generally described as having a chloride concentration in excess of 300 $mg\,l^{-1}$. Seawater has a chloride concentration of about 19 $g\,l^{-1}$ so that only a small volume of seawater is required to cause significant pollution of fresh groundwater.

Previous studies of shallow ground salinity have primarily targeted the root zone and plant zonation, e.g. McNeill (1990) after Wood (1987). Carter (1988) referred to groundwater saline intrusions and briefly described the position of related coastal groundwater interfaces. Carter (1988) stated that severe and problematic local increases in groundwater salinity can occur where coastal wetlands are drained. Thus, the significance of geophysical measurement of groundwater movement is here emphasized for arable areas and conserved natural environments.

An outline of the process of saline intrusion into the fresh-water front is described mathematically by Bear & Verruijt (1987) who stated obvious limitations to any model based on the assumptions of thin and stationary brackish transition zones occurring between seawater and fresh-water. Recent saline groundwater plume mapping using electromagnetic methods by Goldstein *et al.* (1990) and resistivity investigations of salt water intrusion near a major sea-level canal by Hagemeyer & Stewart (1990) and related geophysical work by Rigby-Jones *et al.* (1997), van Overmeeren (1989) and Stesky (1986), show that valuable information for static models of saline intrusion can be obtained using mapped electromagnetic conductivity data and resistivity tomography.

Studies have addressed the delineation of the saline groundwater boundary (Schnoebelen *et al.* 1995); the hydrochemical characteristics of a broad transition zone of seawater intrusion (Xue *et al.* 1993); the effects of salt-water intrusion on a coastal block including rocks of different porosity (Calvache & Pulido-Bosch 1994); and comprehensive evaluations of groundwater flow and transport problems using geostatistics (Poeter & McKenna 1995). A prerequisite of our experiment is that the acquired electromagnetic conductivity and resistivity data can be modelled hydrogeologically for a non-static system (e.g. Taylor 1992).

Location and geological setting of the field area

The study area lies on the northern margin of Langstone Harbour (Fig. 1). The area is within the Palaeogene sediment area of the Hampshire Basin that contains mainly Upper Palaeocene to Middle Oligocene strata. The succession is about 650 m thick and contains a number of transitional marine to non-marine sedimentary facies such as those exposed along the coastal sections of the Isle of Wight (Osborne White 1915; Melville & Freshney 1982; Edwards & Freshney 1987). Controlling factors on the evolution of these rocks are described by Plint (1982, 1988) in his study of the Eocene sequence of Hampshire.

The large-scale geological structure of Hampshire records a Palaeogene tectonic history. The field area rests on the northeastern limb of the asymmetric Hampshire syncline. Bedding generally has a gentle dip, usually 3° towards the south. A large number of long-lived faults have been documented in addition to the local WNW–ESE-trending Portsdown Monocline (Stoneley 1982) which contains an inlier of Upper Chalk at its surface axial trace. The regional unconformity between the Cretaceous Chalk and Palaeogene strata is well-known and is probably related locally to Palaeogene faults (to the inversion of the Mesozoic Channel Basin, a part of the major tectonic event which affected much of Western Europe) as well as a complex depositional and erosional Quaternary history (Osborne White 1915; Melville & Freshney 1982; Edwards & Freshney 1987).

Geological data specific to the study area were acquired from three boreholes drilled to rock-head at Survey Site 1, while invasive investigation was restricted to augured water sample holes at Survey Site 2. Chalk bedrock was found at depths of about 5 to 6 m in the three boreholes (Fig. 2). From the regional geology we anticipated the erosional contact between the veneer of wetland sediment and Chalk to be close to or part of the buried topography of the Palaeogene old land surface upon the Portsdown Monocline.

Consideration of the regional irregularity of the Cretaceous–Palaeogene contact leads to particular attention being paid to the occurrence of any isolated anomalies adjacent to an irregular boundary. Such anomalies could represent preserved pockets of Palaeogene strata, their presence implying that a Holocene or Quaternary land surface is superimposed on or below the old land surface.

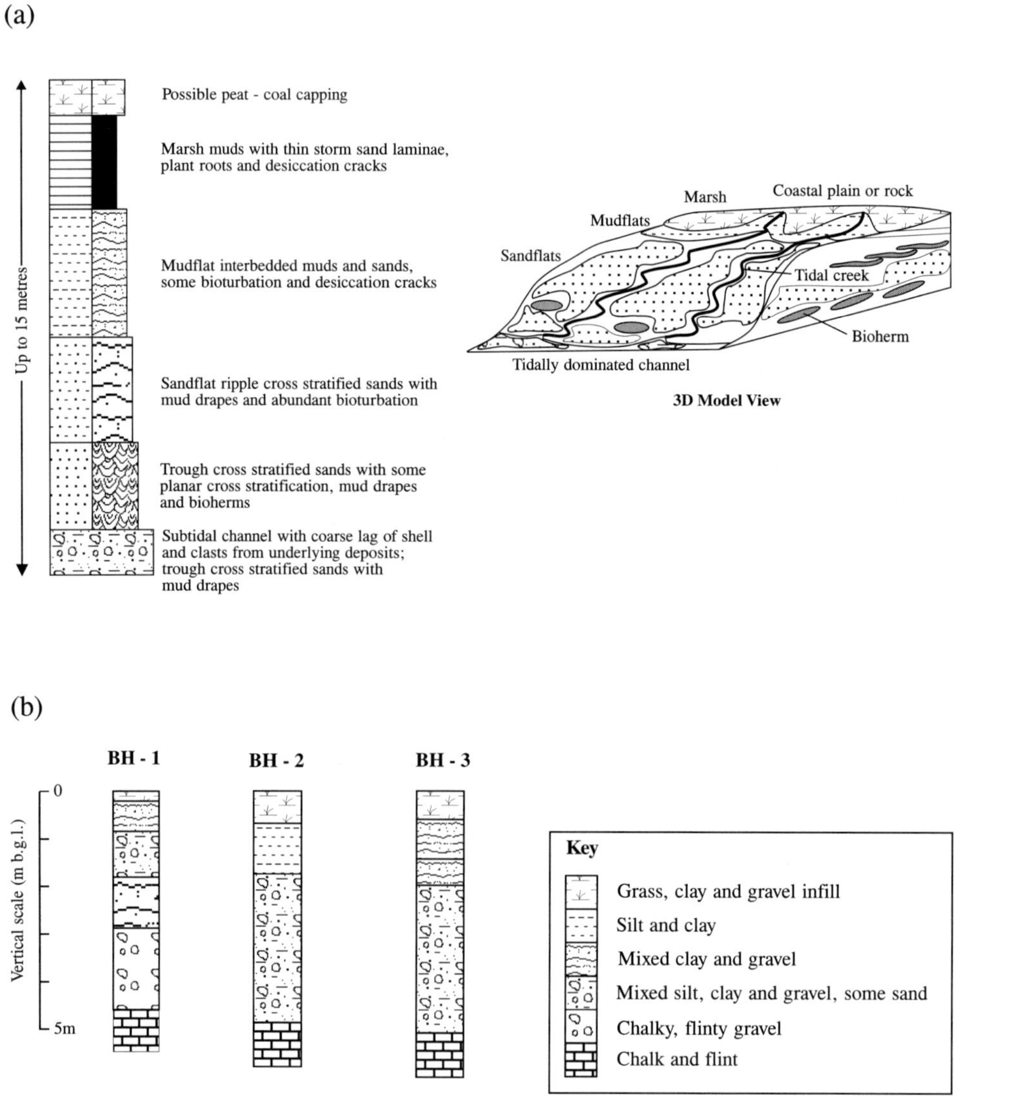

Fig. 2. (**a**) Summary composite log and 3D block model of a typical wetland environment. (**b**) Geological data derived from three shallow boreholes drilled to rockhead at Survey Site 1.

Both survey sites are covered by vegetated soil. The underlying sediments locally comprise mixed wetland subsequences (Fig. 2). The sequences consist of clays, silty clays, mixed clays and gravel, silts, sands, gravels, and a basal gravel containing Chalk and flint clasts. The Chalk–sediment boundary is interpreted to be erosional.

Groundwater levels were monitored for tidal influence and shallow-hole water samples were taken and analysed to assess the conductivity of local waters where geophysical anomalies had been identified.

Geophysical field methods

Survey equipment

Measurements of apparent ground conductivity in units of millisiemens per metre ($mS\,m^{-1}$), were carried out with a Geonics EM34-3 (two-operator instrument) and an EM31 (one-operator instrument). The instruments consist of a dipole transmitter coil (Tx) and a dipole receiver coil (Rx) linked by a lightweight shielded reference cable in the case of the EM34 and contained within a tube in the EM31. The transmitter coil is energized with an alternating current that gives rise to a time-varying primary electromagnetic (EM) field. The primary EM field induces eddy currents within sub-surface conducting bodies in the range of the instrument. The eddy currents in turn generate secondary EM fields which distort the primary field measured at the receiver. The ratio of the secondary to the primary magnetic field is linearly proportional to the ground conductivity. Apparent ground conductivity is read directly via an analogue display on the EM34 and digitally datalogged on the EM31.

Resistivity tomography surveys were carried out using an ABEM LUND resistivity system. Use of the system involves measuring apparent resistivities (Ω m) via a linear array of electrodes linked by multichannel cables to a computer-controlled switching system. Depths of investigation are dependent on selected electrode separation, i.e. to increase depths of penetration greater electrode spacings are required. Results are contoured to produce an apparent resistivity pseudo-section prior to data processing and modelling.

Shallow boreholes were drilled using an Atlas-Copco Cobra vibrocorer which facilitated coring sediments to rockhead with little environmental impact.

Survey methodology

In order to test the effect of tide on local saline intrusion at Survey Site 1, a time-dependent experiment was carried out to target any associated lateral change in ground conductivity. The survey involved taking comparable measurements of apparent ground conductivity on a regular 10 m × 5 m spaced grid at the extremes of the tidal cycle over two close areas. A regular grid (Swan & Sandilands 1995) was chosen, as opposed to a uniform grid, because of topographic favourability and the need for rapid data collection due to the tide dependency. The Geonics EM34 instrument was configured to target depths to 15 m below ground level. The horizontal dipole configuration was chosen as it is more sensitive to shallow conductivity variations, and also to improve the range of proportional linearity between measured apparent conductivity and real ground conductivity. Using horizontal dipoles this linearity remains valid up to true ground conductivities of about 300 mS m^{-1}. Such values were expected for conductive saline intrusion, based on specific conductivity and ion concentration profiles (Cl^{-1}, SO_4^{-2}) given by Hagemeyer & Stewart (1990).

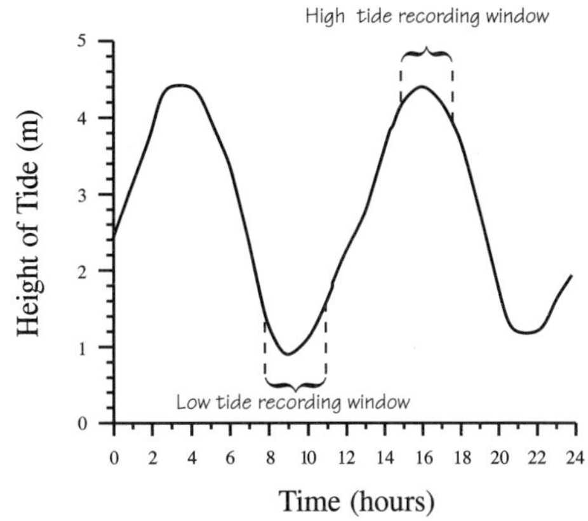

Fig. 3. The average spring tidal cycle during the survey period with 3 h recording windows marked and centred on low and high water.

Measurements of apparent ground conductivity were taken during 3-h time windows centred on high and low tide. These time windows were selected because relatively little variation in sea-level occurs within each window and maximum difference in sea-level occurs between the two separate periods of acquisition (Fig. 3). In order to emphasize tidal influence, the survey was carried out during a period of large tidal range. To ensure data reliability, sufficient distance was maintained between the survey grid and any possible sources of electrical noise.

At Survey Site 2 an experiment was carried out using a Geonics EM31 to locate and map any shallow intrusions of saline water. While the EM31 instrument only penetrates relatively shallow depths (about 6 m), it has the advantages of higher lateral resolution and more rapid acquisition than the EM34 instrument. Restricted access time prevented carrying out a time-dependent experiment; however, the main interest was in targeting long-period contamination of the farmland within a depth range applicable to possible vegetation impact.

Main anomalous features identified from the EM conductivity surveys carried out at both survey sites were targeted by resistivity tomography surveys to investigate their cross-sectional geometry. The ABEM LUND multichannel resistivity system was used because of the need for rapid acquisition to avoid tidal effects on the data. Electrode spacings and instrument settings were varied depending on anticipated target geometry. Smallest spacings of 2 m and 2.5 m were used to investigate depths to 20 m and ensure sufficient resolution of anticipated geological structure.

Data processing and presentation

Apparent conductivity data were processed using GEOSOFT. Data were gridded using a minimum curvature method. Calculations to produce the gridded datasets are based on the inverse distance average of the actual data with iterative recalculations carried out until 99% of the data fit within 1% of the data range. The data were gridded with cell sizes smaller than the Nyquist spatial sampling interval appropriate to the individual dataset.

Three datasets were produced from the time-dependent experiment: (i) apparent conductivity values at high tide; (ii) apparent conductivity values at low tide; and (iii) the difference between high and low tide apparent conductivity values. Maps presenting the data overlain by geographical features were produced from digitized site plans imported to GEOSOFT.

Measured apparent resistivity data were converted to pseudo-section images based on the current and potential electrode separations ('n' level) and the location of the centre of the active electrodes. The pseudo-sections represent measured resistivity values that are apparent in terms of magnitude, location and depth. Modelled resistivity distributions of the sub-surface were then derived from finite-difference forward modelling of the apparent resistivity data using RES2DINV software.

The forward modelling process involves determining a two-dimensional (2D) simulated model of the sub-surface, then comparing this calculated model with the observed data. Convergence between theoretical and observed data is achieved by non-linear, least squares optimization. The procedure is smoothness-constrained to improve stability in the iterative process, the degree of smoothness being determined by a user-specified damping factor. The extent to which the observed and calculated theoretical models agree is an indication of the validity of the interpreted resistivity model indicated by the final root-mean-squared (r.m.s.) error. The final r.m.s. errors achieved for the current resistivity data modelling were all less than 5%.

The interpreted resistivity models are presented as contour plots of the resistivity distribution. To facilitate correlation, most of the resistivity sections have been plotted using the same scaling. The 2D method of presenting resistivity data assumes infinite strike extent and is limited when irregular or complex geological features are present, when the simple cross-section may not be indicative of the true geometry. However, the types of features targeted by this survey, e.g. saline fronts and channels, are often laterally continuous and can be reliably imaged in cross-section.

Results and discussion

Distribution and geometry of saline intrusion and shallow geological structure

Survey Area 1. A contour map of apparent conductivity data collected during the low tidal cycle is shown in Fig. 4a. The map represents the combined effects on ground conductivity of sub-surface saline intrusion, changes in shallow geology and occasional spring-tide flooding. Apparent conductivity ranges from 15 to 180 mS m^{-1} and the following areas can be identified. The highest values of 100 to 180 mS m^{-1} are confined to a zone about 50 m wide along the land side of the coastline. Landwards of this coastline zone is a cuspate band of intermediate apparent conductivity (45 to 90 mS m^{-1}) with a maximum width of about 60 m. Further landwards is an area of low apparent conductivity which represents a relatively stable area.

In general, Fig. 4a shows a landward decrease in apparent conductivity away from coastal features of about 0.75 mS m^{-1} per metre. This conductivity gradient is attributable to the presence of increasingly saline groundwater and possibly increasing clay-rich sediment thickness towards the coastal belt.

A resistivity survey traverse line was positioned to investigate the observed seaward increase in apparent conductivity as mapped by the EM34 survey. The modelled resistivity distribution (Fig. 4d) reveals the geometry of a near-surface wedge-shaped body which thins landwards. The body thickness decreases from about 12 m near the coastline to less than 6 m about 200 m inland and is interpreted as representing the wetland sediments overlying more resistive Chalk bedrock. Its base is less than 12 m below the surface and has a gently undulating morphology. Information from a nearby borehole (Fig. 2b; Borehole 2) indicates that Chalk bedrock is found in the region of 5 m b.g.l. inland.

Interpretation of the combined conductivity and resistivity data suggests that, generally, local shallow saline intrusion is focused through a thin wedge of sediments overlying less permeable Chalk bedrock. There is an apparent non-linear decrease in ground conductivity landwards of the coastal belt which suggests that there is a non-linear, rapid decrease in the salinity of the shallow groundwater. The tide-linked conductivity survey results shown in Fig. 4a–c indicate that seawater incursion is most likely restricted to within about 50 m of the coastline, whereas saline intrusion continues landwards for at least another 50 m (Fig. 5a). However, evidence indicating the presence of channel features within the wetland sediments that have in places incised into the Chalk bedrock (Fig. 5) indicates that saline intrusion can penetrate further inland in places.

Survey Area 2. The results of the ground conductivity survey carried out at Survey Area 2 are shown in Fig. 6a. The site is an apparently uniform area of coastal farmland; however, the survey has revealed the presence of a broad intrusion of saline water. The plan geometry of the intrusion indicates that inland the saline zone is greater than 100 m in width; however, the main connection with the sea is apparently concentrated into a narrower zone. A resistivity traverse positioned to investigate the vertical extent of the conductivity anomaly (Fig. 6b) indicates that the interpreted saline intrusion continues to significant depth and is therefore not likely to be due to surface flooding. Restricted access prevented a follow-up borehole investigation; however, some shallow sample holes were augured and water samples taken and conductivity tested. Water samples from a shallow hole in the interpreted saline intrusion area (Sample Point A) showed values of about 900 $mS\,m^{-1}$ which, using equations provided by Hoekstra *et al.* (1992), suggests a chloride concentration of about 3 $g\,l^{-1}$. Comparing this level of salinity with general lower-limit levels for saline intrusion of about 0.35 $g\,l^{-1}$ (Owen *et al.* 1991) it is evident that the mapped high-conductivity plume is the result of a localized saline intrusion. Measured conductivities of water samples taken at Sample Point B were significantly lower (540 $mS\,m^{-1}$) than at Sample Point A, which provides supporting evidence for the geophysical interpretation.

A resistivity traverse was positioned to investigate the depth extent of the observed ground conductivity anomaly. Without the resistivity tomography data it would be difficult to postulate whether the anomalous conductivities were due to a thin shallow saline sediment horizon, a recent flood-water saline contamination of the soil profile or a more significant intrusion continuing to depth. The modelled resistivity section shown in Fig. 6b suggests that the saline intrusion is a broad deep feature that is fairly well-constrained laterally as indicated by the high resistivity gradient at the edges.

Geophysical properties of saline intrusion

The apparent conductivity maps derived from data recorded at Survey Area 1 (Figs. 4 and 5) indicate that local seawater incursion produces a high apparent conductivity greater than 100 to 120 $mS\,m^{-1}$. More dilute saline intrusion beneath the wetland produces apparent conductivity values in the range of about 45 to 90 $mS\,m^{-1}$. Background conductivities of less than 45 $mS\,m^{-1}$ have been measured over inland areas. Since these conductivity measurements were recorded using an instrument configuration targeting a maximum depth of 15 m it is important to note that the influence of relatively low-conductivity Chalk bedrock residing within this depth range should be considered. Measured conductivities are lowered where bedrock resides at shallow depth and conversely measured conductivities are higher where bedrock is at greater depth due to the relative ratio of high-conductivity sediment to low-conductivity bedrock within the instrument depth range. However, given the known conductivity of seawater (about 5000 $mS\,m^{-1}$) compared to natural waters in sediments (about 350 $mS\,m^{-1}$; Telford *et al.* 1976) it is considered that its presence in the sub-surface has a dominant influence on measured apparent conductivity values in coastal wetlands.

The map of apparent conductivity data recorded at Survey Area 2 (Fig. 6a) shows that greater maximum conductivity values for saline intrusion have been measured, about 300 $mS\,m^{-1}$ compared to 180 $mS\,m^{-1}$ observed at Area 1. This suggests a possibly greater concentration of saline water at shallow depth in Area 2 than in the coastline zone of Area 1.

The apparent resistivities of seawater and saline intrusion are not clearly distinguishable on the resistivity sections (Figs. 4d and 5c) given the limitations of modelling small percentage changes in resitivity due to groundwater salinity in the presence of larger percentage changes at the rockhead boundary. However, model resistivities of anomalies indicated to be saline intrusion within the local wetland sediments are consistently less than about 10 $\Omega\,m$, as anticipated from work by Ward (1990). The resistivity tomography sections indicate that interpreted bedrock resistivities typically exceed 50 $\Omega\,m$ at Survey Area 1. By correlating the known Upper Chalk bedrock porosity of about 38.8% (Bloomfield *et al.* 1995) with measured resistivity values $> 50\ \Omega\,m$ and equations presented by Hoekstra *et al.* (1992) it is possible to infer the possible range of salt concentration in the pore-water within the Chalk bedrock. Using

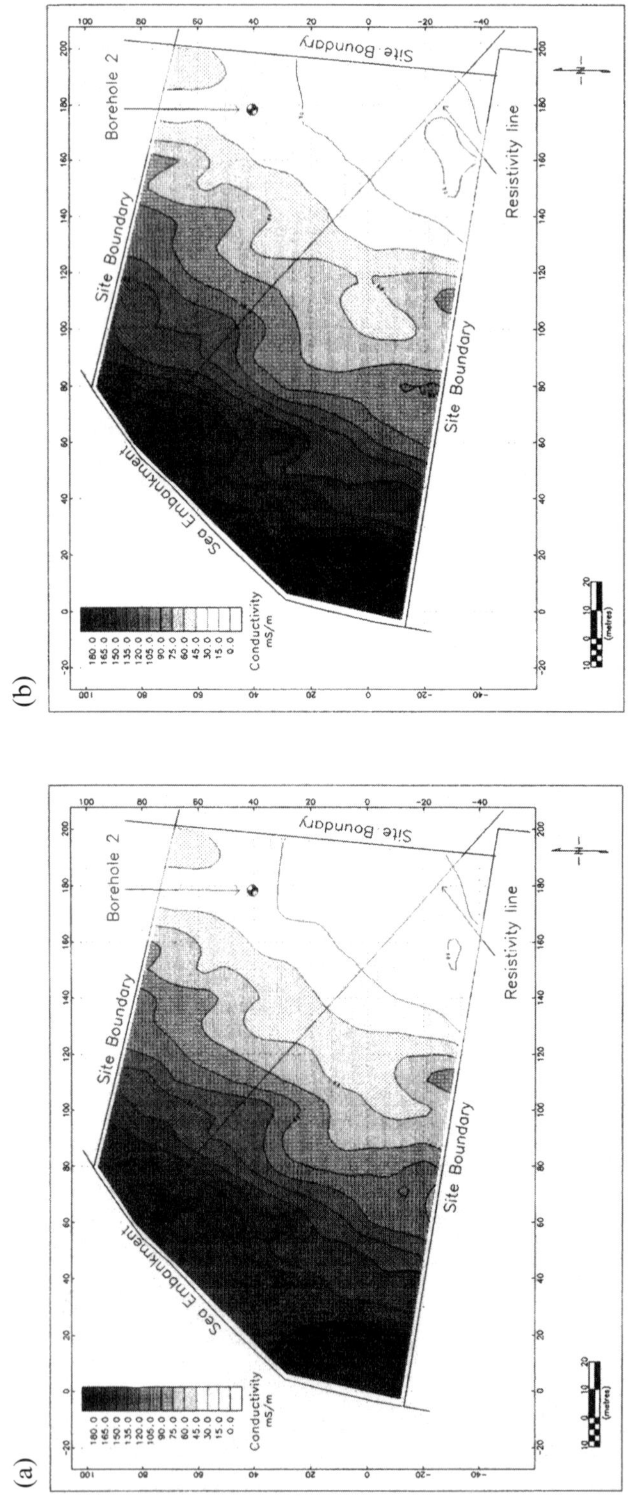

Fig. 4. Contour plans of apparent ground conductivity for an area within Survey Site 1. Plotted data show apparent conductivities at **(a)** low water, **(b)** high water, and **(c)** the difference between high and low water. **(d)** Resistivity tomography section from a traverse perpendicular to the coastline.

(c)

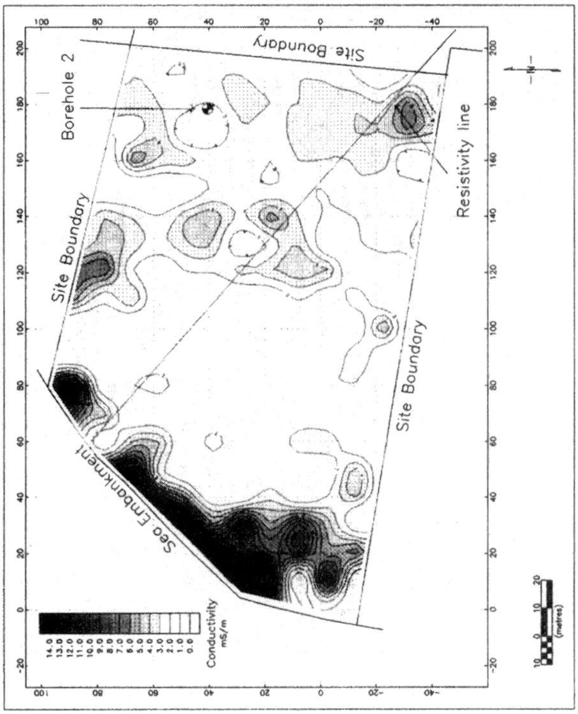
Borehole 2
Site Boundary
Site Boundary
Site Boundary
Resistivity line
Sea Embankment
Conductivity
mS/m
14.0
13.0
12.0
11.0
10.0
9.0
8.0
7.0
6.0
5.0
4.0
3.0
2.0
1.0
0.0
(metres)

(d)

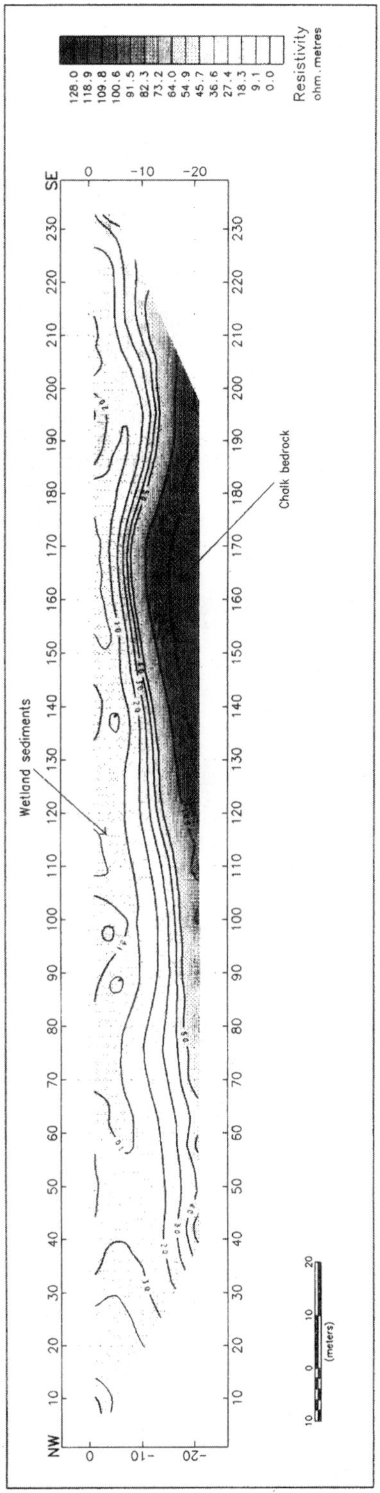
NW
SE
Wetland sediments
Chalk bedrock
Resistivity
ohm.metres
128.0
118.9
109.8
100.6
91.5
82.3
73.2
64.0
54.9
45.7
36.6
27.4
18.3
9.1
0.0
(meters)

(a)

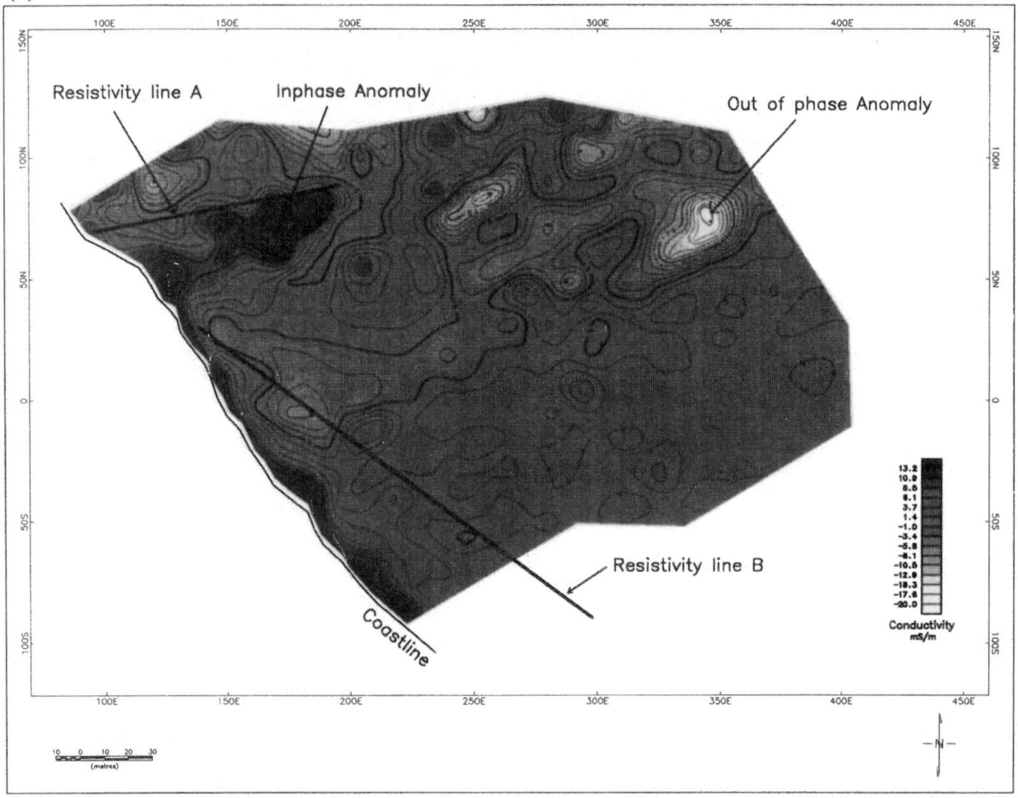

Fig. 5. (**a**) Difference in apparent ground conductivity readings taken at high and low water over a small area within Survey Site 1, where buried channel systems have been identified. Positive residual anomalies represent sites of in-phase tidal groundwater motion and negative anomalies represent out-of-phase groundwater movements.

the possible extreme values for bedrock formation factor it can be shown that the chloride concentration of the groundwater within the local Chalk is between 0 $g l^{-1}$ and 1.7 $g l^{-1}$ while using average values for equation constants a concentration of about 0.48 $g l^{-1}$ is derived. Seawater has a salt concentration of about 19 $g l^{-1}$, water is drinkable at 0.25 $g l^{-1}$ and Owen *et al.* (1991) stated that saline-water intrusions have concentrations exceeding 0.35 $g l^{-1}$. Since modelled resistivity values for the Chalk are typically significantly greater than the minimum value ($\approx$50 Ω m) used to derive possible salt concentrations, it is likely that groundwater within the Chalk at shallow depth is reasonably fresh beneath much of the surveyed area.

Geological control on saline intrusion

In Survey Area 1, bedrock topography has a clear control on the position of localized saline intrusion as evidenced from the identified channel shown in Fig. 5*b*. In places, saline intrusion apparently extends along aquiferous lineaments inland from the coastline. Despite the moderate porosity of the Chalk expected from work by Bloomfield *et al.* (1995), its depth is an apparent major factor restricting the thickness of the saline intrusion at shallow depths. Incisions into the Chalk and their permeable infill material facilitate influx of saline water into inland wetland areas. Wetland sediments show relatively limited permeability to seawater as evidenced by the small tidal influence on conductivity map data (Fig. 5*a*) apart from the exceptional aquiferous tidal channel sediments. The observed phase lag of localized ground water movements in the intermediate conductivity zone also suggests limited permeability.

Tidal control on saline intrusion

The tidal range during the time-dependent component of the survey work was 3.6 m (Fig. 3).

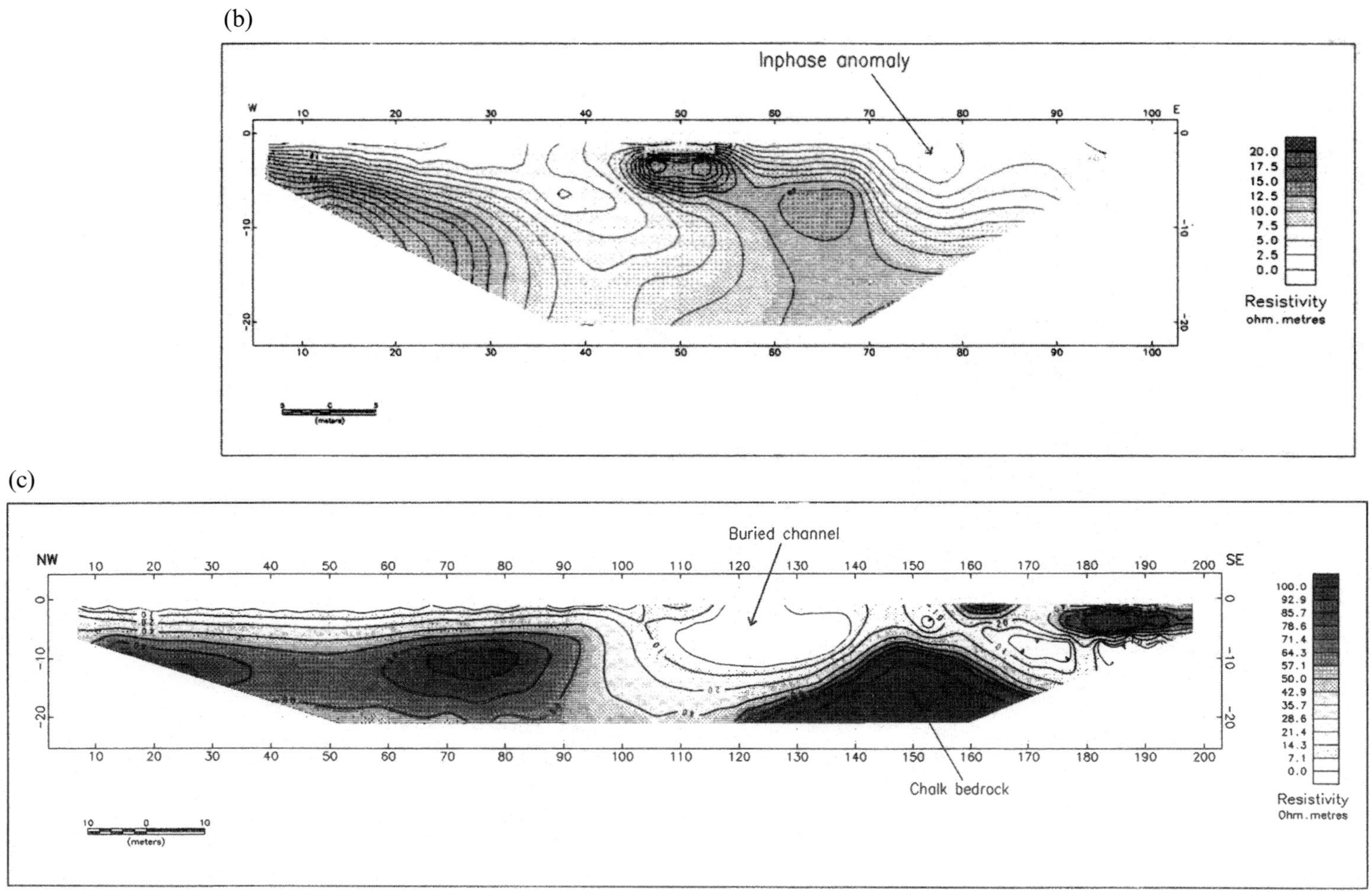

Fig. 5. (**b**) and (**c**) Resistivity tomography sections for traverse lines crossing the identified channels observed in (a).

(a)

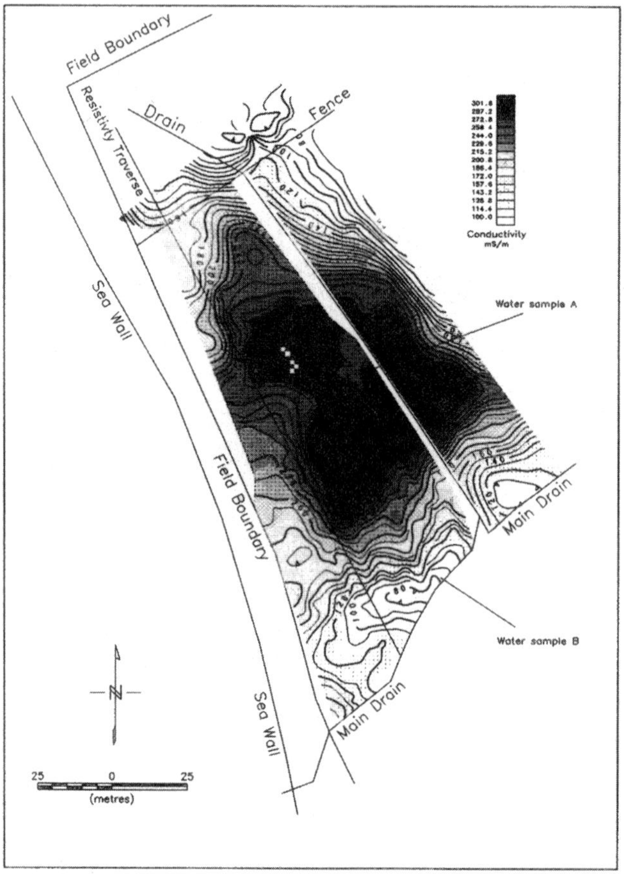

Fig. 6. (**a**) Contour plan of measured apparent ground conductivity for Survey Site 2, showing the location of augured water sample holes.

The associated maximum absolute difference between the high- and low-tide datasets of apparent conductivity was about 20 mS m^{-1} (Fig. 4*c*). Most of the survey area shows little difference between high-tide and low-tide values of apparent conductivity; however, a band of positive anomalous change occurs along the sea boundary. These measured higher apparent conductivities at high tide indicate that coastal sediments are affected by in-phase tidal movements. In addition, differencing high- and low-tide datasets has enabled clear identification of a channel feature that facilitates saline intrusion that is in phase with the tide (Fig. 5*a*). A resistivity traverse that intersects part of this buried tidal channel reveals in part its cross-sectional geometry, Fig. 5*c*.

At several locations in Survey Area 1, zones where higher conductivities are mapped at low tide have been identified, i.e. there is an apparent out-of-phase relationship with tidal movement Fig. 5(a). This has been correlated in places by observed local rise in groundwater at low tide.

A resistivity traverse located approximately parallel to the coastline (Fig. 5*b*) has enabled identification of a channel feature that is not apparently subject to significant tidal groundwater movement since it is not in evidence on the conductivity high tide–low tide difference plot. It is postulated that the channel may contain relatively low-permeability material or may be blocked from the sea.

The wetland in Survey Area 1 can be subdivided into three zones: (i) the in-phase tidal zone near the coastline and along sea-connected buried channels; (ii) an intermediate zone which contains out-of-phase groundwater movements in places; and (iii) a stable zone further inland

(b)

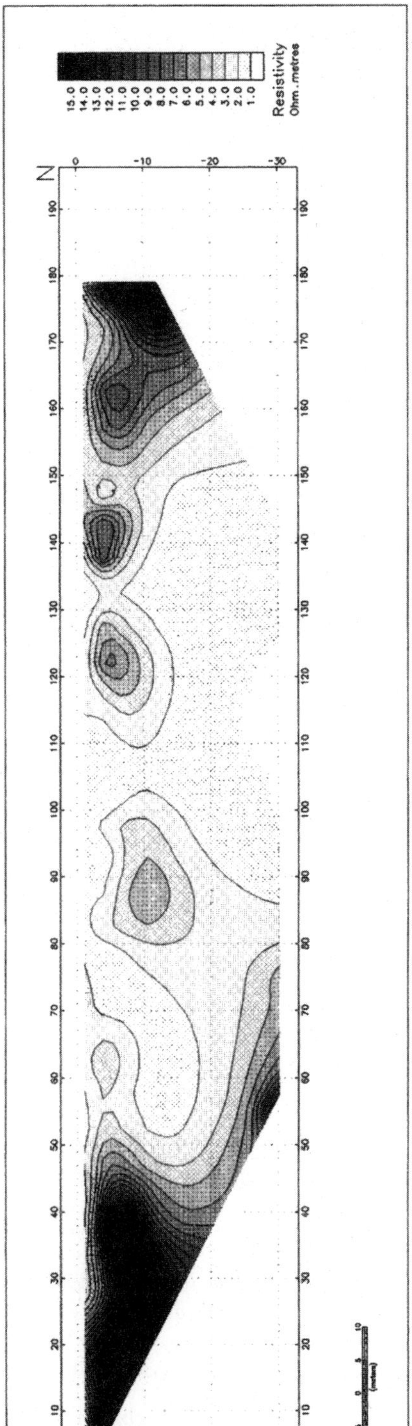

Fig. 6. (**b**) Resistivity tomography section, sub-parallel to the coastline, located across the anomalous zone of high conductivity observed in (a).

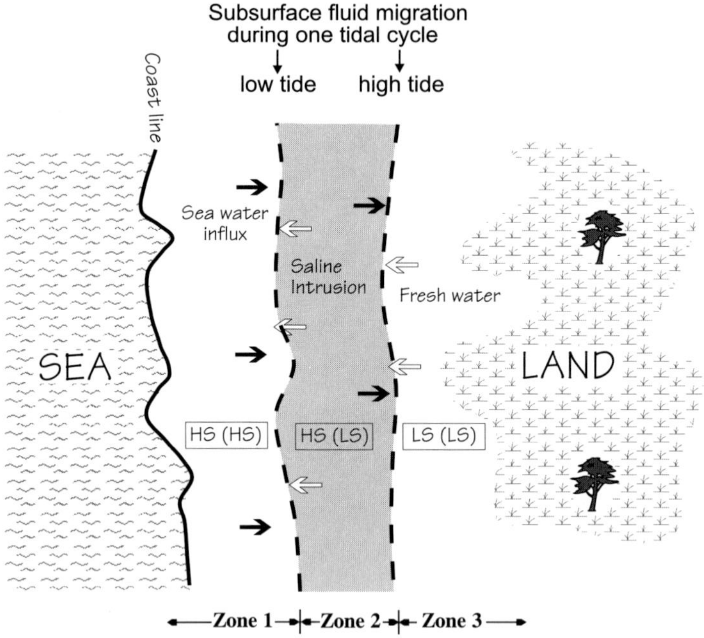

Key:

Zone 1: Tidal seawater incursion
Zone 2: Intermediate zone of mixing (inphase & out of phase)
Zone 3: Stable area of fresh groundwater

HS (HS) Groundwater conditions at high and (in brackets) low water

HS High salinity
LS Low salinity

Influx 'phase factor' removed form tidal cycle

Fig. 7. Schematic map of tide-controlled shallow saline and fresh groundwater movement (simplified geography).

beyond the influence of tidal and long-period saline intrusion (Fig. 7). The flow in zones (1) and (2) is complicated by buried channels and surface inlets (Fig. 8a). The intermediate zone is characterized by out-of-phase groundwater movement and by moderate conductivity values ranging between those of the stable zone and the in-phase zone. Significantly, negative residual anomalies observed on the high tide–low tide conductivity difference map correlate in places with high-conductivity springs appearing close to low-tide times. The out-of-phase relationship between tide and groundwater movement in the intermediate zone is thus evident. Similar out-of-phase groundwater motion is known from work by Stanley (1995) in landfill-covered wetland of the Severn Estuary near Cardiff.

Insufficient access time was available for work in Survey Area 2 to carry out a high tide and low tide conductivity survey. Analysis of a differenced dataset would help establish whether the anomalously conductive saline zone (Fig. 6*a*) was due to present-day tidal saline incursion, or due to long-period intrusion or an abandoned channel independent of tidal activity.

Discussion of general models

There are two empirical models to describe the present geometry of groundwaters beneath the studied coastal wetlands. The first is based on water motion and the permeability of the three sub-areas (Fig. 7). The second model is based on the sub-wetland structure and its effects on the position and geometry of the anomalies (Fig. 8a–c)). A combination of the two models probably gives a more realistic result. Evidence from the conductivity survey supports the first model while selected resistivity traverses reveal

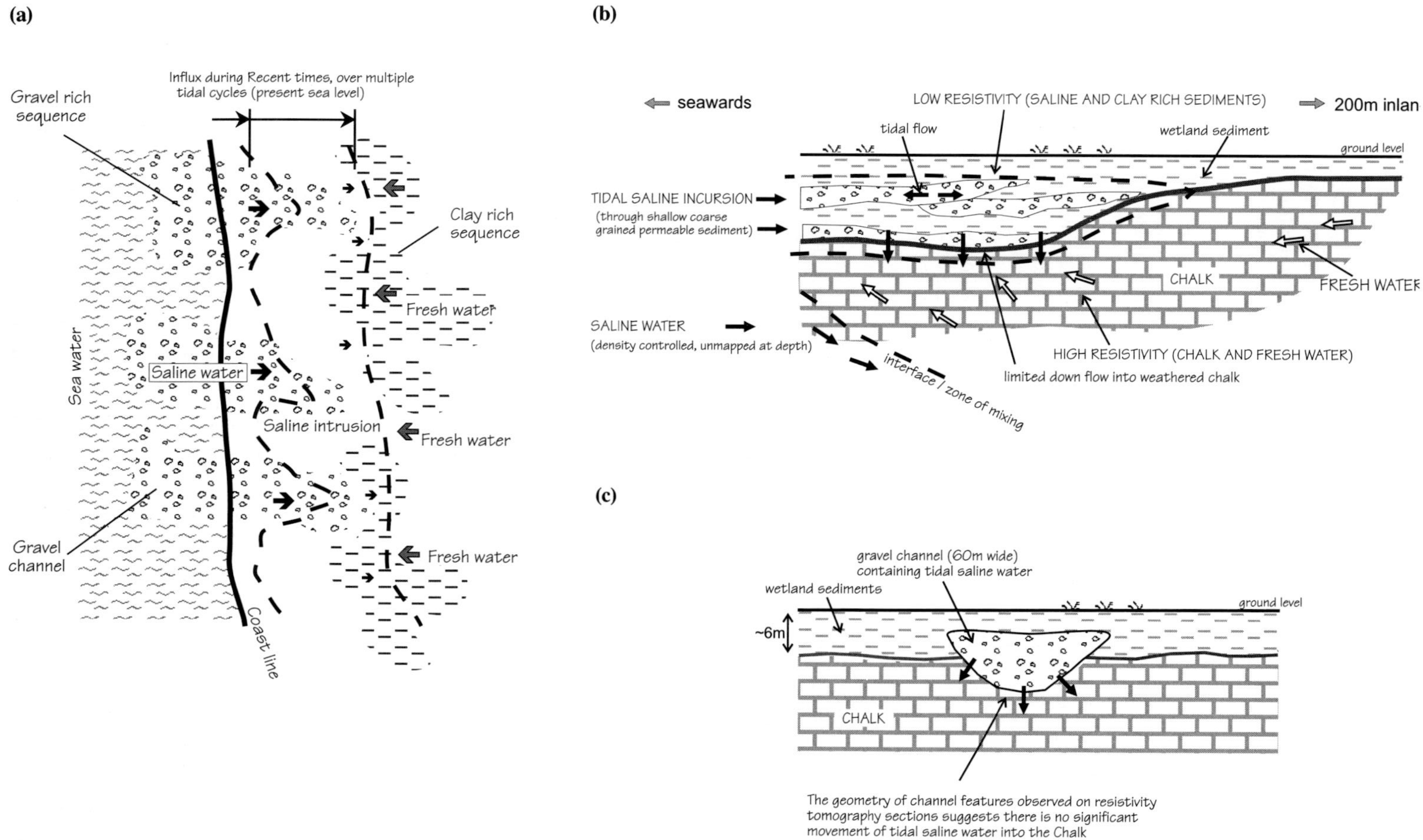

Fig. 8. Simple models for saline intrusion based on observed tidal changes in apparent ground conductivity, resistivity tomography results and field observations. (**a**) Schematic diagram showing the simplified relative plan geometries of the three main zones of shallow groundwater. (**b**) Cross-sectional model for the interpreted shallow-level saline intrusion into the coastal wetland. (**c**) Model illustrating the interpreted channel-limited migration of saline groundwater.

the geometry of the saline intrusion in section and indicate that geological features such as bedrock channels, bedrock–sediment contact, and distance from the coastline are controlling factors.

Standard texts such as Todd (1980) describe the coastal saline–fresh water interface as typically dipping landwards within isotropic coastal sediments and/or bedrock due to the density difference between the two waters. While this model is expected to hold true within the Chalk at depth, it does not incorporate the effects of tidal flow and differential permeability of coastal lithologies at shallow depth. At Survey Area 1, where shallow depths were investigated, high-permeability gravel-rich zones overlie less permeable Chalk; typical values of hydraulic conductivity for gravels are about 8.5 to 850 m day^{-1} (derived from Fetter (1994)) in contrast to the low hydraulic conductivity of Chalk of 0.001 to 0.01m day^{-1} (sample) or inter-well measurements through fractured 46% porosity Chalk of about 2 to 5m day^{-1} (derived from Price (1996)). The Upper Chalk of southern England is known to have an average porosity of 38.8% (Bloomfield *et al.* 1995) which is lower than that tested by Price (1996). The local hydraulic conductivity of the Chalk is expected to be very much lower than that of the overlying gravel-rich channel sediments and hence less susceptible to tidal saline incursion. The model proposed in Fig. 8b illustrates that the tidal saline incursion and related intrusion passes over fresh groundwater present in the underlying Chalk. A thin zone of mixing is inferred at the sediment–bedrock interface due to the rapidity of the saline incursion and regression. This model is further supported by the observed geometry of buried channels subject to saline water flow which indicates that little intrusion of saline water into the local Chalk rockhead has occurred (Fig. 8c).

Conclusions and recommendations

The modelled and observed motion of saline intrusion within an area of coastal wetland has been shown to be out of phase with the tidal cycle in some places and in phase in others. The main areas of in-phase influx are within buried channels, surface inlets and at the coastal front. Out-of-phase influx has been identified in several locations and is attributed to relatively low lateral permeability along shallow conduits.

Non-invasive ground conductivity and resistivity tomography surveys have proved a useful and inexpensive method for investigating environmentally sensitive coastal wetland sites. Narrow features such as buried channels would most likely not have been located with a borehole investigation alone.

General models have been developed that describe an interpreted hydrological regime based on field observations and geophysical survey work carried out to date. The survey results suggest that shallow saline intrusion preferentially moves inland through coarse-grained sediment conduits confined by less permeable wetland sediments and Chalk bedrock which apparently contains fresher groundwater.

Geophysical surveys similar to those described here provide a useful means by which the impact of land neighbouring engineering structures can be monitored or factors affecting possible migration of pollutants within a wetland environment can be assessed. Target areas such as buried inlets, channels and springs can be identified with a suitable survey and subsequently monitored for coastal management and protection.

Four forms of additional work that would provide data to enhance the hydrogeological model of the wetland sites are: (i) water-level monitoring through the tidal cycle in boreholes positioned where anomalous areas have been identified from the geophysical surveys; (ii) carrying out seismic refraction surveys over interpreted channel features to provide more accurate local depth to bedrock information (e.g. Corsmit *et al.* 1989; Goforth & Hayward 1992; McDonald 1994, Thomas 1995); (iii) recording self-potential (SP) data via a network of non-polarizing electrodes to enable observation of time-varying anomalies associated with groundwater movement; and (iv) an extended period of 3D resistivity tomography surveying to further characterize the sub-surface.

Environmental implications

Observations relating to the phase variations of tidal water levels within wetland sediments have implication for engineering or environmental projects that require investigation and/or monitoring of a coastal wetland. Where local water resources are exploited or there is a landfill adjacent to a wetland environment, potential hydrological links with the wetland sediments often require consideration. The possible influx, movement and retention of pollutants, whether saline or waste-derived, may require investigation. A geophysical survey combined with a follow-up borehole investigation can provide the necessary data to enable management of the wetland environment and facilitate monitoring of temporal changes due to over-pumping of resources or migration of pollutants.

The authors wish to thank the anonymous referees for their endeavours, and their suggestions were much appreciated. Field data acquisition was undertaken with the helpful assistance of A. R. Lewis of TerraDat (UK) Ltd. The efforts of Dr P. J. Brabham and Dr R. Gayer in reviewing the manuscript are also appreciated.

References

ALLEN, J. R. L. 1992. Tidally influenced marshes in the Severn Estuary, south west Britain. *In*: ALLEN, J. R. L. & PYE, K. (eds) *Saltmarshes: Morphodynamics, Conservation and Engineering Significance*. Cambridge University Press, 123–147.

—— 1994. A continuity-based Sedimentological model for temperate-zone tidal salt marshes. *Journal of the Geological Society, London*, **151**(1), 41–49.

ANDERTON, R., BRIDGES, P. H., LEEDER, M. R. & SELLWOOD, B. W., 1979. Cainozoic basins of Northern Europe. *In*: *A Dynamic Stratigraphy of the British Isles. A study in Crustal Evolution*. Allen and Unwin, London, 251–254.

AQRAWI, A. A. M. & EVANS, G. 1994. Sedimentation in the lakes and marshes (Ahwar) of the Tigris-Euphrates Delta, southern Mesopotamia. *Sedimentology*, **41**, 755–776.

BARKER, R. D. 1990. Investigation of groundwater salinity by geophysical methods. *In*: WARD, S. H. (ed.) *Geotechnical and Environmental Geophysics*, Volume 2: *Environmental and Groundwater*. Investigations in Geophysics, No. 5, 201–211.

—— 1997. Electrical imaging and its application in engineering investigations. *In*: MCCANN, D. M. (ed.) *Modern Geophysics in Engineering Geology*. Geological Society, London, Engineering Geology Special Publication, **12**, 37–43.

BEAR, J. & VERRUIJT, A. 1987. *Modelling Groundwater Flow and Pollution*. Reidel, Dordrecht, 196–215.

BLOOMFIELD, J. P., BREWERTON, L. J. & ALLEN, D. J. 1995. Regional trends in matrix porosity and bulk density of the Chalk of England. *Quarterly Journal of Engineering Geology*, **28**, S131–S142.

CALVACHE, M. L. & PULIDO-BOSCH, A. 1994. Modelling the effects of salt–water intrusion dynamics for a coastal karstified block connected to a detrital aquifer. *Ground Water*, **32**(5), 767.

CARTER, R. W. G. 1988. *Coastal Environments. An Introduction to the Physical, Ecological and Cultural Systems of Coastlines*. Academic, London, 335–346.

CORSMIT, J., VERSTEEG, W. H., BROUWER, J. H. & HELBIG, K. 1989. High-resolution 3D reflection seismics on a tidal flat: acquisition processing and interpretation. *First Break*, **6**(1), 9–23.

CURRY, D. 1965. The Palaeogene beds of South-east England. *Proceedings of the Geologists' Association*, **76**, 151–173.

EDWARDS, R. A. & FRESHNEY, E. C. 1987. *Geology of the country around Southampton*. Memoir for 1:50,000 Geological sheet 315 (England and Wales). HMSO, London.

EVANS, G. 1965. Intertidal flat sediments and their environments of deposition in the Wash. *Quarterly Journal of the Geological Society*, **121**, 209–245.

—— & COLLINS, M. B. 1975. The transport and deposition of suspended sediment over the intertidal flats of the Wash. *In*: HAILS & CARR (eds) *Nearshore Sediment Dynamics and Sedimentation*. Wiley, Chichester, 273–306.

FETTER, C. W. 1994. *Applied Hydrology*. Macmillan, New York.

GOFORTH, T. & HAYWARD, C. 1992. Seismic reflection investigations of a bedrock surface buried under alluvium. *Geophysics*, **57**(9), 1217–1227.

GOLDSTEIN, N. E., BENSON, S. M. & ALUMBAUGH, D. 1990. Saline groundwater plume mapping and electromagnetics. *In*: WARD, S. H. (ed.) *Geotechnical and Environmental Geophysics*, Volume 2: *Environmental and Groundwater*. Investigations in Geophysics, No. 5, 17–25.

GREER, S. A. 1975. Sand body geometry and sedimentary facies at the estuary-marine transition zone. Ossabaw Sound, Georgia: a stratigraphic model. *Senckenberg. Mar.*, **7**, 105–135.

HAGEMEYER, R. T. & STEWART, M. 1990. Resistivity investigation of salt–water intrusion near a major sea-level canal. *In*: WARD, S. H. (ed.) *Geotechnical and Environmental Geophysics*, Volume 2: *Environmental and Groundwater*. Investigations in Geophysics, No. 5, 67–77.

HAYES, M. O. 1980. General morphology and sediment patterns in tidal inlets. *Sedimentary Geology*, **26**, 139–156.

HOEKSTRA, P., HILD, J. & TOTH, D. 1992. Time domain electromagnetic measurements to determine water quality in the Floridan Aquifer. *In*: BELL, R. S. (ed.) *Proceedings of the Symposium on the Application of Geophysics to Engineering and Environmental Problems*, Society of Engineering and Mineral Exploration Geophysicists, Golden, Colorado, Vol. 1, 125.

KAUAHIKAUA, J. 1989. Can electrical geophysical methods detect the salinity transition zone beneath ground water lenses? *In*: *Proceedings of the Symposium on the Application of Geophysics to Engineering and Environmental Problems*, Society of Engineering and Mineral Exploration Geophysicists, Golden, Colorado, 424.

MCDONALD, R. J. 1994. *The application of shallow seismic reflection to onshore quaternary studies*. PhD Thesis, University of Wales, Cardiff.

MCNEILL, J. D. 1988. Advances in electromagnetic methods for groundwater studies. *In*: *Proceedings of the Symposium on the Application of Geophysics to Engineering and Environmental Problems*, Society of Engineering and Mineral Exploration Geophysicists, Golden, Colorado, Vol. 1, 251–348.

—— 1990. Use of electromagnetic methods for groundwater studies. *In*: WARD, S. H. (ed.) *Geotechnical and Environmental Geophysics*, Volume 1: *Review and Tutorial*. Investigations in Geophysics, No. 5, 191–218.

MELVILLE, R. V. & FRESHNEY, E. C. 1982. *British Regional Geology. The Hampshire Basin and Adjoining Areas*. 4th edn, HMSO, London.

MURRAY, J. W. & WRIGHT, C. A. 1974. Palaeogene foraminiferida and palaeoecology, Hampshire and Paris Basins and the English Channel. *Palaeontology*, **14**, 171.

OSBORNE WHITE, H. J. 1915. *The Geology of the country near Lymington and Portsmouth*. Memoirs of the Geological Survey. England and Wales. Explanation of sheets 330 and 331 (mainland). HMSO, London.

OWEN, M., HEADWORTH, H. G., & MORGAN-JONES, M. 1991. Groundwater in basin management. *In*: DOWNING, R. A. & WILKINSON, W. B. (eds) *Applied Groundwater Hydrology. A British Perspective. Clarendon, Oxford*, **3**, 29.

PLINT, A. G. 1982. Eocene sedimentation and tectonics in the Hampshire Basin. *Journal of the Geological Society, London*, **139**, 249–254.

—— 1988. Global eustasy and the Eocene sequence in the Hampshire Basin, England. *Basin Research*, **1**, 11–22.

POETER, E. P. & MCKENNA, S. A. 1995. Reducing uncertainty associated with ground-water flow and transport predictions. *Ground Water*, **33**(6), 899.

PRICE, M. 1996. *Introducing Groundwater*, 2nd edn. Chapman and Hall, London.

PYE, K. 1992. Saltmarshes on the barrier coastline of North Norfolk. *In*: ALLEN, J. R. L. & PYE, K. (eds) *Saltmarshes: Morphodynamics, Conservation and Engineering Significance*. Cambridge University Press, 148–178.

REED, D. J., 1989. Patterns of sediment deposition in subsiding coastal salt marshes, Terrebonne Bay, Louisiana: The role of winter storms. *Estuaries*, **12**(4), 222–227.

RICHARDS, M. T. 1994. Transgression of an estuarine channel and tidal flat complex: the Lower Triassic of Barles, Alpes de Haute Provence, France. *Sedimentology*, **41**(1), 55–82.

RIGBY-JONES, J., MATTHEWS, M. C. & MCDOWELL, P. W. 1997. Electrical resistivity imaging systems for ground investigations, with particular reference to dissolution features in Chalk areas. *In*: MCCANN, D. M. (ed.) *Modern Geophysics in Engineering Geology*, Geological Society, London, Engineering Geology Special Publication, **12**, 235–245.

RUST, I. C. 1977. Evidence of shallow marine and tidal sedimentation in the Ordovician Graafwater Formation. Cape Province, South Africa. *Sedimentary Geology*, **18**, 123–133.

SCHNOEBELEN, D. J., BUGLIOSI, E. F. & KROTHE, N. C. 1995. Delineation of a saline ground-water boundary from borehole geophysical data. *Ground Water*, **33**(6), 965–976.

STANLEY, P. C. 1995. *The potential environmental risk to developments adjacent to, and the remediation of, the former Ferry Road Municipal Waste Disposal Site*. MSc Thesis, University of Wales, Cardiff, 25–26.

STESKY, R. M., 1986. Electrical conductivity of brine-saturated fractured rock. *Geophysics*, **51**(8), 1585–1593.

STODDART, D. R., REED, D. J. & FRENCH, J. R. 1989. Understanding salt-marsh accretion, Scolt Head Island, Norfolk, England. *Estuaries*, **12**(4), 228–236.

STONELEY, R. 1982. The structural development of the Wessex Basin. *Journal of the Geological Society, London*, **139**, 543–554.

SWAN, A. R. H. & SANDILANDS, M. 1995. *Introduction to Geological Data Analysis*. Blackwell, Oxford, 266–267.

TARNOCAI, C. 1979. Canadian wetland registry. *In*: RUBEC, C. D. A. & POLLETT, F. C. (eds) *Proceedings of a Workshop on Canadian Wetlands Environments*. Canadian Land Directorate, Ecological Land Classification Series, **12**, 9–38.

TAYLOR, R. W. 1992. Continuous electrical resistivity surveys along the Lake Michigan and Green Bay coastlines of Wisconsin. *In*: BELL, R. S. (ed.) *Proceedings of the Symposium on the Application of Geophysics to Engineering and Environmental Problems, Society of Engineering and Mineral Exploration Geophysicists, Golden, Colorado, Vol. 1, 129–134*.

TELFORD, W. M., GELDART, L. P., SHERIFF, R. E. & KEYS, D. A. 1976. *Applied Geophysics*, Cambridge University Press, 524–542.

THOMAS, J. W. 1995. *Estuarine channel response to natural and man-made influences: a detailed study from the Burry Inlet, South Wales*. PhD Thesis, University of Wales, Cardiff.

TODD, D. K. 1980. *Groundwater Hydrology*, 2nd edn. Wiley, New York.

VAN OVERMEEREN, R. A. 1989. Aquifer boundaries explored by geoelectrical measurements in the coastal plain of Yemen. A case of equivalence. *Geophysics*, **54**(1), 38–48.

WARD, S. H. 1990. Resistivity and induced polarization methods. *In*: WARD, S. H. (ed.) *Geotechnical and Environmental Geophysics*, Volume 1: *Review and Tutorial*. Investigations in Geophysics, No. 5, 147–189.

WOOD, R. C. 1987. *Uses of the EM31 terrain conductivity meter for mapping superficial geology and groundwater features in dryland salinity investigations – a discussion paper*. Internal Report, Prairie Farm Rehabilitation Administration, Lethbridge, Alberta.

WRIGHT, V. P. & PLATT, N. H. 1995. Seasonal wetland carbonate sequences and dynamic catenas: areappraisal of Palustrine limestones. *Sedimentary Geology*, **99**(2), 65–71.

XUE, Y., WU, J., LIU P., WANG, J., JIANG, Q. & SHI, H., 1993. Sea-water intrusion in the coastal area of Laizhou Bay, China: 1. Distribution of sea-water intrusion and its hydrochemical characteristics. *Ground Water*, **31**(4), 532.

Evidence for rapid groundwater flow and karst-type behaviour in the Chalk of southern England

A. M. MacDONALD[1], L. J. BREWERTON[2] & D. J. ALLEN[2]

[1]*British Geological Survey, Murchison House, West Mains Road, Edinbugh, EH9 3LA, UK*
[2]*British Geological Survey, Maclean Building, Crowmarsh Gifford, Wallingford, Oxfordshire, OX10 8BB, UK*

Abstract: With the growing importance of groundwater protection, there is increasing concern about the possibility of rapid groundwater flow in the Chalk of southern England and therefore in the frequency and distribution of 'karstic' features. Pumping test data, although useful in quantifying groundwater resources and regional flow, give little information on groundwater flow at a local scale. Evidence for rapid groundwater flow is gathered from other, less quantifiable methods. Nine different strands of evidence are drawn together: tracer tests; observations from Chalk caves; Chalk boreholes that pump sand; descriptions of adits; the nature of water-level fluctuations; the Chichester flood; the nature of the surface drainage; geomorphological features; and the presence of indicator bacteria in Chalk boreholes. Although the evidence does not prove the widespread existence of karstic features, it does suggest that rapid groundwater flow should be considered seriously throughout the Chalk. Rapid groundwater flow is generally more frequent close to Palaeogene cover and may also be associated with other forms of cover and valley bottoms.

Recently, emphasis has been placed on protecting both groundwater sources and resources from contamination (National Rivers Authority (NRA) 1992). As a consequence it has become important to identify flow routes from the ground surface, through the unsaturated zone and the aquifer, to water supply boreholes or wells. As the most important of British aquifers, the Chalk has come under particular scrutiny. Chalk groundwater accounts for more than half of the groundwater used in the country (55% in 1988 (Downing 1993)) and 18% of the total water used in England and Wales (Water Authorities Association 1986). The large outcrop area of the Chalk, over 21 500 km^2, combined with the high population density of southern England make protection of both the resource and individual sources a large and expensive task.

The hydraulic properties of the Chalk result from a combination of matrix and fracture properties (Price 1987; Price *et al.* 1993). Groundwater flow in the saturated zone is primarily through fractures; usable groundwater storage comprises storage within macro- and microfractures and a little from within the matrix. Although complex, these general hydraulic properties still allow the aquifer to be approximated at a regional scale as having intergranular flow: this permits groundwater head distributions to be modelled relatively simply. However, at a local scale, extremely rapid groundwater flow also occurs within the Chalk. Groundwater sources can be linked to the ground surface via rapid transport routes in which groundwater can flow through the unsaturated zone and the aquifer in a matter of days. The frequency and distribution of these rapid groundwater flow features in the Chalk are largely unknown.

Various strands of evidence suggesting the existence of rapid groundwater flow and karst-type behaviour are presented here for the Chalk of southern England. Taken on their own, each piece of evidence is not sufficiently convincing to draw any conclusions; however, together, they suggest that rapid groundwater flow may be widespread throughout the Chalk of southern England. Before presenting the evidence, the limitations of pumping test data must first be examined.

Pumping test data

As part of a joint-funded project between the British Geological Survey (BGS) and the Environment Agency, pumping test data were collected for the major aquifers in England and Wales (Allen *et al.* 1997). Data on aquifer properties for the Chalk are available from 2000 pumping tests at approximately 1300 locations throughout England. These data are of varying quality, having been calculated from pumping tests of different lengths analysed using both observation and production borehole measurements.

MacDonald, A. M., Brewerton, L. J.. & Allen, D. J. 1998. Evidence for rapid groundwater flow and karst-type behaviour in the Chalk of southern England. *In*: Robins, N. S. (ed.) *Groundwater Pollution, Aquifer Recharge and Vulnerability*. Geological Society, London, Special Publications, **130**, 95–106.

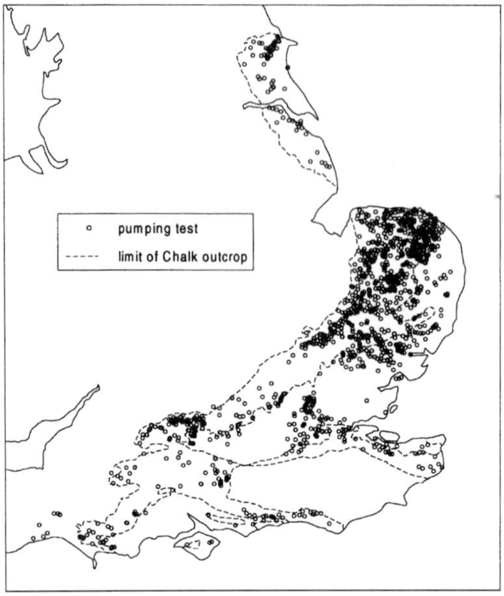

Fig. 1. The distribution of pumping test data for the Chalk of England (data from Allen *et al.* 1997).

Figure 1 shows the distribution of all the available data. The data are not evenly spread over the Chalk outcrop: most are clustered within East Anglia, with a much lower density over the Hampshire and Thames Basins and to the north in Yorkshire and Lincolnshire. At a larger scale, the data are even less randomly distributed. Data are calculated from pumping tests undertaken in boreholes that have generally been drilled for production and are therefore in areas known to be high-yielding. Throughout the Chalk outcrop, therefore, data are clustered within valleys, where the transmissivity and storage tend to be high. This bias towards high value data points creates a significantly unbalanced dataset. For example, the geometric mean of the transmissivity values calculated for the 2000 tests in the Chalk is approximately 460 $m^2 day^{-1}$; which is unreasonably high for large areas of Chalk. Therefore, the data should not be treated as indicative of the aquifer properties of the entire Chalk, but only of the measured sites, which are by their nature highly biased.

Pumping tests integrate the hydraulic properties of the aquifer throughout the depth of the borehole. Similar aquifer properties might be estimated from two boreholes with radically different flow regimes; e.g. a borehole that intersects one highly permeable fracture can give the same transmissivity and storage coefficient estimates as a network of small fractures with moderate permeability (e.g. Foster & Milton 1974). Consequently, groundwater models that use transmissivity and storage, derived from pumping tests, to characterize the aquifer, although adequately describing regional groundwater flow and resources, give little indication of groundwater transport (and in particular, groundwater flow velocities) around a source. The potential for rapid groundwater flow is not easily identified from pumping tests; evidence must be gathered from other, less quantifiable sources.

Evidence for rapid groundwater flow and karst

The term 'karst' is a geomorphological term associated with terrain having distinctive landforms and hydrology. In hydrogeological terms, the importance of karst is that groundwater is concentrated in, and flows rapidly through, a network of fractures, conduits (significantly enlarged fractures) and caves (conduits that are large enough to be explored physically by man). Karst characteristics are more developed in some terrains (e.g. Carboniferous limestone) than others (e.g. Jurassic limestone, Chalk). This prompted Atkinson & Smart (1981) to suggest that carbonate aquifers should be considered as possessing varying degrees of karst, rather than being wholly karstic or wholly non- karstic. Characterizing the Chalk as having 'karstic' properties can lead to emotive discussion, and secondary debates on the precise definition of terms. However, for understanding and managing the Chalk aquifer, it is not the precise terminology that is of primary concern; rather, it is the velocity with which groundwater can flow from the ground surface to the aquifer, and within the aquifer to a borehole. In this discussion 'karst' is used to describe features of any diameter that allow rapid groundwater flow and also, where appropriate, geomorphological dissolution features.

Evidence of hydrogeologically significant karstic features in the Chalk can be gained in a variety of ways.

Tracer tests

Tracer tests in the Chalk may be divided into two broad categories: those associated with observed karstic features, and those carried out without reference to such features. Tests in the first category have generally been carried out in the Chalk of southern England, while the second category of tests has a more widespread geographical distribution.

'Karstic' tracer tests

Sub-surface karstic-type flow has been investigated at several sites in the Chalk by means of tracer tests, some of the most prominent examples of which are given below.

One of the best-known sites of swallow-hole activity in the Chalk is at Water End in Hertfordshire, where the Mimmshall Brook, a headwater of the River Colne, disappears at certain times of the year into a swallow-hole complex. The swallow holes are located at the edge of the boundary between the Chalk and Eocene cover. Three tracer tests, carried out in 1927, 1928 and 1932 using fluorescein dye, showed that water recharging the swallow holes travelled northeast towards the River Lea Catchment. Tracers entering the swallow holes were detected in a series of springs and wells up to 16 km from the swallow holes. The breakthrough times gave flow velocities ranging up to 5.5 km day^{-1}, with tracers being detected over a 30° arc from the swallow holes (Harold 1937). The velocities between the swallow holes and a particular detection point in the Lea Valley varied by up to 37% between tests, and the fastest route also varied between tests. A test carried out in 1935 in swallow holes at South Mimms (3.5 km to the southwest of Water End) also showed flows towards the Lea Valley over distances up to 19 km, with velocities ranging up to 3.2 km day^{-1}.

There are several implications of this work. It showed that rapid (several kilometres per day) natural flows can occur in the Chalk in the Water End–Lea Valley area over distances of about 15 to 20 km. The conduit system through which flow occurs is complex, and the individual pathlines followed by flow are not constant and probably vary with water level. The system of conduits through which the flow occurs is widespread, as is shown by the dispersion of the detected tracer along a section of the Lea Valley several kilometres in length.

Bacteriological examination of groundwater samples taken from the Lea Valley in 1935–36 showed a strong correlation between rainfall and the level of bacteria (Harold 1937). The increase in bacteria numbers coincided with periods when streamflows known to contain bacteria entered the swallow holes at North and South Mimms. Given the connection between the swallow holes and the sampling points in the Lea Valley, this suggested that particles of bacterial size could travel rapidly through fractures in this area of Chalk.

Atkinson & Smith (1974) reported a tracer experiment in the Havant–Bedhampton area. Here, swallow holes occur near to the northern margins of the Eocene outcrop, with groundwater flows occurring in Upper Chalk. Rhodamine WT dye was pumped into a swallow hole for three days. The travel time to its emergence at the Bedhampton spring (a distance of 5.75 km) was 62.5 hours, corresponding to a velocity of 2.2 km day^{-1} (peak concentration). From calculations involving the quantities of flow and the hydraulic gradient, Atkinson & Smith suggested that the fracture was the equivalent of a pipe 0.74 m in diameter (although in reality several features would probably be involved). Price (1987) suggested that in an ideal case (a plane parallel fracture with no roughness or channelling) a fracture of only 4.5 mm width, with a transmissivity of 5000 m^2 day^{-1}, could carry the observed flow.

Banks *et al.* (1995) reported a tracer test in Berkshire between a swallow hole and a spring 4.7 km away (both less than 1 km from the Chalk/Eocene boundary) which gave the highest velocities yet observed in the Chalk: 5.8 km day^{-1} for peak concentration and 6.8 km day^{-1} for breakthrough. The authors suggest that little attenuation occurred as the tracer moved from the sinkhole to the spring. By using Price's (1987) method, the authors calculated that a single fracture 5.4 mm in width could theoretically be sufficient to represent the fracture system.

The pollution of a well at Addington, Croydon, led to a series of tracer tests being carried out in the early 1900s (Richards & Brinckner 1908) to establish whether a waste-disposal site located in a depression 3.2 km from the well could be the source. The well is close (around 1 km) to the edge of Palaeogene cover, with local deposits of Clay-with-Flints. Four tests were carried out, two by washing salt into the depression, and two using bacteria. All tests showed a connection between the depression and the well, with breakthrough times giving velocities of between 2.6 and 3.4 km day^{-1} for the salt, and 1.0 to 1.1 km day^{-1} for the bacteria. The well collected water from headings and an interesting observation was that only certain fractures produced flow with significant chloride, indicating discrete flow routes within the fracture system.

'Non-karstic' tracer tests

The above tracer tests, and others, show that fast preferential flow routes (or karstic flows) do occur in the Chalk and are commonly associated with the proximity of Palaeogene cover. However, what is not known is how common such behaviour is. Are these special cases on which interest has been focused because of obvious features such as sinking streams and resurgences, or do they represent only the most obvious examples of a more widespread phenomenon?

In southern England there are few examples of tracer tests carried out in the Chalk that are unconnected with obvious karstic features.

A tracer experiment was undertaken at the M1/M25 motorway intersection and was reported by Price *et al.* (1992). This area is close to the Chalk/Palaeogene boundary and karstic features are common, but the studies were carried out in soakaways that were apparently unassociated with karstic features. Tracer studies carried out in soakaways at the motorway intersection showed that some tracer travelled rapidly to a pumping station (a distance of 3 km), with maximum recorded velocities in excess of 2.4 km day^{-1}, but tracer recovery was very low, and it was thought that significant amounts were moving through fine fractures.

At Arish Mell, in the Lulworth area of Dorset, a tracer experiment was carried out between a borehole, 400 m inland from the coast, and coastal springs (Alexander 1981). Dye injected into the borehole arrived at the coastal springs, giving a sharp peak after four days (implying a velocity for the peak of 0.1 km day^{-1}). The shape of the peak and the high hydraulic conductivity implied by a locally low hydraulic gradient were taken to suggest karstic flow in the area.

A number of tracer tests have been undertaken in East Anglia by researchers of the University of East Anglia. These are chiefly borehole dilution tests, or radial tests from an injection borehole to a pumped borehole (e.g. Kachi 1987; Ward 1989). The main conclusion of the extensive testing and modelling carried out by Ward was that the flow was dominated by microfractures with a range of sizes, rather than by a few discrete, high-permeability conduits.

A tracer test at Kilham in North Yorkshire suggested flow velocities along a 3.1 km section of the Broachdale Valley of 0.28–0.44 km day^{-1} which in turn implies a network of interconnected fractures extending over several kilometres (Ward & Williams 1995). Very few tracer tests from the northern Chalk province are reported in the literature.

In summary, the tracer test data support the contention that there can be rapid flow in the Chalk, particularly in southern England and particularly close to Palaeogene cover. Tracer information is, at present, insufficient to suggest whether rapid flows may occur away from such cover. In East Anglia, available tracer information does not support the existence of rapid flow through well- developed conduit systems. Further north, a test at Kilham in Yorkshire did indicate rapid flows.

Caves

Speleologists have long recognized and explored one aspect of the karstic nature of the Chalk (e.g. Reeve 1976, 1977, 1981, 1982; Proctor 1984; Fogg 1984; Lowe 1992). Explorable caves are seen in the Chalk at many locations (e.g. Sussex, Devon, Kent), mostly around the coast. Although some are of marine origin, a large proportion are connected to palaeosolution features of fresh-water origin. Often, these caves are now dry, but some extend down into the saturated zone where they are less easily explored.

One such cave system has been identified and examined at Beachy Head (Reeve 1981). The cave system is developed along a series of faults and joints and although generally sub-horizontal, has occasional vertical steps. The retreat of the cliff line at Beachy Head intersected and exposed the system allowing the cave to be examined (Fig. 2). The main cave is several hundreds of metres long and has been followed until it becomes filled with water. A tabular flint layer appears to be associated with the cave development and was observed to be followed by the caves (including vertical displacements). It is possible that the perturbation in groundwater flow caused by the presence of the flint layer initiated the dissolution and enlargement of the fractures, which were then further enlarged by the concentration of flow through the fractures. As the caves are now mainly above the water-table they may have developed when water levels were higher during interglacial periods of the Devensian period.

Chalk boreholes pumping sand

Chalk boreholes in southern England have been known to pump sand. In one such borehole in the South Downs, sand was found within large, solution-enhanced bedding-plane fractures to a depth of 70 m below the ground surface (Southern Science 1992). Groundwater flow induced by abstraction disturbs the sand within the fractures

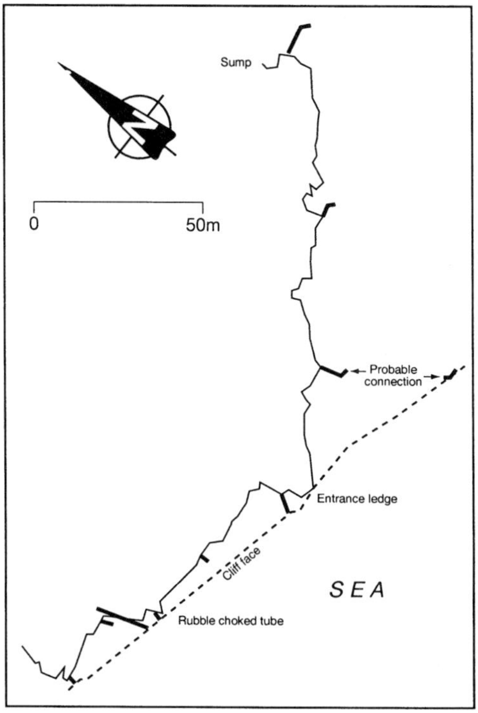

Fig. 2. A sketch map of the Beachy Head Chalk cave (after Reeve 1981).

and transports it into the borehole. The origin of sand and gravel in bedding-plane fractures can be identified from cliff sections: in cliff sections in France (Dieppe West, Beauville, Tilleul Plage) overlying sands and gravels have been washed down large sub-vertical solution pipes and deposited in horizontal bedding planes. The fractures are observed to have significant solution enhancement. Similar features are observed throughout the Chalk in England; e.g. in cliff sections along the south coast and also in excavations in Caversham (V. Robinson, pers. comm.).

Therefore, the presence of sand and gravel in Chalk boreholes may indicate the presence of large solution features in the surrounding Chalk (see Fig. 3). In addition, sand and gravel may be present in solution pipes that provide access for aggressive recharge deep into the Chalk and solution enhancement of bedding-plane fractures.

Adits

Extensive adit systems were built in different parts of the Chalk aquifer during the second half of the 19th century and the early part of the 20th century. The notes made during their construction are extremely interesting and useful for understanding the aquifer properties of the Chalk. During construction, engineers often estimated the location and flow from individual fractures. Unfortunately, many of the construction details have been lost over the years and few documentary records now exist, apart from the Brighton water supply (Mustchin 1974). This adit system extends for many kilometres and is generally located between -10 and $+10$ m OD. The yield of each adit is recorded as coming from a limited number of fractures, possibly as few as one every 30 to 50 m, each having large inflows (up to 50 $l\,s^{-1}$). This information implies that at this particular depth in the Brighton area, there exist widely spaced sub-vertical fractures with substantial groundwater flow.

The information is highly subjective, however, and depends on what the original engineer thought constituted a productive fracture. Also, adits are often deep. Therefore the majority of groundwater flow may occur in a fracture network above the adit with only a few vertical fractures extending to the adit. Nevertheless, if studied and mapped carefully, adits can provide additional insight into the behaviour of the aquifer not available from boreholes. Boreholes can only give information on horizontal and sub-horizontal fractures; since adits and tunnels are horizontal, information can be gained on the distribution and yield of sub-vertical fractures.

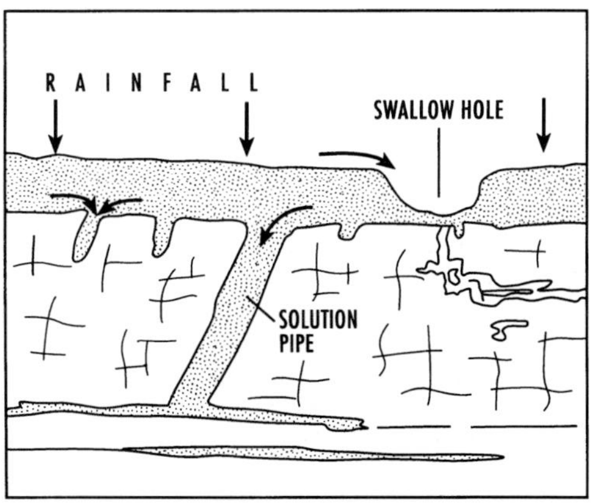

Fig. 3. A cross-section through the Chalk aquifer where overlain by thin cover (from Allen *et al.* 1997). Dissolution features can develop which lead to rapid groundwater flow and also the infilling of sediment.

Water levels

Significantly different water levels are often recorded in Chalk boreholes in close proximity. This change in rest water level is often due to boreholes intersecting different fracture sets that are not necessarily connected. In addition, water-level variations during pumping tests can also indicate the presence of discrete fracture sets: pumping can rapidly affect water levels in a borehole or spring several kilometres away but have little significant effect on much closer boreholes. Examples of these phenomena are observed throughout the Chalk, e.g. Swanbourne Lake, South Downs (Southern Science 1994) and South Dorset (Alexander 1981; Houston *et al* 1986). Such effects could be attributed to discrete, highly directional fracture zones that allow rapid flow.

Some boreholes respond quickly to rainfall. Water levels can be observed to rise rapidly in some areas of the Chalk within a matter of days following heavy rainfall (Institute of Hydrology and British Geological Survey 1996). One reason for such behaviour is rapid by-pass flow through the unsaturated zone connecting to a fracture network within the aquifer. This behaviour, which gives rise to 'flashy' hydrographs, is a common feature in aquifers that are thought of as karst (European Commission 1995).

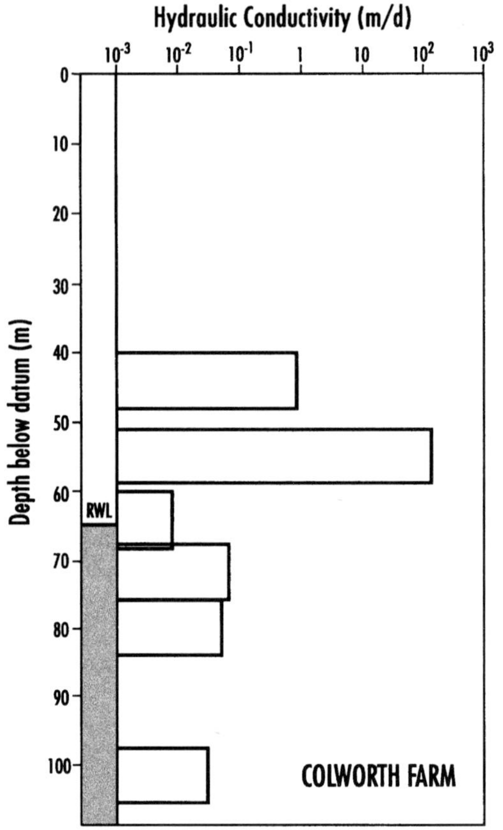

Fig. 4. The permeability distribution from a packer test undertaken at Colworth Farm borehole, situated on an interfluve in the South Downs (from Allen *et al.* 1997: data from NRA 1993).

The Chichester flood

In early January 1994, Chichester was subjected to severe flooding that lasted for nearly a month (Posford Duvivier 1994; Taylor 1995). Groundwater played a significant role in this episode. Groundwater levels were low at the end of the 1993 recession but during the early part of the winter the area received above average rainfall, with about 350 mm falling in approximately six weeks, 40% of which fell in just six days (Taylor 1995). Groundwater levels rose extremely rapidly and springs appeared throughout the catchment. Consequently, flow within the River Lavant increased rapidly and the resulting flood caused much damage around the Lavant Valley and in the city of Chichester.

In analysing groundwater-level data from the Chilgrove borehole, Posford Duvivier (1994) identified a critical level within the Chalk aquifer. Once water levels exceeded this level, river flows in the area increased markedly. It appears that the capacity for the Chalk aquifer to store significant quantities of recharge, and therefore buffer the intense rainfall, had been overcome and the catchment underwent a significant change, becoming 'flashy'. It has been suggested that this critical level corresponds to a zone of highly permeable Chalk that provides a rapid flowpath connecting groundwater in the interfluves to springs within the valley (Taylor 1995). Packer testing (NRA 1993) in the area corroborates this theory by showing the existence of high permeability within the unsaturated zone (see Fig. 4).

Surface drainage

The nature of streams and rivers on the Chalk outcrop is also indicative of a karstic aquifer. In an unpublished study, the NRA, Wessex Region, recorded flow during two seasons in ten winterbournes throughout the Chalk of Wessex and Dorset (W. Stanton, pers. comm.). Each winterbourne (i.e. the seasonally flowing upper reaches of streams) dried up in different but well-defined stages, with small springs and

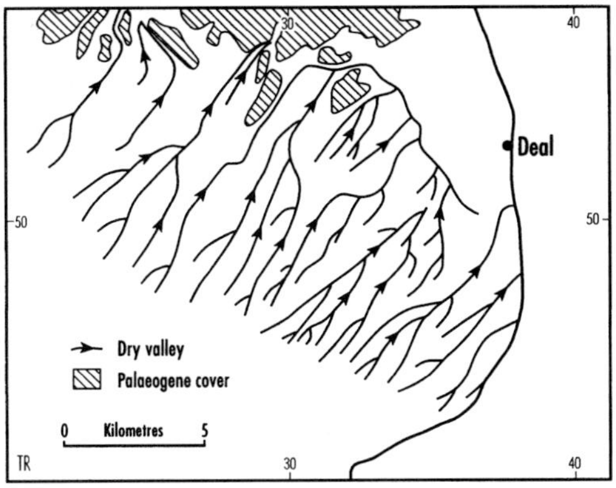

Fig. 5. Distribution of dry valleys within the Chalk of the Dover/Deal area (after Allen *et al.* 1997)

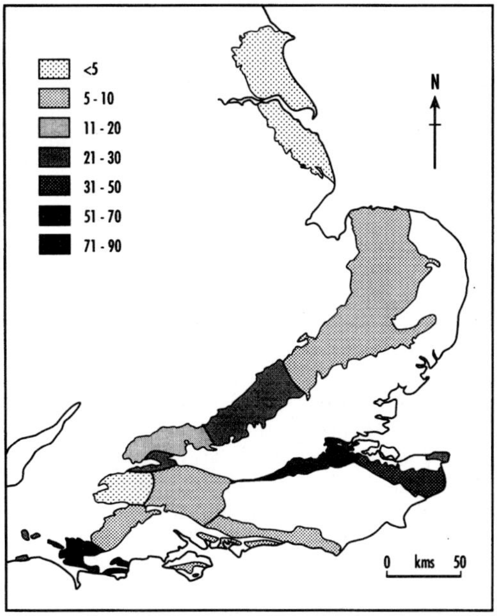

Fig. 6. The apparent frequency of dolines per 100 km^2 on the Chalk outcrop (based on Edmonds 1983).

swallow holes defining the start and end points of each stage. Some winterbournes had reaches high up the valley that flowed almost continuously throughout the year, while reaches downstream dried up quickly. Such disappearance and re-emergence of streams implies that a discrete network of fractures is present in the valley bottom.

Surface drainage on the Chalk outcrop is poorly developed. The dendritic river patterns that are seen on most other geological formations are not observed; instead the valley patterns tend to be orthogonal, possibly defined by the fracture directions (e.g. see Fig. 5). The lack of surface drainage illustrates the high permeability of Chalk strata – most water flows through the catchment as groundwater. Rapid groundwater flow can also result in groundwater flowing across surface catchment boundaries, e.g. the Alre in Hampshire (Giles and Lowings 1990), the Frome valley and Lulworth Cove in Dorset (Houston *et al.* 1986).

Geomorphological features

There is significant geomorphological evidence of surface dissolution features on the Chalk outcrop (e.g. Fagg 1958; Docherty 1971; West & Dumbleton 1972; Sperling *et al.* 1977; Edmonds 1983; Goudie 1990). Dolines, solution pipes and swallow holes are all observed in the Chalk. Although the regional frequency of these features is apparently lower than in other limestones, locally the frequency can be comparable. The highest density is found in Dorset (e.g. > 150 per km^2 at Puddletown Heath) with other important areas within the Chilterns, Pewsey area, Kent Downs, and the Surrey and West Kent Downs (Fig. 6). The lowest density areas are the Salisbury Plain and Yorkshire and Lincolnshire (Edmonds 1983). Swallow holes and dolines are found both on recharge areas (i.e. interfluves) and discharge areas (i.e. valleys). In addition, the surface of the Chalk, where overlain by some sort of cover, has undulations quite similar to the clints and grykes observed in harder limestones (as in the Carboniferous Limestone). These can be exposed in quarries and excavations (e.g. West & Dumbleton 1972). It appears that the Chalk is too soft to maintain them at outcrop.

Active swallow holes are also observed in the Chalk. For example, a tributary of the River Colne flows down a series of swallow holes at Water End in Hertfordshire and re-emerges in a different catchment (Harold 1937). At a series of swallow holes in Little Bedwyn, Berkshire [SU 306 650] surface water streams were observed to flow through the bottom of the swallow holes, directly into the Chalk: the recharge rate into the Chalk would require a well-developed fracture system to accommodate the water. Swallow holes have also been identified in the bed of the River Mole in Kent (Fagg 1958) where river water was observed to mix freely with groundwater through a series of swallow holes and springs.

There is little information on the distribution of active swallow holes throughout the Chalk outcrop. Since swallow holes are often active for only a few weeks in the year, it is difficult to distinguish them from dolines and collapse features. The link between surface karstic features and rapid groundwater flow is uncertain. Further investigations, perhaps small excavations or tracer tests, would help bridge the knowledge gap.

Indicator bacteria

Escherichia (E.) coli bacteria, which are indicators of faecal contamination, have been detected in untreated water from boreholes where the unsaturated zone of the Chalk is several metres thick (in some cases in excess of 30 m thick). The survival time of these indicator bacteria in either the unsaturated or saturated zones of the Chalk is not known, but times reported in the literature are generally of the order of four days to four weeks, depending on the environmental conditions (Lewis *et al.* 1980). *E. coli* are too large to enter the pores in the Chalk and therefore cannot reach the water table by a piston flow mechanism through the unsaturated zone matrix. This implies that bacteria must travel through fractures, especially those enlarged by solution. Therefore, the presence of these bacteria can be taken as an indicator of rapid groundwater flow from the ground surface, through the unsaturated zone and the aquifer, to the borehole. However, although the indicator bacteria are widespread, they are not universal, therefore it is possible for karstic flow to exist in the vicinity of a borehole but for no bacteria to be detected.

In a study of the occurrence of indicator bacteria in abstraction boreholes, groundwater samples taken in the vicinity of the Palaeogene deposits contained indicator bacteria more frequently than groundwater sampled at a distance from the Palaeogene deposits (Fig. 7). Indicator bacteria were detected in four out of five boreholes situated within 1 km of the Palaeogene cover but in only two out of five boreholes where the Palaeogene cover is greater than 1 km distant. This implies that the bacteria were transported more easily through the Chalk in the vicinity of the Palaeogene cover and, therefore, that rapid groundwater flow, particularly in the unsaturated zone, may be common near the Palaeogene deposits.

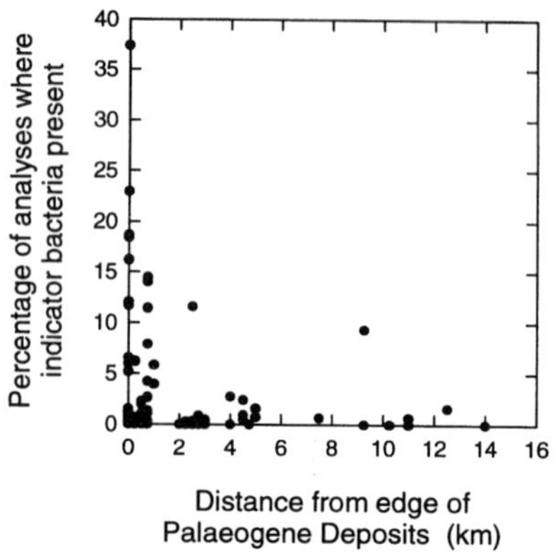

Fig. 7. Frequency of groundwater samples in southern England showing indicator bacteria related to the distance from Palaeogene deposits.

Emerging patterns

From the evidence given above, it is apparent that the Chalk exhibits many karst-like features and associated rapid groundwater flow. Geomorphologists have noted various surface expressions of karst on the Chalk outcrop. Limited data from the saturated zone show the importance of discrete, solution enhanced fractures with channelled, rapid groundwater flow. From the above evidence, various patterns begin to emerge: in particular, rapid groundwater flow in the Chalk of southern England is generally associated with cover, often Palaeogene cover, and to a lesser extent with valleys. These are examined below.

Palaeogene deposits or other forms of cover

Surface expressions of karst – dolines, solution pipes and swallow holes – have long been associated with cover. The high density of dolines found in south Dorset, for example, are formed beneath Palaeogene cover (Sperling *et al.* 1977); dolines and swallow holes in the Chilterns are formed beneath both Palaeogene deposits and Clay-with-Flints (Edmonds 1983). Figure 8 illustrates the distribution of the Clay-with-Flints in southern England. Areas with little Clay-with-Flints cover (Salisbury Plain, South Downs) also have fewer surface karst features (cf. Fig. 6). Although there has been little work, apart from occasional tracer tests in swallow holes, directly linking rapid groundwater flow with surface karst, much of the evidence for rapid groundwater flow is from areas on the Chalk outcrop close to Palaeogene deposits, the same area that surface karst is most evident.

All of the rapid flows proved by the tracer studies in southern England, discussed above, have been in catchments with, or close to,

Palaeogene deposits, i.e. Water End, Bedhampton, Addington and Arish Mell. The sand observed in solution pipes and fractures deep within the aquifer also had its origin from deposits overlying the Chalk. This suggests that these solution features developed at a time when the Chalk was overlain by Palaeogene deposits. This cover has subsequently been removed by erosion.

Perhaps the most convincing evidence of the link between cover and rapid sub-surface groundwater flow is from a study of indicator bacteria in untreated groundwater samples. The occurrence of these bacteria in groundwater samples is correlated to the presence of Palaeogene deposits (Fig. 7). It has already been suggested that the presence of bacteria in groundwater samples is an indicator of rapid flow, particularly in the unsaturated zone, since the bacteria must have travelled from the ground surface to the borehole in a matter of weeks. Unlike the tracer tests above, this evidence is not biased by being taken from visible karst features – the samples were taken from boreholes throughout a large area.

Many transmissivity data are available for the same area as the indicator bacteria. There was no correlation found between high transmissivity and distance from the Palaeogene deposits. Also, sites with high transmissivity did not necessarily have a high frequency of indicator bacteria in untreated samples. The occurrence of bacteria (and therefore rapid groundwater flow) may therefore be predicted more accurately by the distance from Palaeogene deposits than by the measured transmissivity value. This can be explained by returning to the pumping test. Often pumping tests are carried out when water levels are low, to estimate the available resources. Pumping tests do not test the unsaturated zone and also integrate the effect of individual fractures. Transmissivity calculated from pumping tests is therefore not a good indicator of the possibility of rapid groundwater flow.

Several factors could account for the high degree of solution activity associated with cover:

1. soils associated with Palaeogene deposits and Clay-with-Flints tend to be quite acidic (Edmunds *et al.* 1992);
2. chalk soils are generally permeable, but those associated with cover can be quite clayey and therefore concentrate runoff to discrete points;
3. as recharge drains through the cover, it remains undersaturated with respect to calcite, thus allowing the acidic recharge to reach the Chalk surface.

In summary, whatever the mechanism, rapid groundwater flow appears to be correlated with cover. Most evidence highlights the important role of the Palaeogene deposits; more work will

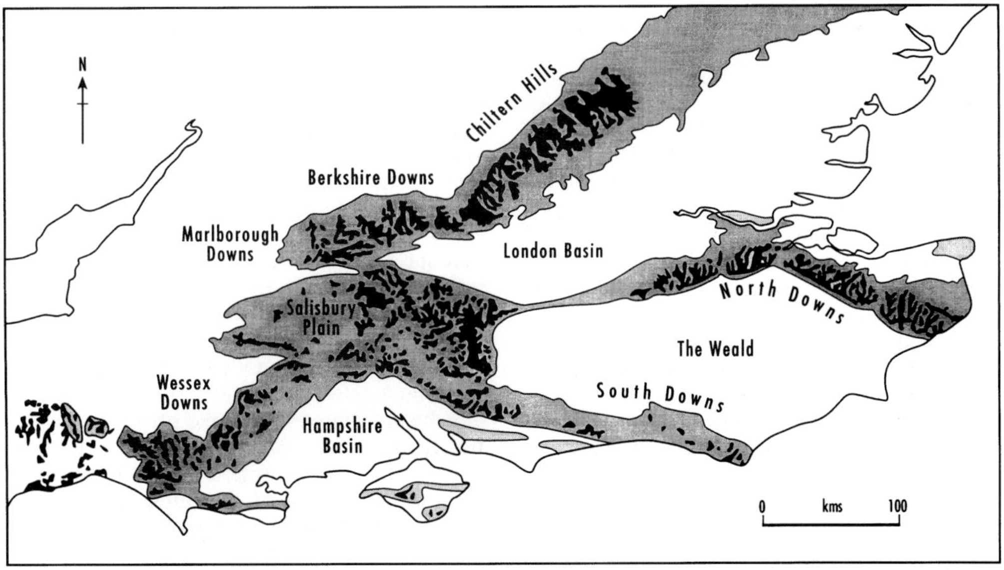

Fig. 8. The distribution of Clay-with-Flints on the Chalk outcrop of southern England (based on Goudie 1990, taken from Allen *et al.* in press).

need to be undertaken to find out if, like surface karst features, rapid groundwater flow is associated to the same degree with other forms of cover.

Valleys

There is evidence, albeit rather limited, that suggests rapid groundwater flow may also be found in major valleys. Some of the tracer tests, although close to Palaeogene deposits, were also in major valleys, e.g. the Blue Pool in the River Pang (Banks *et al.* 1995) and the Water End swallow holes in Hertfordshire (Harold 1937). Also, the large number of swallow holes found in the bed of the River Mole have formed because of the presence of the river (Fagg 1958).

The distinct pattern observed in the ephemeral sections of the Wessex Chalk streams indirectly suggests the presence of swallow holes and springs in valleys. Discrete features, such as individual springs and sinks in the river-bed, are probably responsible for the streams drying up in distinct stages. Below the stream-bed there may be a series of fractures joining the springs and sinks, although no direct evidence exists for this. Although not part of the evidence presented above, geophysical borehole logging in valleys often highlights a high degree of fracturing at the top of the Chalk (e.g. Owen & Robinson 1978); where this is connected to the ground surface by dissolution features, rapid groundwater flow could occur between ground surface and borehole.

Chalk valley bottoms have undergone a complex evolution and any number or combination of factors may be responsible for the development of rapid groundwater flow features. For example, periglaciation (Younger 1989), high historic rainfall, changes in sea-level, and the tectonic history may all have contributed to the development of valleys and high-permeability zones within the valleys.

These valley 'karst'-type features may also be developed in today's environment. Surface water flowing within streams has different chemistry from the groundwater; mixing with groundwater perturbs the chemistry and can produce water with a higher dissolution potential. This aggressive water, coupled with the high flux through the valleys and the dynamics of surface water flow, could produce sink holes and springs in stream-beds, possibly linked with conduits (Fig. 9). Docherty (1971), when reviewing the work of Fagg in the Mole Gap, suggested that such karstic features might exist in many valleys that used to contain flowing streams, although the surface expressions might now be concealed by cover deposits.

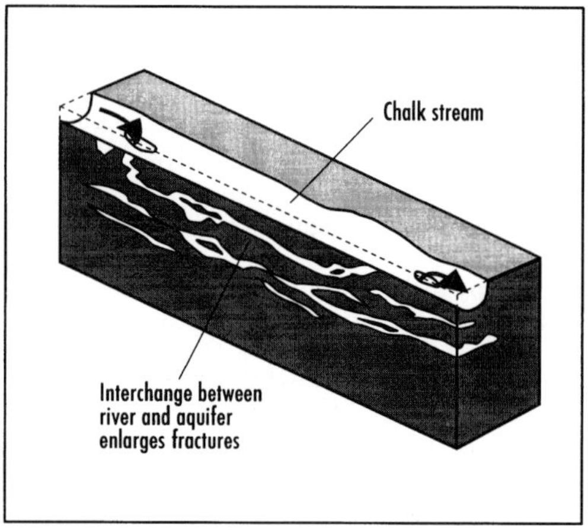

Fig. 9. The development of discrete dissolution-enhanced fractures beneath Chalk streams (from Allen *et al.* 1997).

There is presently not as much evidence for rapid groundwater flow in valleys as there is close to Palaeogene cover. Direct study of valley bottoms for surface karst features, possibly below cover, would help to establish the significance of rapid flow. Tracer tests could be used to help prove the existence of rapid links between valley bottoms and the underlying aquifer.

Conclusions

Evidence for rapid groundwater flow is mainly qualitative, with tracer tests being the only quantitative method of gaining data. Taking all the various strands of evidence together, however, allows the hydrogeologist a slightly different view of the Chalk aquifer. The potential for rapid groundwater flow in the Chalk may be widespread, but rapid flow might only occur infrequently, when the features are saturated. There is a large amount of evidence that suggests rapid groundwater flow is more frequent close to Palaeogene deposits. Other forms of cover, surface karstic features, and valley bottoms may also be associated with rapid groundwater flow, but further investigation is needed to establish a definite link. Examining the Chalk aquifer with tools other than the pumping test might expose more of the karstic nature of the Chalk aquifer; for example, interdisciplinary geomorphological studies, packer tests, tracer tests and examination of adits and tunnels would provide valuable additional hydrogeological data.

This paper is published with the permission of the Director, British Geological Survey. The work on which the paper is based was carried out as part of the Aquifer Properties Review funded jointly by the British Geological Survey and the Environment Agency. Specific elements of the research were supported by Thames Water Utilities. The study could not have been carried out without the enthusiasm and cooperation of staff from these organizations.

References

ALEXANDER, L. S. 1981. *The hydrogeology of the Chalk of South Dorset*. PhD Thesis, University of Bristol, Bristol.

ALLEN, D. J., BREWERTON, L. J., COLEBY, L. M., GIBBS, B. R., LEWIS, M. A., MACDONALD, A. M., WAGSTAFF, S. J. & WILLIAMS, A. T. 1997. *The physical properties of major aquifers in England and Wales*. British Geological Survey Technical Report WD/97/34.

ATKINSON, T. C. & SMART, P. L. 1981. Artificial tracers in hydrogeology. *In*: *A Survey of British Hydrogeology, 1980*. The Royal Society, London, 173–190.

—— & SMITH, D. I. 1974. Rapid groundwater flow in fissures in the Chalk: an example from south Hampshire. *Quarterly Journal of Engineering Geology*, **7**, 197–205.

BANKS, D., DAVIES, C. & DAVIES, W. 1995. The Chalk as a karstic aquifer: evidence from a tracer test at Stanford Dingley, Berkshire, UK, *Quarterly Journal of Engineering Geology*, **28**, S31–S38.

DOCHERTY, J. 1971. Chalk Karst, *Proceedings of the Croydon Natural History and Scientific Society*, **15**(2) 21–34.

DOWNING, R. A. 1993. Groundwater resources, their development and management in the UK: an historical perspective. *Quarterly Journal of Engineering Geology*, **26**, 335–358.

EDMONDS, C. N. 1983. Towards the prediction of subsidence risk upon the Chalk outcrop, *Quarterly Journal of Engineering Geology*, **16**, 261–266.

EDMUNDS, W. M., DARLING, W. G., KINNIBURGH, D. G., DEVER, L. & VACHIER, P. 1992. *Chalk groundwater in England and France: hydrogeochemistry and water quality*. British Geological Survey Research Report **SD/92/2**.

EUROPEAN COMMISSION. 1995. *Cost action 65. Hydrogeological aspects of groundwater protection in karstic areas. Final report*. Office for Official Publications of the European Communities, Luxembourg, **EUR 16547**.

FAGG, C. C. 1958. Swallow holes in the Mole gap. *South East Naturalist and Antiquary*, **62**, 1–13.

FOGG, T. 1984. Surprise find in Irish Chalk, *Caves and Caving*, **26**, 26.

FOSTER, S. S. D. & MILTON, V. A. 1974. The permeability and storage of an unconfined chalk aquifer. *Hydrological Sciences Bulletin*, **19-4**, 485–500.

GILES, D. M. & LOWINGS, V. A. 1990. Variation in the character of the Chalk aquifer in east Hampshire. *In*: BURLAND, J. B., MORTIMORE, R. N., ROBERTS, T. S., JONES, D. L. & CORBETT, B. O. (eds) *Chalk*. Thomas Telford, London, 619–626.

GOUDIE, A. S. 1990. *The Geomorphology of England and Wales*. Basil Blackwell, Oxford.

HAROLD, C. H. H. 1937. *Thirty-second annual report on the results of the chemical and bacteriological examination of the London waters for the twelve months ended 31st December, 1937*. Metropolitan Water Board.

HOUSTON, J. F. T., EASTWOOD, J. C. & COSGROVE, T. K. P. 1986. Locating potential borehole sites in a discordant flow regime in the Chalk aquifer at Lulworth using integrated geophysical surveys, *Quarterly Journal of Engineering Geology*, **19**, 271–282

INSTITUTE OF HYDROLOGY AND BRITISH GEOLOGICAL SURVEY. 1996. *Hydrological Data United Kingdom: 1995 Handbook*. Wallingford, Oxfordshire.

KACHI, S. 1987. *Tracer studies in the Chalk aquifer near Cambridge*. MPhil, University of East Anglia.

LEWIS, W. J., FOSTER, S. S. D. & DRASAR, B. S. 1980. *The risk of groundwater pollution by on-site sanitation in developing countries: a literature review*. International Research Centre for Waste Disposal (IRCWD), Duebendorf, Switzerland, Report No. **01/82**.

LOWE, D. J. 1992. Chalk caves revisited. *Cave Science*, **19**, 55–58.

MUSTCHIN, C. J. 1974. *Brighton's Water Supply from the Chalk 1834–1956: A history and description of the heading systems*. Brighton Corporation Water Department, Brighton.

NATIONAL RIVERS AUTHORITY 1992. *Policy and Practice for the Protection of Groundwater*. National Rivers Authority, Bristol.

—— 1993. *Chichester Chalk Investigation, Double Packer Testing*. National Rivers Authority, Southern Region.

OWEN, M. & ROBINSON, V. K. 1978. Characteristics and yield in fissured Chalk. *Symposium on Thames Groundwater Scheme*, Institution of Civil Engineers. Paper 2, 33–49

POSFORD DUVIVIER. 1994. *River Lavant Flood Investigation*. Posford Duvivier, Haywards Heath.

PRICE, M. 1987. Fluid flow in the Chalk of England. *In*: GOFF, J. C. & WILLIAMS, B. P. J. (eds) *Fluid Flow in Sedimentary Basins and Aquifers*, Geological Society, London, Special Publication, **34**, 141–156.

——, ATKINSON, T. C., BARKER, J. A., WHEELER, D. & MONKHOUSE, R. A. 1992. A tracer study of the danger posed to a Chalk aquifer by contaminated highway run-off. *Proceedings of the Institution of Civil Engineers, Water, Maritime & Energy*, **96**, 9–18.

——, DOWNING, R. A. & EDMUNDS, W. M. 1993. The Chalk as an aquifer. *In*: DOWNING, R. S., PRICE, M. & JONES, G. P. (eds) *The Hydrogeology of the Chalk of North-west Europe*. Clarendon, Oxford, 14–34.

PROCTOR, C. 1984. Chalk Caves in Devon. *Caves and Caving*, **26**, 31.

REEVE, T. J. 1976. Cave development in Chalk at St Margarets Bay, Kent. *British Cave Research Association*, **11**, 10–12.

—— 1977. Chalk caves in Sussex. *British Cave Research Association*, **18**, 3.

—— 1981. Beachy head cave. *Caves and Caving*, **12**, 2–5.

—— 1982. Flamborough Head. *Caves and Caving*, **17**, 2–3.

Richards, H. M. & Brincker, J. A. H. 1908. The potential dangers of water derived from wells in the Chalk. *Proceedings of the Royal Society of Medicine (Epidemiological Section)*, **1**, 191–203.

Southern science 1992. *A historical review of the Warningcamp borehole and recommendations for future work*. Southern Science, Worthing, Report No. **92/6/451**.

—— 1994. *The effect of pumping from Tortington and Madehurst in 1993 on water levels in the vicinity of Swanbourne Lake*. Southern Science, Worthing, Report No. **94/7/800**.

Sperling, C. H. B., Goudie, A. S., Stoddart, D. R. & Poole, G. G. 1977. Dolines of the Dorset Chalklands and other areas in southern Britain. *Transactions of the Institute British Geographers*, **NS2**, 205–223.

Taylor, S. M. 1995. The Chichester flood, January 1994. *In*: *Hydrological data UK, 1994 yearbook*. The Institute of Hydrology, Wallingford.

Ward, R. S. 1989. *Artificial tracer and natural* ^{222}*Rn studies of the East Anglian Chalk aquifer*. PhD Thesis, School of Environmental Sciences, University of East Anglia.

—— & Williams, A. T. 1995. *A tracer test in the Chalk near Kilham, North Yorkshire*. British Geological Survey, Technical Report **WD/95/7**.

Water Authorities Association. 1986. *Waterfacts*. Water Authorities Association, London.

West, G. & Dumbleton, M. J. 1972. Some observations on swallow holes and mines in the Chalk, *Quarterly Journal of Engineering Geology*, **5**, 171–177

Younger, P. L. 1989. Devensian periglacial influences on the development of spatially variable permeability in the Chalk of southeast England. *Quarterly Journal of Engineering Geology*, **22**, 343–345.

Groundwater recharge: an overview of estimation 'problems' and recent developments

I. SIMMERS

Faculty of Earth Sciences, Vrije Universiteit, De Boelelaan 1085, 1081 HV Amsterdam, The Netherlands

Abstract: Since the mid-1980s there has been a relative explosion of recharge studies for (semi-)arid regions reported in the scientific literature. It is therefore relevant to assess what we now know and to offer guidance to the water resources development practitioner. This paper summarizes recurring recharge estimation 'problems' and reviews some recent advances in (*inter alia*) estimation techniques and recharge time/space variability.

The determination of recharge fluxes in (semi-)arid regions remains fraught with uncertainty; multiple tracer approaches appear to offer the best potential for reliable results in local studies which require 'at-point' information. It is also clear that sufficient advances have been made in recent years to show that the value of water balance and Darcian approaches should not be under-estimated. The frequently studied issues of localized recharge and spatial variability need not be a problem if concern is with regional estimates. The key for the practitioner is the project objective; this will dictate whether multiple 'at-point' or area-based estimation methods are appropriate. For the latter, a combination of reliable local data, remote sensing, GIS and geostatistical techniques offers considerable promise for a better understanding and determination of recharge over extended areas.

Groundwater use is of fundamental importance to meet the rapidly expanding urban, industrial and agricultural water requirements in (semi-)-arid areas. To quantify the current rate of groundwater recharge is thus a basic prerequisite for efficient groundwater resource management in these regions, where such resources are often the key to economic development. It is indeed unfortunate that of all the factors in the evaluation of groundwater resources, this rate of aquifer replenishment is one of the most difficult to derive with confidence.

Attention in this discussion focuses on recharge of phreatic aquifers, often the most readily available and affordable source of water in (semi-)arid regions. These aquifers are also the most susceptible to contamination, with the recharge rate determining their level of vulnerability.

Historical framework

The above aspects serve to illustrate the growing international demand for reliable quantitative information on (semi-)arid zone groundwater recharge estimation and the determination of its variability in both time and space.

Early responses to these demands are typified by the *ad hoc* norms for estimating groundwater recharge introduced by the Government of India (1984). Such norms, despite their (now) evident shortcomings, have proved of value to India. However, the data are the result of a pioneering study based largely on empirical information and are country-specific; the international community requires guidelines for wider application.

Since the mid-1980s there has been a relative explosion of recharge studies reported in the scientific literature. Recognition of the growing need for reliable recharge estimation has also been reflected in the active support by international agencies and non-government organizations (NGOs), and the publications which have emerged from various international meetings (e.g. Simmers 1988, 1997; Sharma 1989; Lerner *et al.* 1990; Bredenkamp *et al.* 1995).

Recharge processes

The various sources of recharge to a groundwater system are of course well-known. Principal recharge mechanisms from these sources have been defined by Lerner *et al.* (1990) as direct, localized or indirect.

- *Direct recharge*: water added to the groundwater reservoir in excess of soil moisture deficits and evapotranspiration, by direct vertical percolation through the unsaturated zone.
- *Localized recharge*: an intermediate form of groundwater recharge resulting from the horizontal (near-)surface concentration of water in the absence of well-defined channels.
- *Indirect recharge*: percolation to the watertable through the beds of surface water courses.

SIMMERS, I. 1998. Groundwater recharge: an overview of estimation 'problems' and recent developments. *In*: ROBINS, N. S. (ed.) *Groundwater Pollution, Aquifer Recharge and Vulnerability*. Geological Society, London, Special Publications, **130**, 107–115.

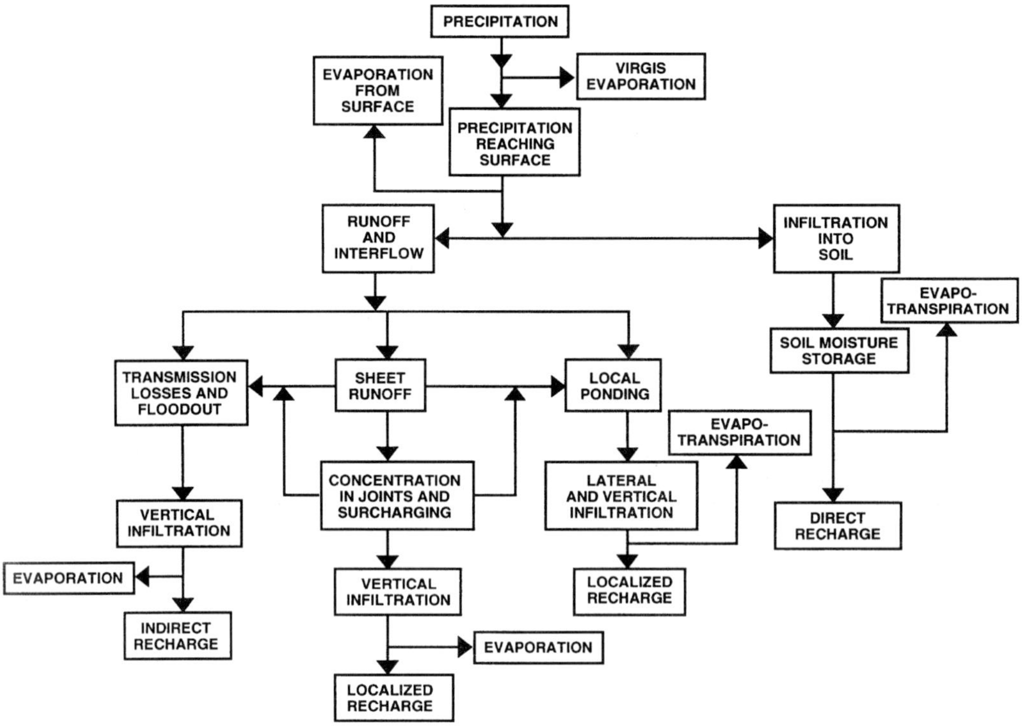

Fig. 1. The various elements of recharge in a (semi-)arid area (after Lloyd 1986).

Figure 1, from Lloyd (1986), presents a conceptual illustration. The definitions are of course a simplification of reality, since in many locations combinations of the various types will occur. It is also clear from the literature that:

(1) recharge occurs to some extent in even the most arid regions;
(2) as aridity increases, direct recharge is likely to become less important than localized and indirect recharge in terms of total aquifer replenishment;
(3) estimates of direct recharge are likely to be more readily derived than those of either localized or indirect recharge.

Recharge estimation

Estimation methods. Differences in sources and processes of groundwater recharge will mean that the applicable value of available estimation techniques will vary. Although direct recharge is known to be of decreasing significance with increasing aridity, the processes are conceptually the easiest to define and still form the basis of numerous recharge estimation techniques in common (but at times inappropriate) use. Further, the actual frequency of recharge events and the transit time until recharge takes place are also important.

The procedures which may be used to quantify recharge from the various sources (i.e. direct measurement, water balance methods, Darcian approaches, tracer techniques, empirical methods) and the problems encountered with each, are addressed in detail by, *inter alia*, Gee & Hillel (1988) and Lerner *et al.* (1990).

Since groundwater recharge estimation is an iterative process, with hopefully increasing precision following successive iterations, the water resources practitioner should particularly note that:

- realistic estimation depends on first identifying the probable flow mechanisms and important features influencing recharge for a given locality;
- it is desirable to apply and compare a number of independent approaches; Lerner *et al.* (1990) offer practical advice on the well-known methods and their limitations;

- model development is too often seen as an alternative to fieldwork and the model applied may be inappropriate to actual field conditions.

Challenges in estimating recharge

(Semi-)arid zone recharge, while generally low, can be highly variable; the greater the aridity, the smaller and potentially more variable the natural flux. In addition to the perennial difficulties associated with sparse information, numerous recharge estimation 'problems' recur in a scan of the literature to the early 1990s. These include:

- the assessment and hydrological consequences of *localized* and *indirect* recharge;
- recharge time and space variability (*inter alia*: effects of climatic and land-use changes on tracer profiles/mass balances; the spatial extrapolation of 'at-point' data; determination of reliable (representative) water balance and soil hydraulic parameters);
- the impacts of urban development on groundwater recharge.

A review of the more recent literature shows that not all these 'problems' have been satisfactorily resolved. Equally evident, however, is that considerable progress has been made. This paper briefly addresses the above and other challenges identified, outlines recent developments and, where possible, offers guidance for potential solutions.

Recharge from precipitation

Direct recharge estimation. In principle, one can determine recharge through the vadose zone by direct measurements of water fluxes or by establishing a vadose zone water balance on the basis of soil, climate and vegetation data. However, in arid and semi-arid climates such methods are not straightforward; comprehensive recent reviews are provided by Stephens (1994) and Allison *et al.* (1994).

In their evaluation of the range of techniques available for estimating recharge from precipitation the latter authors indicate that mutually supporting, multiple tracer approaches, or a combination of tracer and physical methods, are now the norm rather than an exception (e.g., Cook *et al.* 1994; Phillips 1994; Tyler & Walker 1994; Leaney & Herczeg 1995; Liu *et al.* 1995; Nativ *et al.* 1995; Wood & Sanford 1995).

The fundamental difficulty with Darcy's law solutions for water movement in the unsaturated zone is that they require knowledge of the soil water retention and unsaturated hydraulic conductivity curves; direct determination of these is difficult. However, indirect methods for estimating soil hydraulic properties have now been developed from easily measured soil data (Rawls *et al.* 1991; Van Genuchten *et al.* 1992), which often yield sufficiently precise information for many practical applications without increasing costs.

A further technique which is re-emerging, with reported success, uses upper vadose zone changes in soil temperature. As an example, Taniguchi & Sharma (1993) derived recharge rates from the method which corresponded closely to those estimated using natural Cl and artificial Br tracers.

Recharge time and space variability. Variations in groundwater recharge with time are well-documented; Fig. 2 offers a further example from work recently completed in Botswana (Gieske 1992). Although the figure illustrates model output using measured daily rainfalls, comparable results were also obtained from an 80–year stochastically generated rainfall sequence. An obvious implication from these studies is that when significant recharge results only from infrequent large events, it is highly misleading to talk of mean annual recharge, or of recharge as a proportion of mean annual rainfall. Barnes *et al.* (1994) further propose that one implication of this episodic mechanism is that a climatic variation which results in an increased mean rainfall could be offset by a decrease in variability, leading to a reduction in mean annual recharge. Such speculation assumes, however, that the decreased variability is accompanied by fewer major (recharge-causing) events.

Of the two variability aspects, it has often been stated that the more important problem yet to be overcome for practitioners is the assessment and prediction of recharge spatial variability. The issue has obvious practical importance; if recharge estimation is based on point measurements, water resource managers will need to know how or whether these relate to values over a specified area of interest. However, the degree of 'problem' is directly related to project objectives (and hence scale). If concern is with acquiring good recharge estimates over a limited area (e.g. for waste disposal or local water supply purposes), then the need for detailed information is evident. In this situation, multiple 'at-point' investigations are appropriate, with identification of the preferential flow contribution a prere-

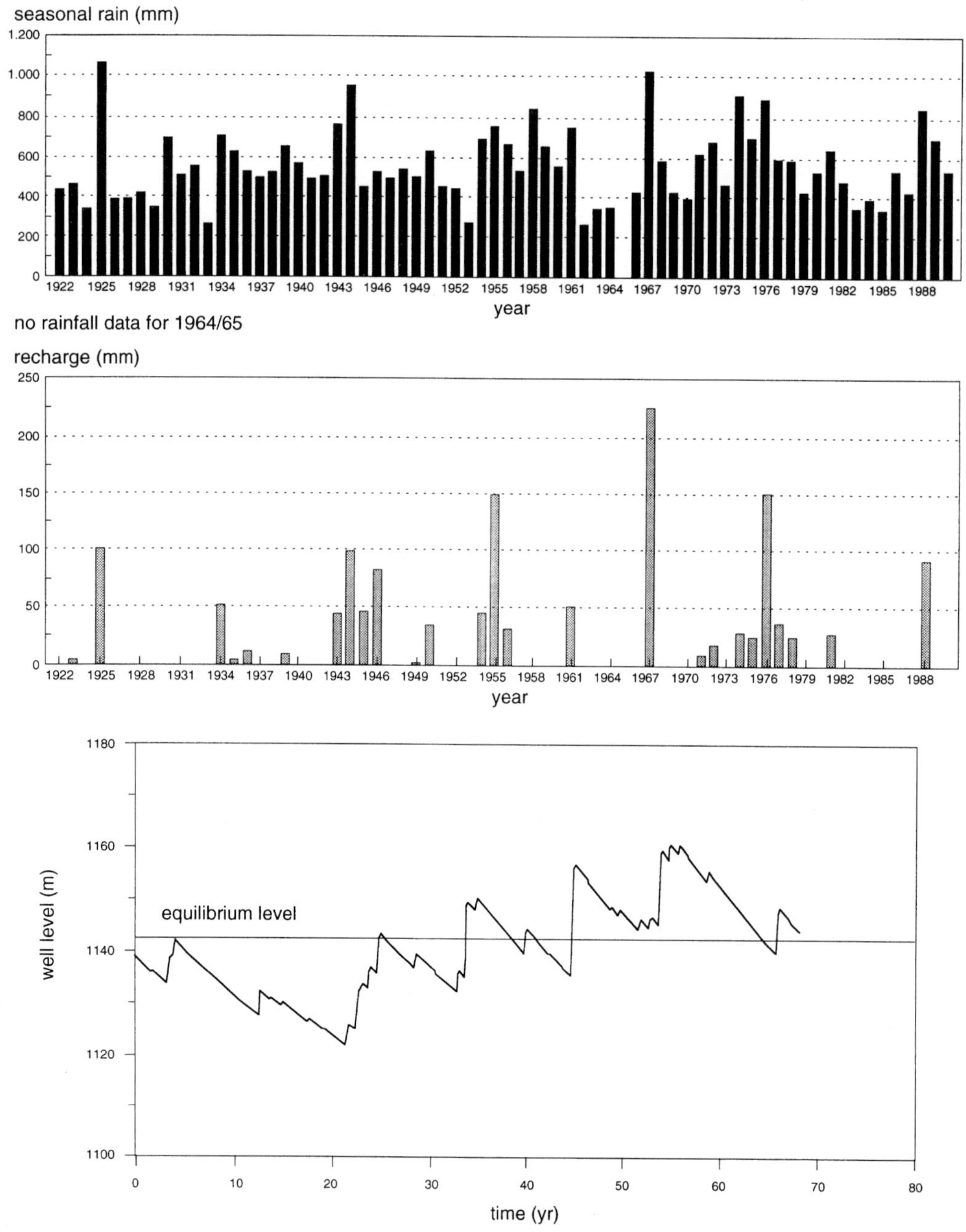

Fig. 2. Results from a 68-year recharge simulation using the model EARTH and Lobatse (Botswana) daily rainfalls (Gieske 1992).

quisite. Conversely, for projects on a regional scale, or those requiring only preliminary recharge estimates, area- or groundwater-based methods are relevant and small-scale variability in local recharge ceases to be an issue.

To illustrate the latter, several studies have shown that it is quite possible to derive what appear to be very reasonable initial estimates of recharge over extended areas, using readily derived low-cost field data, without considering the complicating aspects of small-scale (local) variability; good examples are given by Athavale *et al.* (1992), Edmunds & Gaye (1994) and Leaney & Herczeg (1995). Other recent methods used to estimate recharge over extended areas reflect, to some degree, a proposal in Lerner *et al.* (1990) that 'some potential lies in a combination of field measurements, remote sensing and geostatistics'. Examples of this approach are afforded by Sophocleous (1992), Kennett-Smith *et al.* (1994) and Salama *et al.* (1994*b*). Results indicate (*inter alia*) that despite its present shortcomings, the areal water balance method can be a powerful tool to understand the main features of recharge processes if short time periods are considered and the variability of the water balance components are taken into account. To this end, it is expected that procedures arising from, for example, the ongoing FIFE programme (Sellers *et al.* 1992) will make a valuable contribution, particularly with respect to determining evapotranspiration and soil moisture status over an extended target area.

Recharge and land-use change. With regard to the hydrological consequences of land-use change, many earlier attempts to estimate recharge under such situations have generally assumed that evapotranspiration and precipitation are the only time-variant factors affecting the various processes. The obviously related issue, however, is when transit time until recharge is long and land-use changes are known to have occurred in an area prior to current data collection programmes. In this situation the hydrological system is in a state of dynamic evolution; numerous examples are cited in Allison *et al.* (1994) and Phillips (1994). Solutions are usually sought from 'at-point' Cl profiles which, under non-steady-state conditions, typically display complex shapes. An example of this approach is the study by Walker *et al.* (1991) using a combination of tracer (Cl) and physical (suction profile) methods. The described procedure remains valid even when the chloride profiles become distorted; it also allows an estimate to be made of probable time delay before the land-use change effects reach the water-table.

Localized recharge estimation. Recharge along preferential flowpaths has received considerable attention in the past decade, with an increasing number of studies verifying the significance of the process and attempting to quantify its relative contribution to the groundwater body. The literature also clearly indicates that local groundwater recharge models which ignore the possibility of such phenomena may be highly misleading. An overview of the topic is given by Stephens (1994).

Gee & Hillel (1988) visualize localized recharge as occurring on three spatial scales:

(a) 'micro-scale' pathways, several centimetres or decimetres apart, such as those formed by shrinkage cracks, roots and burrowing animals;
(b) 'meso-scale' flow paths, with a spacing of several metres or tens of metres, initiated by local topographic or lithological variations;
(c) 'macro-scale' flow paths, spaced hundreds (or more) metres apart, caused by major landscape features such as karst sinks or playa basins.

Such variations in local recharge have been demonstrated and quantified for many soil types. Typical examples are presented by Gieske (1992), Scanlon (1992), Nativ *et al.* (1995) and Wood & Sanford (1995); all used isotopic and/or chemical tracer techniques to solve their specific problem. The Wood & Sanford (1995) example is particularly interesting, since it addresses both surface run-off and spatially variable vertical fluxes. Although the research has a regional perspective, much of the effort focused on recharge processes in and around an ephemeral playa lake. The results show that nearly 50% of total recharge on the Texas and New Mexico High Plains occurs as piston flow through playa basin floors occupying only 6% of the area. The remaining recharge, from 94% of the area, comes from macropore flux near the basin floors and diffuse infiltration over the region.

An alternative approach to solving the variability problems inherent in localized recharge estimation is the combination of frequency-domain electromagnetic (EM) measurements and 'at-point' Cl mass balances. The method is not new conceptually, but more recent studies (e.g. Cook *et al.* 1992; Salama *et al.* 1994*a*) have shown promise.

Recharge from intermittent flow

The two principal mechanisms of natural groundwater recharge under conditions of ephemeral flow are mountain front and channel recharge; in (semi-)arid regions both types frequently occur in the same drainage basin and are difficult to separate in practice. Descriptions of their respective physical and hydrological characteristics are given by Lerner *et al.* (1990).

Mountain front recharge. With respect to mountain front recharge, estimates of this component are vital for management purposes in areas where total recharge is small and development may lead to overdraft conditions. Determination by water balance calculations is unreliable, and methods based on either hydrochemical mass balances or environmental isotopes are shown to be associated with large uncertainties. An alternative approach has recently been proposed by Chavez *et al.* (1994*a,b*), who developed and tested an analytical seasonal streamflow model which includes mountain front recharge as one of the parameters. The model is formulated in terms of parameters with physical significance, makes use of normally recorded climatic / hydrometric data and favours remotely sensed input of basin characteristics.

Recharge from ephemeral channels. The most common approaches to determining recharge from wadis have been by water balance methods, empirical formulae, isotope tracer studies and Darcian approaches. None are straightforward and the empirical formulae, in particular, are site-specific. Examples of recent detailed studies are afforded by Abdulrazzak *et al.* (1989), Parissopoulos & Wheater (1992), Hughes & Sami (1992), Sorman & Abdulrazzak (1993) and Scanlon (1994). All target the unsaturated response of wadi alluvium under field conditions and require an intensive data collection programme; with the possible exception of the Tabalah (Saudi Arabia) study, none addressed the issue of spatial heterogeneity in bed material. It is clear that the relationship between 'at-point' hydraulic properties and reach transmission losses demands further investigation.

Recharge from permanent water bodies

Typical examples of seepage recharge from permanent water-bodies are losses from rivers or lakes and irrigation losses from canals or through the bunds of flooded fields. In urban situations recharge occurs from water mains, sewers and drainage ditches. Since the processes involved and the numerous techniques used to estimate recharge from these line sources are well-documented, the present discussion emphasizes aspects which are either new or innovative.

Seepage recharge from rivers, canals and flooded rice irrigation. Realistic estimates of seepage recharge from rivers rely primarily on field measurements of flow losses. Recent examples of this approach are detailed by Knighton & Nanson (1994) and by Zellweger (1994); the latter study made use of multiple tracer (lithium, sodium, chloride and bromide) continuous injection.

With respect to canals, earlier guidelines such as those proposed by the Government of India (1984) have assumed minimal losses from lined canals. However, more recent ponding tests and water balance studies have demonstrated that large losses can occur which are rarely less than 25% of water supplied. It has also now been shown that time-varying losses in excess of 50% are typical from unlined canals and form an important component of total recharge in areas where there are canal irrigation schemes.

A similar situation has been demonstrated for flooded rice irrigation, with field and modelling studies indicating typical recharge water losses of 10–15 mm day^{-1}: 2 mm day^{-1} as vertical flow through the plough layer and 8–13 mm day^{-1} through the bunds. This last result is particularly surprising, but has been confirmed by Tuong *et al.* (1994) during a comprehensive study in the Philippines. Two practical recommendations from the latter work, with regard to reducing water losses, were to lower ponding water depths and to seal bund walls with mud from the plough layer.

Recharge in urban areas. Urban development can have a profound impact on the water balance of an area. The changes are even more marked when major water imports from outside the urban limits are involved, particularly when combined with a lack of (sub-)surface drainage; in such cases, as much as 90% of the imported water may be recharged. The effects vary with the type and stage of urbanization and between advanced and developing countries, but can give similar problems irrespective of climatic region.

Although urbanization can induce an increase in surface run-off and a reduction in direct recharge, new sources of recharge water are introduced. The most common net effect of these is to increase recharge to at least pre-urbanization rates (Foster *et al.* 1994; Fig. 3). To consider briefly each of the new recharge sources:

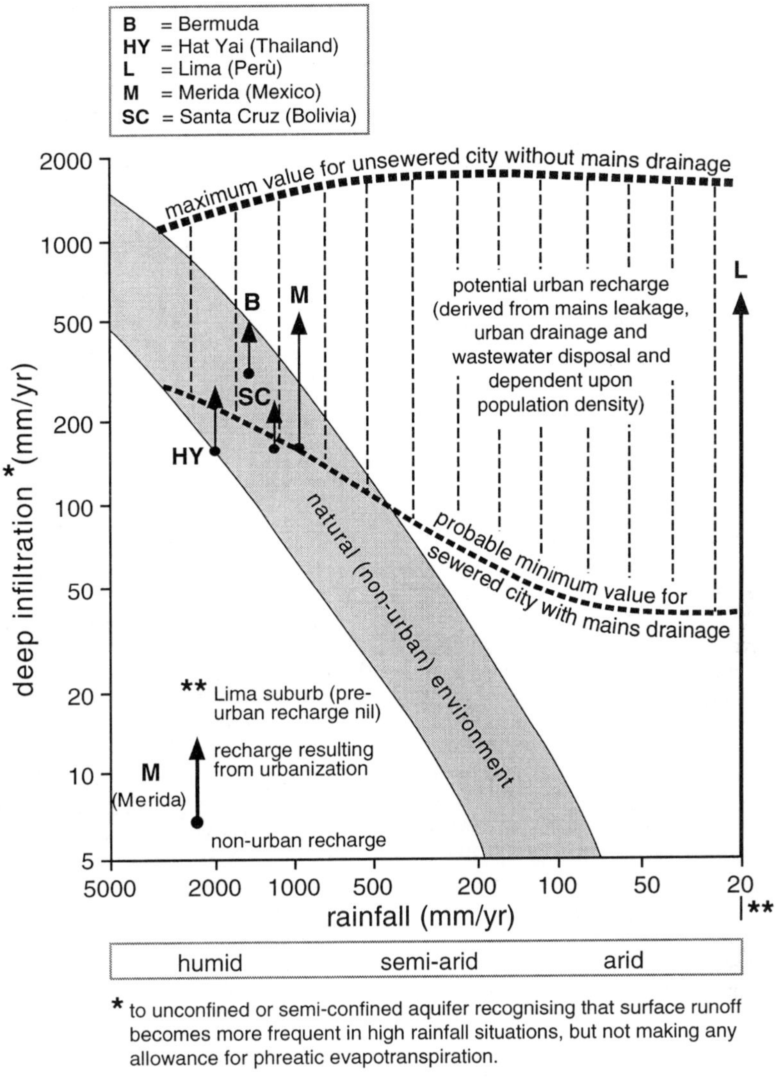

Fig. 3. Potential range of increase in recharge due to urbanization (Foster *et al.* 1994).

1. *Recharge from leaking water mains.* Two facts are clear from literature: (a) the amount of water in distribution systems can be large in relation to excess rainfall; and (b) since water mains are normally pressurized, leakage will occur from any location where there is a weakness, even if pipes are below the water-table.
2. Few authorities claim to be able to reduce leakage below 10% of water supplied and rates of 65% have been reported. This can create a potential recharge which is orders of magnitude greater than natural rates. Results from several recently reported studies show that mains leakage is rarely less than 20% and more typically in the order of 30% of supply. Published figures also indicate that this component can account for up to 45% of total urban recharge, though approximately 30% is again more usual.
3. *Recharge from septic tanks and leaking sewers.* Septic tanks, latrines and cesspits are an obvious source of polluted recharge, particularly in developing countries where the provision of mains sewage lags behind

urban population growth. For example, such sources are thought to contribute almost 40% of total urban recharge in Riyadh, 30% in Bermuda and 20% in Doha.

4. *Recharge from over-irrigation of parks and gardens.* Contributions from this source vary widely, but typically total 20–40% of total urban recharge. As illustration, an application rate of about 23 mm day^{-1} is reported for Riyadh, with ET demand at 8 mm day^{-1}. Such over-irrigation is particularly important in cities where high rates of urban recharge are associated with rising groundwater levels.

Many detailed data are needed to determine the effects of urbanization on recharge. The most direct method is to carry out field measurements for each of the component processes. However, to consider each individually, with their associated errors, can lead to a large error in the final estimate. Some alternatives are discussed by Lerner *et al.* (1990).

Concluding remarks

Although there are evident limitations associated with the well-established recharge estimation methods, Lerner *et al.* (1990) is a useful source of initial information. If necessary, as dictated by project requirements, the practitioner is encouraged to explore recent techniques identified in the present paper. A note of caution, however: many of the studies reported in recent articles are in response to specific local issues; it should therefore not be concluded that all the described procedures will prove equally reliable for general application.

Groundwater recharge estimation should be treated as an iterative process, allowing progressive aquifer response data collection and resource evaluation. There is also a need to use more than one technique to verify results; field measurements are a necessary component of a recharge investigation, at any scale, as they form the only means to realistically investigate recharge processes. Further, the principal criterion for all models is that they must represent the essential features of likely flow mechanisms.

Multiple tracer approaches appear to offer the best potential for reliable recharge results in local studies which require 'at-point' information. It is also clear that sufficient advances have been made in recent years to show that the value of water balance and Darcian methods should not be under-estimated.

The frequently investigated issues of localized recharge and spatial variability need not be a problem if concern is with regional estimates. The key for the practitioner, therefore, is the project objective; this will dictate whether multiple 'at-point' or area-based estimation methods are appropriate. It is also concluded that the combination of reliable local data, remote sensing, GIS technology and geostatistical techniques offers considerable promise for a better understanding and quantification of recharge over extended areas.

References

ABDULRAZZAK, M. J., SORMAN, A. U. & ALHAMES, A. S. 1989. Water balance approach under extreme arid conditions – a case study of Tabalah basin, Saudi Arabia. *Hydrological Processes*, **3**, 107–122.

ALLISON, G. B., GEE, G. W. & TYLER, S. W. 1994. Vadose-zone techniques for estimating groundwater recharge in arid and semiarid regions. *Soil Science Society of America Journal*, **58**, 6–14.

ATHAVALE, R. N., RANGARAJAN, R. & MURALIDHARAN, D. 1992. Measurement of natural recharge in India. *Journal of the Geological Society of India*, **39**, 235–244.

BARNES, C. J., JACOBSON, G. & SMITH, G. D. 1994. The distributed recharge mechanism in the Australian arid zone. *Soil Science Society of America Journal*, **58**, 31–40.

BREDENKAMP, D. B., BOTHA, L. J., VAN TONDER, G. J. & VAN RENSBURG, H. J. 1995. *Manual on quantitative estimation of groundwater recharge and aquifer storativity*. Water Research Commission, Pretoria, Report TT **73/95**.

CHAVEZ, A., DAVIS, S. N. & SOROOSHIAN, S. 1994*a*. Estimation of mountain front recharge to regional aquifers. 1. Development of an analytical hydroclimatic model. *Water Resources Research*, **30**(7), 2157–2167.

—— 1994*b*. Estimation of mountain front recharge to regional aquifers. 2. A maximum likelihood approach incorporating prior information. *Water Resources Research*, **30**(7), 2169–2181.

COOK, P. G., JOLLY, I. D., LEANEY, F. W., WALKER, G. R., ALLAN, G. L., FIFIELD, L. K. & ALLISON, G. B. 1994. Unsaturated zone tritium and chlorine-36 profiles from southern Australia: their use as tracers of soil water movement. *Water Resources Research*, **30**(6), 1709–1719.

——, WALKER, G. R., BUSELLI, G., POTTS, I. & DODD, A. R. 1992. The application of electromagnetic techniques to groundwater recharge investigations. *Journal of Hydrology*, **130**, 201–229.

EDMUNDS, W. M. & GAYE, C. B. 1994. Estimating the variability of groundwater recharge in the Sahel using chloride. *Journal of Hydrology*, **156**, 47–59.

FOSTER, S. S. D., MORRIS, B. L. & LAWRENCE, A. R. 1994. Effects of urbanization on groundwater recharge. *In*: *Proceedings of ICE International Conference on Groundwater Problems in Urban Areas*, London, 43–63.

GEE, G. W. & HILLEL, D. 1988. Groundwater recharge in arid regions: review and critique of estimation methods. *Hydrological Processes*, **2**, 255–266.

GIESKE, A. S. M. 1992. *Dynamics of groundwater recharge: a case study in semi-arid eastern Botswana*. PhD thesis, Vrije Universiteit, Amsterdam.

GOVERNMENT OF INDIA. 1984. *Ground water estimation methodology*. Report of the Ground Water Estimation Committee, Ministry of Irrigation, Government of India, Delhi.

HUGHES, D. A. & SAMI, K. 1992. Transmission losses to alluvium and associated moisture dynamics in a semiarid ephemeral channel system in southern Africa. *Hydrological Processes*, **6**, 45–53.

KENNETT-SMITH, A., COOK, P. G. & WALKER, G. R. 1994. Factors affecting groundwater recharge following clearing in the south western Murray Basin. *Journal of Hydrology*, **154**, 85–105.

KNIGHTON, A. D. & NANSON, G. C. 1994. Flow transmission along an arid zone anastomosing river, Cooper Creek, Australia. *Hydrological Processes*, **8**, 137–154.

LEANEY, F. W. & HERCZEG, A. L. 1995. Regional recharge to a karst aquifer estimated from chemical and isotopic composition of diffuse and localised recharge. *Journal of Hydrology*, **164**, 363–387.

LERNER, D. N., ISSAR, A. S. & SIMMERS, I. 1990. *Groundwater recharge. A guide to understanding and estimating natural recharge*. Heinz Heise, International Contributions to Hydrogeology, **8**.

LIU, B., PHILLIPS, F., HOINES, S., CAMPBELL, A. R. & SHARMA, P. 1995. Water movement in desert soil traced by hydrogen and oxygen isotopes, chloride, and chlorine-36, southern Arizona. *Journal of Hydrology*, **168**, 91–110.

LLOYD, J. W. 1986. A review of aridity and groundwater. *Hydrological Processes*, **1**(1), 63–78.

NATIV, R., ADAR, E., DAHAN, O. & GEYH, M. 1995. Water recharge and solute transport through the vadose zone of fractured chalk under desert conditions. *Water Resources Research*, **31**(2), 253–261.

PARISSOPOULOS, G. A. & WHEATER, H. S. 1992. Experimental and numerical infiltration studies in a wadi stream bed. *Hydrological Sciences Journal*, **37**(1), 27–37.

PHILLIPS, F. M. 1994. Environmental tracers for water movement in desert soils of the American southwest. *Soil Science Society of America Journal*, **58**, 15–24.

RAWLS, W. J., GISH, T. J. & BRAKENSIEK, D. L. 1991. Estimating soil water retention from soil physical properties and characteristics. *In*: STEWART, B. A. (ed.) *Advances in Soil Science*, Springer, **16**, 213–234.

SALAMA, R. B., BARTLE, G., FARRINGTON, P. & WILSON, V. 1994*a*. Basin geomorphological controls on the mechanism of recharge and discharge and its effect on salt storage and mobilization - a comparative study using geophysical surveys. *Journal of Hydrology*, **155**, 1–26.

——, TAPLEY, I., ISHII, T. & HAWKES, G. 1994*b*. Identification of areas of recharge and discharge using Landsat-TM satellite imagery and aerial photography mapping techniques. *Journal of Hydrology*, **162**, 119–141.

SCANLON, B. R. 1992. Moisture and solute flux along preferred pathways characterized by fissured sediments in desert soils. *J. Contaminant Hydrology*, **10**, 19–46.

—— 1994. Water and heat fluxes in desert soils. 1. Field studies. *Water Resources Research*, **30**, 709–719.

SELLERS, P. J., HALL, F. G., ASRAR, G., STREBEL, D. E. & MURPHY, R. E. 1992. An overview of the First Internat. Satellite Land Surface Climatology Project (ISLSCP) Field Experiment (FIFE). *Journal of Geophysical Research*, **97**(D17), 18 345–18 371.

SHARMA, M. L. (ed.) 1989. *Groundwater recharge*. Balkema, Rotterdam.

SIMMERS, I. (ed.) 1988. *Estimation of Natural Groundwater Recharge*. NATO ASI Series C, Reidel, Vol. 222.

—— (ed.) 1997. *Recharge of phreatic aquifers in (semi-)arid areas*. Balkema, Rotterdam, International Contributions to Hydrogeology, **19**.

SOPHOCLEOUS, M. 1992. Groundwater recharge estimation and regionalization: the Great Bend Prairie of central Kansas and its recharge statistics. *Journal of Hydrology*, **137**, 113–140.

SORMAN, A. U. & ABDULRAZZAK, M. J. 1993. Infiltration recharge through wadi beds in arid regions. *Hydrological Sciences Journal*, **38**(3), 173–186.

STEPHENS, D. B. 1994. A perspective on diffuse natural recharge mechanisms in areas of low precipitation. *Soil Science Society of America Journal*, **58**, 40–48.

TANIGUCHI, M. & SHARMA, M. L. 1993. Determination of groundwater recharge using the change in soil temperature. *Journal of Hydrology*, **148**, 219–229.

TUONG, T. P., WOPEREIS, M. C. S., MARQUEZ, J. A. & KROPFF, M. J. 1994. Mechanisms and control of percolation losses in irrigated puddled rice fields. *Soil Science Society of America Journal*, **58**, 1794–1803.

TYLER, S. W. & WALKER, G. R. 1994. Root zone effects on tracer migration in arid zones. *Soil Science Society of America Journal*, **58**, 25–31.

VAN GENUCHTEN, M. Th., LEIJ, F. J. & LUND, L. J. (eds) 1992. *Indirect methods for estimating the hydraulic properties of unsaturated soils*. Proceedings of 1989 International Workshop, California.

WALKER, G. R., JOLLY, I. D. & COOK, P. G. 1991. A new chloride leaching approach to the estimation of diffuse recharge following a change in land use. *Journal of Hydrology*, **128**, 49–67.

WOOD, W. W. & SANFORD, W. E. 1995. Chemical and isotopic methods for quantifying ground-water recharge in a regional, semiarid environment. *Ground Water*, **33**(3), 458–468.

ZELLWEGER, G. W. 1994. Testing and comparison of four ionic tracers to measure stream flow loss by multiple tracer injection. *Hydrological Processes*, **8**, 155–165.

Water transport through the unsaturated zone of the Middle Chalk: a case study from Fleam Dyke lysimeter

H. K. JONES[1] & J. D. COOPER[2]

[1]*British Geological Survey, Maclean Building, Crowmarsh Gifford, Wallingford, Oxfordshire, OX10 8BB, UK*

[2]*Institute of Hydrology, Maclean Building, Crowmarsh Gifford, Wallingford, Oxfordshire, OX10 8BB, UK*

Abstract: The role of fractures in transporting water in the unsaturated zone of the Chalk has been a source of controversy for many years. There is clearly a potential for fractures to transport water and contaminants rapidly to the saturated zone, although direct observations are rare. The evidence from well hydrographs, solute profiles and soil water measurements is ambiguous concerning the importance of by-pass flow.

This paper examines drainage data from several years' operation of a 5 m cube undisturbed monolith lysimeter at Fleam Dyke, Cambridgeshire, for evidence of by-pass flow and investigates the factors responsible for its onset. Evidence from measurements of unsaturated hydraulic conductivity of the Chalk at the site and *in situ* measurements of soil water are also presented. Both the lysimeter and soil water observations agree that flow through fractures accounts for approximately 30% of the total drainage from this site, significantly higher than has been inferred at other Chalk sites. By-pass flow was observed to occur above a threshold of approximately −50 cm H_2O matric potential, or above a flow rate of approximately 1 mm day^{-1}. During periods of no soil water deficit, about 50% of drainage occurred via fractures. In general, there was a delay of three or four days between rainfall and drainage at the base of the lysimeter. On once occasion, intense summer rainfall satisfied the existing soil moisture deficit and initiated fracture flow, the peak in drainage through the lysimeter occurring within 24 hours. Drainage from the lysimeter was consistently higher than MORECS estimates of hydrologically effective precipitation during the period after the onset of soil water deficits.

Mechanisms of recharge are, in general, poorly understood. Although it is generally accepted that water movement through the unsaturated Chalk is a combination of matrix flow (via larger pores and possibly microfractures) and fracture (or by-pass) flow, many questions still surround the mechanisms of recharge, and the relative contributions from fractures and matrix. Such issues have implications for the timing of recharge, as well as for the movement of solutes through the unsaturated zone.

The Fleam Dyke lysimeter was initially constructed to obtain direct measurements of natural recharge, without the need for rainfall and evaporation estimates (Kitching & Shearer 1982). The lysimeter is located at the Fleam Dyke pumping station near Fulbourn, Cambridgeshire (TL 539549), and is constructed within the lower beds of the Middle Chalk. It is essentially a 5 m cube of undisturbed Chalk, contained such that infiltrating water is collected at the base. Details of the construction may be found in Kitching & Shearer (1982).

Drainage, meteorological and soil physics data were collected over a period of years. The data were originally compared with the soil physics estimates of drainage at the site. They were later analysed to gain a better understanding of the processes operating in the unsaturated zone, the results of which are presented here. The paper initially reviews existing evidence for the occurrence of by-pass flow through the unsaturated Chalk and the attempts made to reconcile this with observed solute profiles. It then examines the available data for the Fleam Dyke lysimeter to assess the main controls on by-pass flow, and to estimate the proportion of water recharging the aquifer which by-passes the matrix. The data are also compared with MORECS estimates of hydrologically effective precipitation to assess the accuracy of the estimates for this site.

Terminology

There is some confusion over terminology, particularly that applied to fractures. Where authors have used a particular term, it is used in reference to their work. Otherwise the following terminology is used.

JONES H. K. & COOPER, J. D. 1998. Water transport through the unsaturated zone of the Middle Chalk: a case study from Fleam Dyke lysimeter. *In*: ROBINS, N. S. (ed.) *Groundwater Pollution, Aquifer Recharge and Vulnerability*. Geological Society, London, Special Publications, **130**, 117–128.

- *Fracture* – any discontinuity within the Chalk, unless otherwise specified. This term describes all discontinuities, including faults, joints and bedding plane discontinuities.
- *By-pass flow* – the process by which water moves through only a small volume of the total porosity. This term, rather than fracture flow, is used to describe rapid flow through the unsaturated zone.
- *Piston flow* – water movement resulting from the addition of water at the top of the unsaturated zone; water moves downwards displacing the entire water column.
- *Recharge event* – taken to be a period of recharge when rapid by-pass flow is occurring.
- *Matric potential* – negative pressure potential exerted by the matrix.
- *Matric hydraulic conductivity* – hydraulic conductivity of the matrix.

Previous work

Much work has been carried out to explore movement of water and solutes through the Chalk unsaturated zone.

Smith *et al.* (1970) investigated tritium movement through the unsaturated zone of the Upper Chalk at a site in Berkshire. They calculated that if transport were via intergranular seepage, tritium would not be expected below 13 m. However, tritium below this depth accounted for 15% of the total tritium found. The amount of tritium in the profile indicated that the classic concept that groundwater recharge was equivalent to rainfall minus actual evapotranspiration, was incorrect for the Chalk aquifer. Reduction of actual evaporation by 25% of that calculated by Penman's approach combined with allowing 15% of recharge to occur as by-pass flow produced a good fit to the data. A second model assumed rapid transport of recharging water through the weathered layer close to the surface, allowing 30% of the monthly rainfall to recharge by this means. This also provided a reasonable match with the tritium concentration profile. Smith *et al.* (1970) concluded that intergranular seepage dominated the downward movement of water, although a significant proportion of additional water must have moved through crack systems.

Foster (1975) attempted to reconcile the apparently small proportion of fracture flow suggested by this work. He assumed that water movement occurs mainly in the fracture systems, and proposed that the retardation of tritium movement through the unsaturated zone is due to diffusion between fractures and pore-water. If the equilibrium of tritium concentration between fracture and pore-water is sufficiently rapid, the concentration profiles appear almost identical to what would be expected if water movement were mainly through the pores.

Based on tritium profiles and other evidence (e.g. pollution of Chalk boreholes following periods of heavy rainfall), Downing *et al.* (1978) suggested that rapid infiltration through fractures could represent 10% of total recharge, with flow velocities of up to 50 m day^{-1} during and after periods of heavy rain. They estimated that water flowing through the microfissures and larger pores represented recharge velocities of between 1 and 4 m day^{-1}, although solutes moved at significantly slower rates through the unsaturated zone than would be expected as a consequence of piston displacement because of diffusion between fracture and pore-water. Citing the example of the Chilgrove borehole in West Sussex, they suggested that an infiltration rate of greater than 4 mm day^{-1} was required to initiate rapid by-pass flow.

Reeves (1979) reviewed existing work on the Chalk. He defined microfissures as having apertures up to 20 μm, and macrofissures as up to 10 mm aperture. He estimated that 80 to 90% of flow through the unsaturated zone takes place at rates of less than 1 mm day^{-1}, implying that only the microfissure system is permanently saturated during the infiltration period. If infiltration rates increase and exceed the microfissure capacity, the macrofissure systems will become active. Vertical rates of flow in macrofissures were estimated to be in the range 1 to 10 m day^{-1}, and could be several metres per hour through enlarged macrofissures.

Workers at the Institute of Hydrology used soil physics approaches in the late 1970s and early 1980s, initially to measure recharge to the aquifer beneath deep unsaturated zones of the Chalk (Cooper 1981; Wellings & Bell 1980; Wellings 1984*a*; Wellings & Cooper 1983; Gardner *et al.* 1991; Cooper *et al.* 1990), and later to measure nitrate flux to the water-table (Wellings 1984*b* Barraclough *et al.* 1994). Measurement of the unsaturated hydraulic conductivity of the *in situ* profiles at a number of Upper and Middle Chalk sites in southern England (Gardner *et al.* 1991) provided a range of matric hydraulic conductivities between 1 and 10 mm day^{-1} in the upper 3 m of the profile. This was then plotted against matric potential and inferred that by-pass flow becomes increasingly important as matric potentials rise above −50 cm H_2O. They showed that the hydraulic conductivity of the bulk Chalk remained almost constant from matric potentials of −600 cm H_2O and below to

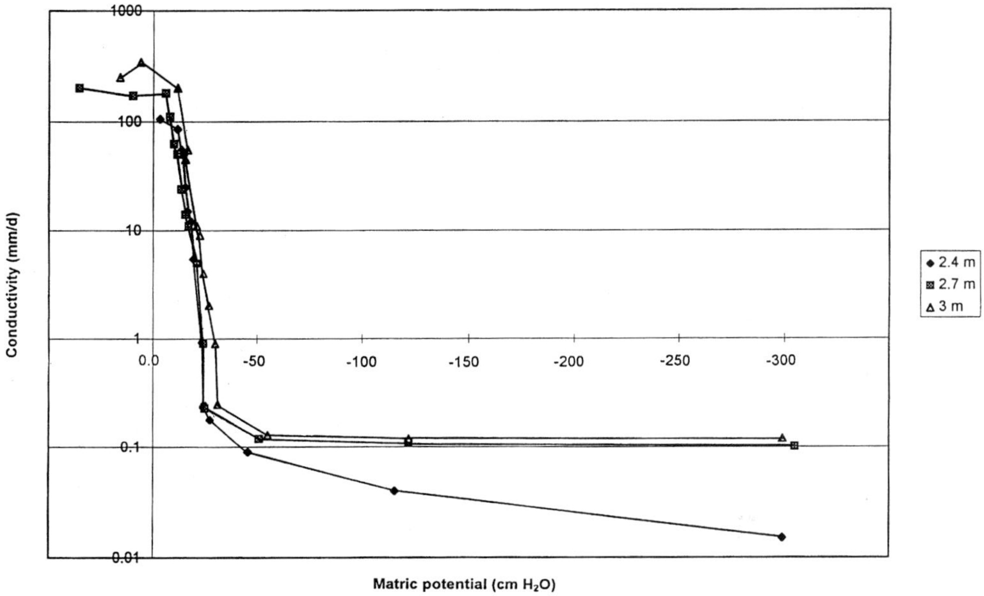

Fig. 1. Unsaturated hydraulic conductivity against matric potential for the Fleam Dyke site, Cambridgeshire.

-50 cm H_2O. Above this matric potential, the total hydraulic conductivity rose rapidly to values of up to 1 m day^{-1} (this is illustrated for Fleam Dyke in Fig. 1). this was taken to indicate that the fracture system contributes negligibly to the hydraulic conductivity of the Chalk up to matric potentials of about -50 cm H_2O, by-pass flow would be expected to occur.

The relationship between matric potential and hydraulic conductivity allowed episodes when fracture flow was occurring to be identified easily from the available tensiometer records. At Bridget's Farm (Hampshire; Upper Chalk), with a thick weathered zone and high matric conductivity (6 to 8 mm day^{-1}), matric potential throughout the upper 3 m of the profile during the winter months was in the region of -200 cm H_2O for most of the winter, well below the range that would be expected for the occurrence of fracture flow. Wellings (1984*a*) concluded that, at this site, only once in a five-year period was fracture flow observed, although there may have been occasions between the twice-weekly reading dates when such conditions occurred. In contrast, at Fleam Dyke (Cambridgeshire; Middle Chalk), where the soil is thin and matric hydraulic conductivity is low, Wellings & Cooper (1983) found that winter matric potentials were often only around -20 cm H_2O, suggesting that fracture flow was a common phenomenon. In fact, of the sites described by Gardner *et al.* (1991), by-pass flow would be likely to be initiated most frequently at Fleam Dyke.

The transport of nitrate and other solutes was estimated by multiplying the calculated water flux by the observed concentration of solute in cores recovered from the unsaturated zone down to 6 m depth. At both Bridget's Farm and at West Ilsley (Berkshire), Wellings (1984*b*) and Barraclough *et al.* (1994) found little evidence of missing solute when constructing solute balances. These would have been expected if substantial amounts of by-pass flow had occurred at these sites. The likely errors in determination of the balances, however, leave considerable room for doubt on this point.

It can be seen that, in spite of extensive work on mechanisms of recharge through the unsaturated zone of the Chalk, there is considerable discord about the processes involved, and the proportion of water transmitted via different routes. Owing to the variation of Chalk hydraulic properties, laterally, vertically and regionally, this is perhaps not surprising.

Site description

The Fleam Dyke lysimeter is located approximately 100 m north of the Fleam Dyke Pumping Station (Cambridge Water Company), near Fulbourn, Cambridgeshire (National Grid Reference TL539549). The site is approximately 30 m

AOD, and is relatively flat. Rest water level is about 18 m below ground level, with the pumping water level at 32 m (at a pumping rate of approximately 63 $m^3 h^{-1}$) (Kitching & Shearer 1982).

The lysimeter is constructed in the lower beds of the Middle Chalk. The uppermost 1 to 1.5 m of the section consists of unstratified, weathered khaki-buff, rubbly Chalk (Kitching & Shearer 1982). Below this, the Chalk is off-white, becoming harder with depth. No flints or marl seams were observed. Fragments of Inoceramus sp. indicate the M. labiatus s.l. Zone of the Middle Chalk.

The dip of the Chalk was estimated to be approximately 20° SE. A large number of high-angle joints, together with a smaller number of high-angle, small-scale faults were observed. Major joints could be followed for the length of a face of the construction pit (approximately 20 m), or from one face to another, and were all stained with a black or greenish-black mineral, assumed to be manganese. These joints were described as being open, in comparison with the smaller, high-angled joints which were impersistent vertically. The effect of jointing and faulting was to give a very broken aspect to the Chalk, such that individual blocks rarely exceeded 20 cm.

The lysimeter was constructed between February and April 1977, and records of drainage kept for almost the complete period between 1977 and 1986. It has recently (end of 1995) been brought back into operation by the Anglian Region of the Environment Agency (then National Rivers Authority).

Lysimetric approach to recharge estimation

Lysimeters provide a means of measuring recharge directly, without recourse to indirect measurements and estimates. However, as described below, they may have their own drawbacks. The Fleam Dyke lysimeter is basically a 5 m cube of undisturbed Chalk, contained within sheet piling on all four sides and beneath the base, in such a way that water percolating through it is collected from the base (Kitching & Shearer 1982). Drainage was measured both by recording the water level in a collection tank which was emptied automatically when full, and by tipping bucket. Recharge was estimated by dividing the volume of throughput by the surface area of the lysimeter.

The lysimeter was constructed to overcome the problems typical of lysimetry (that is, small surface area, insufficient depth, and repacking of fill), being essentially a block of undisturbed Chalk, with large surface area (25 m^2), and sufficient depth (5 m) such that the base was likely to have negligible effect on the water balance of the soil profile. It is necessary for a water-table to exist at the base of the lysimeter in order for drainage to occur. Therefore ideally the lysimeter should extend into the saturated zone, with the water-table within it maintained at the same level as outside (Kitching & Shearer 1982). However, this was impractical with the deep water-table at this site, and the deepest practical achievable depth was used instead. As the primary controlling influences on recharge are exerted at or near the ground surface, it was considered that the lysimeter should be able to quantify the amount of recharge with good accuracy. However, the timing of recharge at a water-table several metres below the base of the lysimeter would obviously be affected.

Soil physics approach to recharge estimation

In parallel with the lysimeter measurements, a soil physics approach was used both on the lysimeter, and in the undisturbed profile about 20 m away. This arrangement was designed not only to assess recharge, but also to investigate the variability due to different instrument sets. The approach was based on the use of the zero flux plane (ZFP) technique for the part of the year corresponding approximately to that when soil water deficits are experienced. For the rest of the year, when errors in evaporation estimates are likely to be small, a simple estimate based on water balance derived from measured rainfall, estimated evaporation (the Penman potential value) and measured change in soil water storage was used.

The ZFP method requires no meteorological input other than rainfall. Water in the unsaturated zone moves under the influence of potential gradients. The ZFP method assumes all water flow in the upper part of the soil profile is upwards, either within plant roots or the soil, and in the lower part of the profile it is downwards within the soil or porous rock. Separating the upward from the downward moving water is a depth at which there is no water movement – the zero flux plane. Since all water above the ZFP is moving upwards and there is no flow through the ZFP (by definition), any water content changes within this part of the profile must be due to water transfer across the upper boundary, i.e. to evaporation and rainfall infiltration. Similarly all water content changes within the lower part of the profile are due to water transfer across the lower boundary, i.e. drainage through the deepest depth at which measurements are made. The method requires, therefore, hydraulic potential measurements through the profile (to define the ZFP depth), water content measurements at a series of depths at regular intervals, and measurements of rainfall.

The theoretical basis of the ZFP method, and a more rigorous discussion of the assumptions inherent in the method, are given by Cooper *et al.* (1990). It should, however, be noted that the method relies on there being a unique matric potential at each depth and that different hydraulic conditions in fractures and in the Chalk matrix would violate these assumptions, although whether it would invalidate any water balance estimates made using the ZFP technique is more open to question. The major advantage of the ZFP technique is that it is capable of estimating both evapotranspiration and drainage from the unsaturated zone during the part of the year when there is maximum uncertainty surrounding the response of transpiration to weather and soil conditions.

The lysimeter was equipped with three neutron probe access tubes, and an array of pressure-transducer tensiometers. The original pressure transducer tensiometers were found to be unsatisfactory, and were removed in April 1980. They were replaced in January 1982 by a set of tensiometers of improved design, installed at depths of 0.6, 1.2. 2.1, 3.3 and 4.5 m. Two further plots were set up approximately 20 m from the lysimeter. Each consisted of four neutron probe access tubes, one of which extended to about 12 m depth, the others to 3.9 m, and two sets of 12 mercury manometer tensiometers at depths over the range to 3 m. The tensiometers and one neutron probe access tube per plot were read five days per week, whilst the other access tubes were read once per week. These test sites confirmed that the infiltration characteristics of the lysimeter were likely to be representative of the natural state, except very close to the base. Estimates of potential evapotranspiration (Penman 1948) and rainfall were obtained from an automatic weather station (AWS) installed on-site. All the data necessary for calculation of recharge estimates using both the ZFP and water balance techniques were therefore available.

Variables controlling by-pass flow

From the review of previous work, it was determined that the onset and continuation of by-pass flow was likely to be controlled by a number of variables, operating in combination.

Infiltration flux

The flux of water through the soil zone is likely to be the principal factor determining whether by-pass flow is initiated. This, in turn, is controlled by rainfall and buffering effects in the soil. If the water flux arriving at the top of the Chalk exceeds the capacity of the matrix to conduct the water, they by-pass flow is likely to occur. If the unit potential gradient is assumed (i.e. gravity flow), then by-pass flow would be expected to be initiated once the flux exceeded the hydraulic conductivity of the matrix.

The infiltration flux will therefore be affected by the following.

(a) Thickness of soil cover and weathered Chalk: a thick soil and/or weathered zone above the Chalk would buffer the effects of heavy rainfall events, damping the release of water to the underlying unweathered Chalk.
(b) Vegetation cover: the type of vegetation affects interception of rainfall, evaporation, and infiltration into the soil. This affects not only the total quantity of recharge, but also the rate at which water is conducted down the profile by, for instance, opening up preferential flowpaths down old root channels or encouraging earthworm activity.
(c) Rainfall intensity and duration: depending on other controlling variables, at a certain level of input, the Chalk matrix would no longer be able to transmit infiltrating water.

Antecedent water content of the soil and underlying Chalk

This has a significant control over the infiltration capacity of the soil (e.g. Kutilek & Nielsen 1994). As a general rule, whilst a soil moisture deficit exists, any precipitation (with the exception of intense storms) would be used to restore the soil to field capacity, before any drainage to the underlying Chalk occurred.

Antecedent matric potential of the Chalk

At low matric potentials, the matrix exerts a high suction. Therefore, even if the level of input to the unsaturated zone exceeded the matric hydraulic conductivity and by-pass flow was initiated, any water being transported through the fractures would be rapidly absorbed into the matrix. This would raise the matric potential at that level, and additional input would be transported further along the fracture.

Hydraulic conductivity of the matrix

Measurement of the unsaturated hydraulic conductivity of the profile *in situ* at Fleam Dyke indicated a matric hydraulic conductivity of around 1 mm day^{-1}. A particularly low hydraulic conductivity band from 1.5 m to 2.2 m depth had a matric hydraulic conductivity of about 0.1 mm day^{-1}.

Methodology

Daily drainage and rainfall data were available for the period from 1977 to 1986. Pressure transducer tensiometer data were also available for 1982 to 1983. Weekly MORECS data (rainfall, and MORECS estimates of soil moisture deficit (SMD), actual evapotranspiration (EVT), and hydrologically effective precipitation (HEP)), were available for 1981 to 1983, and monthly for all other periods.

The data were analysed to assess recharge mechanisms at this site. In particular:

- identification of the controls on by-pass flow;
- calculation of the proportion of water transported rapidly through the fracture system;
- estimation of the frequency and duration of recharge events, that is, drainage events caused by rapid by-pass flow;
- comparison of measured drainage from the lysimeter and HEP as calculated by MORECS.

Initially, interpretation was carried out by visual examination of the data. Thus, on a daily basis, lysimeter drainage was plotted against daily rainfall for non-SMD periods. Additionally, periods when matric potential throughout the profile was above -50 cm H_2O (i.e. matric potential at every tensiometer depth was above -50 cm H_2O) were shown. However, it was found that daily fluctuations in rainfall masked any underlying trends in the data, and the three-day moving average was therefore calculated and used in place of the actual daily rainfall.

Data were extracted for a 2.5 month period (1 November 1982 to 14 January 1983), and correlation coefficients between various measures of rainfall, matric potential and drainage were examined to assess which variables accounted for the variation in drainage on a daily basis. This period was chosen as one where a lengthy non-SMD period was recorded, and during which there were a number of recharge events. Furthermore, at this time of year it was assumed that EVT would be very low and there was no SMD. It was assumed that the soil and vegetation parameters and the hydraulic conductivity of the matrix did not vary during the time period. With these variables removed, the relationship between rainfall parameters, matric potential and drainage could be investigated. The variables used included:

- daily rainfall
- three-day average rain
- three-day average with a time lag of two, three and four days
- matric potential at 0.6 m depth
- matric potential at 0.6 m depth with a time lag of two, three and four days.

Finally, to allow evaluation of MORECS estimates of HEP at this site, drainage amounts were calculated on a weekly basis, and compared with HEP for two years of data (1982 and 1983).

Results

Comparison of soil moisture deficit (SMD) data and drainage showed that by-pass flow did not occur (with one exception) during periods when an SMD existed. Therefore, most of the following is restricted to periods with no SMD.

Figure 2 shows lysimeter drainage and three-day average rainfall for three periods when there was minimal or no SMD. Periods when matric potential was greater than -50 cm H_2O throughout the profile are indicated. The figure illustrates peaks in drainage that occur three to four days after peaks in the three-day average rainfall. The general coincidence between periods of high matric potential and drainage peaks can also be seen, with drainage greater than 1 mm day^{-1} occurring during periods of high matric potential. The rapid increase in hydraulic conductivity at matric potentials greater than -50 cm H_2O has been shown in Fig. 1; as described earlier, this increase was taken to imply that matrix flow carried virtually all recharge at potentials less than -50 cm H_2O. This theory is further verified by the fact that drainage greater than 1 mm day^{-1} occurs when matric potential at all measured depths through the profile was greater than -50 cm H_2O (see Fig. 2).

At the beginning of recharge events, matric potential throughout the profile generally increased to above -50 cm H_2O shortly before, or at the same time as, drainage exceeded 1 mm day^{-1}. At the end of recharge events, matric potential, although remaining high

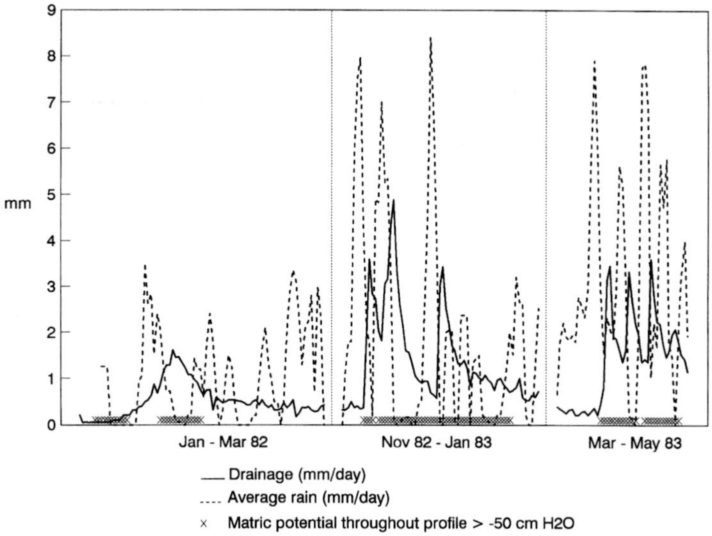

Fig. 2. Drainage through the lysimeter and three-day average rainfall, for periods when soil moisture deficit was non-existent.

throughout much of the profile, tended to decrease to less than -50 cm H_2O at either 0.6 m or 3.3 m depth. In some cases, drainage greater than mm day^{-1} continued for several days after potentials at these horizons had decreased below -50 cm H_2O. This may indicate that the matric potential at which fracture flow ceases in these layers is actually slightly greater than -50 cm H_2O.

The decrease of matric potential at 3.3 m may be explained by the presence of a low conductivity band between 1.5 and 2.2 m. Thus the matric potential at 3.3 m may decrease as the fractures below the band drain, while the matric potential above and within the band remains high. It was observed that during dry spells, matric potentials could drop to -250 cm H_2O at depths greater than 2.4 m, while those higher in the profile were typically -20 to -25 cm H_2O (Fig. 3).

Steady drainage through the matrix continued well after the onset of each SMD period. However, by-pass flow did not occur whilst a significant SMD existed, even after storms of greater than 20 mm day^{-1}. Generally, these periods of heavy rainfall had no major effect on matric potential (as recorded at 0.6 m and below).

By-pass flow was observed only once during an SMD period; a rainfall event of 42.5 mm on 19 June 1985 resulted in a peak of 13.3 mm drainage the following day. Daily SMD was estimated from daily rainfall and the MORECS estimate of SMD at the beginning of June, assuming that evaporation was equal to the average daily potential value for the month. This indicated that the SMD actually disappeared for a period of approximately 16 days. After the initial storm, subsequent rainfall of less then 6 mm day^{-1} on average was sufficient to maintain by-pass flow through the lysimeter in the 16 days following the storm.

As outlined earlier, various parameters were correlated with drainage from the lysimeter, including daily rainfall, three-day average rainfall, delayed three-day average rainfall, maximum rainfall intensity over a 24 hour period, and matric potential. Of these, the three-day moving average rainfall was most closely correlated with drainage, with a time lag of four days ($r = 0.71$). Matric potential with a four-day time delay had a correlation coefficient of -0.51 (see Table 1). Other time delays were used, but did not improve the correlation between rainfall or matric potential and drainage.

It was inferred that significant recharge events and, by inference, by-pass flow, are influenced by conditions existing four days previously. It is the combination of rainfall and matric potential at the time of the rainfall event that determine whether or not by-pass flow occurs. Again, the main exception to this statement was the rainfall event in June 1985. On this occasion exceptionally high drainage of 13.3 m was recorded on the day following rainfall of 42.5 mm.

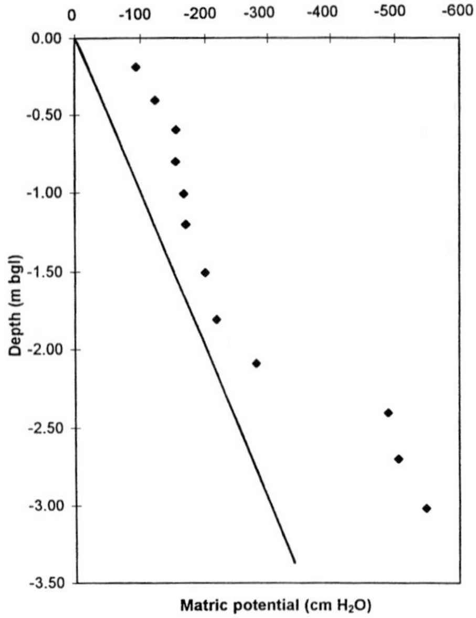

Fig. 3. Variation of matric potential with depth through the profile.

Table 1. *Correlation matrix for recharge, rainfall and matric potential*

	Recharge	Rain	Three-day av. rain	Lag rain	MP	Lag MP
Recharge	1.00					
Rain	−0.8	1.00				
Three-day av. rain	−0.12	0.60	1.00			
Lag rain (four day)	0.71	−0.06	−0.09	1.00		
MP	−0.44	0.11	−0.15	−0.29	1.00	
Lag MP (four day)	−0.51	0.10	0.28	−0.17	0.405	1.00

MP = matric potential

Proportions of matrix and by-pass flow

It it is assumed that any drainage in excess of 1 mm day^{-1} is by-pass flow and any less than this is all matrix flow, then the drainage can be partitioned as shown below:

For the period 1981–1986:
% drainage through fractures = 34.3%
% drainage through matrix = 65.7%

Of total matrix flow:

32.4% (of total drainage) occurred during by-pass flow periods
33.3% (of total drainage) occurred at other times.

Drainage greater than 1 mm day^{-1} was also calculated only for periods when the matric potential throughout the profile was greater than −50 cm H_2O. This gave a figure of 31%, i.e. only 3.3% of the drainage greater than 1 mm day^{-1} occurred when matric potentials at some (or all depths) were less than −50 cm H_2O. This reinforces the assumption that drainage greater than 1 mm day^{-1} represents by-pass flow through the fracture network.

Thus, at Fleam Dyke, to a depth of 5 m, rapid by-pass flow contributes about one-third of total drainage. Moreover, during those periods when by-pass flow occurs, on average about half of the drainage flow occurs as by-pass flow, the other half being via the matrix. Since flow through the matrix has been assumed to be equal to 1 mm day^{-1}, this implies that, during by-pass flow events, the total flow rate, on the average, was about 2 mm day^{-1}.

Timing and duration of recharge events

The data were also examined to determine the timing and duration of recharge events. It was found that during the period from 1981 to 1985, some occurrences of by-pass flow were observed in all months of the year except for August and September. By-pass flow was observed most frequently between November and June, and was possible, although less likely, during July and October.

Once initiated, by-pass flow continued for between 1 and 28 days, dependent on the rainfall input.

Comparison with MORECS data

Hydrologically effective precipitation (HEP) calculated by MORECS was compared with drainage from the lysimeter. As defined by the Meteorological Office:

$$\text{HEP} \equiv \text{Rainfall} - \text{Actual ET (when } SMD = 0)$$

It should be noted that MORECS HEP calculations do not use site-specific rainfall, and that, being spatially variable, this may vary significantly across the 40 km square for which it is calculated. Furthermore, actual ET refers to an estimate made by the MORECS procedure, which may differ from the evapotranspiration that actually occurred.

Cumulative drainage and HEP were calculated for 1982 and 1983, and the results are compared (Figs 4 and 5). Although only two years' data were available for comparison, certain differences are worth comment.

Timing. Differences in timing would be expected as, by definition, HEP is the difference between rainfall and actual (i.e. MORECS calculated) evaporation, whilst the lysimeter drainage is subject to a delay, as the water percolates through the upper 5 m of the profile. Hence, in the short term, peaks in HEP tend to occur in the week preceding peaks in drainage. In the longer term, during winter and before the start

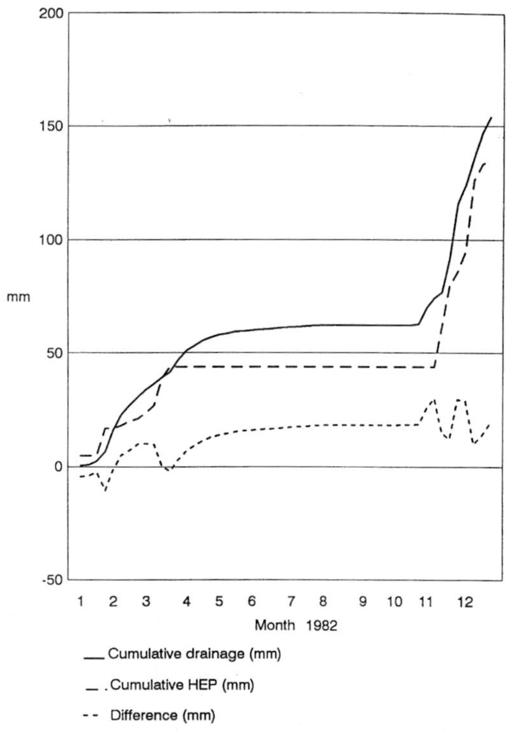

Fig. 4. Comparison of cumulative lysimeter drainage and MORECS estimates of hydrologically effective precipitation, 1982.

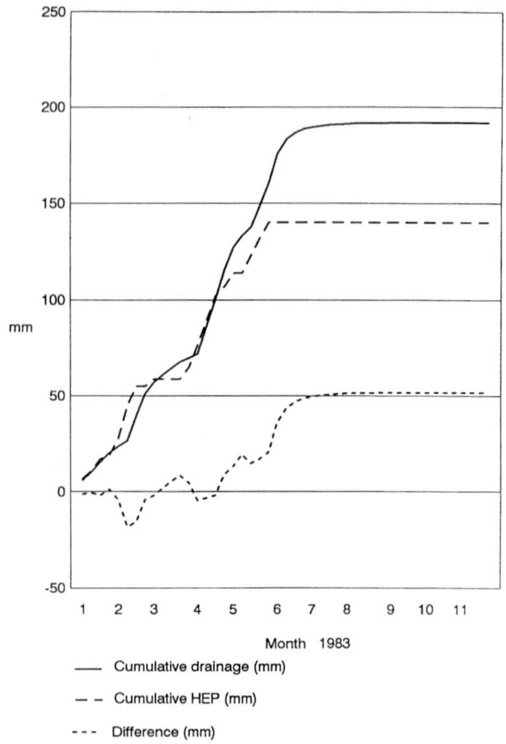

Fig. 5. Comparison of cumulative lysimeter drainage and MORECS estimates of hydrologically effective

of the SMD period, short-term differences between cumulative drainage and cumulative HEP tend to cancel out, and there is reasonable agreement overall. At the beginning of the SMD period, according to the MORECS definition, HEP becomes zero. However, drainage through the lysimeter continues well after the onset of the SMD period and therefore the difference between HEP and drainage increases. It should be noted that drainage from the lysimeter itself ceased before that calculated using the soil physics approach as a consequence of the artificial discontinuity at the bottom of the lysimeter.

Magnitude. Drainage from the lysimeter (cumulative) over a year was greater than HEP; the difference in 1983 was 47 mm. Long-term differences between the two are likely to be caused by rainfall at the site being different from the average across the 40 km MORECS square and/or the evaporation occurring at the site being different from that calculated by MORECS. These data suggested that the later the SMD developed, the greater the difference between cumulative drainage and HEP. It has been suggested (Cooper *et al.* 1990) that this is due to over-estimation of actual EVT in spring, the differences being greatest when there is a wet spring and the SMD does not develop until late spring or early summer.

Discussion

Previous work (e.g. Cooper *et al.* 1990) had suggested that at matric potentials greater than −50 cm H_2O, the fracture system becomes partially saturated. With increasing matric potential, a greater proportion of the fracture network becomes saturated, with the infiltrating water occupying fractures of greater aperture. The tortuosity of the flowpaths also decreases, and drainage volume and rate increases.

Fracture aperture can be related to matric potential using the following equation:

$$d = |2\gamma/\psi_m|$$

where d is the fracture aperture, γ is surface tension, and ψ_m is matric potential. Using this equation, the fracture aperture which would be saturated at a matric potential of −50 cm H_2O is 0.12 mm. This implies that fracture apertures greater than 0.12 mm are responsible for rapid by-pass flow.

Intuitively, it seems likely that the fractures of the Chalk unsaturated zone contain a continuous range of apertures and spacings. Above the -50 cm H_2O threshold, increasing matric potentials represent the increasing saturation of the fracture network, and hence increasing amounts of by-pass flow. However, evidence provided in this paper suggests a fairly distinct boundary between matrix and rapid by-pass flow.

It is probable that there is some water flow in smaller fractures, but that this cannot be resolved from flow through the matrix. It is also likely that, at a low level of saturation of the fracture network, the patches of water-filled fracture are likely to be discontinuous and hence unable to contribute to flow substantially. Also, the smaller fractures may not be well-connected and hence contribute little to by-pass flow.

This work has confirmed that the onset of by-pass flow is controlled by antecedent matric potential and rainfall amount. The time lag of four days observed between rainfall and the occurrence of a recharge event recorded at a depth of 5 m indicates that, in the absence of an SMD (i.e. the whole profile is already at relatively high potential), rainfall gradually wets up the layers of the profile, increasing the matric potential until, provided that the infiltration rate is greater than the saturated hydraulic conductivity of the matrix, by-pass flow commences. If the input to the system is sufficient, matric potentials remain high and by-pass flow may continue for several days. The time for drainage to respond to rainfall would be expected to vary according to the matric potential throughout the profile at the time of the rainfall event.

It was observed that by-pass flow only occurs during an SMD period under exceptional circumstances. At rainfall intensities greater than a certain threshold, it appears that the infiltrating water wets up the soil/weathered Chalk and unweathered Chalk very quickly. Excess water then enters the fractures, causing by-pass flow. If the rainfall amount is sufficiently high, this process occurs very rapidly and continues down through the profile. This was observed during June 1985; the rainfall during a single day appears to have been enough to overcome the SMD and matrix suction, with the response being seen as a drainage peak the following day. The drainage and rainfall over the following weeks indicate that by-pass flow continued for a further 16 days. The observation that recharge is very rare during SMD periods was also noted by Headworth (1972), who stated that during periods of groundwater recession, there were no fluctuations in groundwater level due to the infiltration of rain, except in the case of the 'heaviest storm'.

Wellings and Cooper (1983) had identified this site as one which was likely to have a significant proportion of by-pass flow. Despite this, only about one-third of drainage occurred via the fracture network, and this occurred almost entirely when there was no antecedent SMD and matric potentials were already high.

At this site, two-thirds of the flow is through the matrix, and one-third is by-pass flow through the fractures. By-pass flow is normally through a relatively small proportion of the fracture network, and in parallel with (on average) an equal flux of water in the matrix. Other Chalk sites with thin soils, at which soil physics measurements have been made (West Ilsley, Bridget's Farm, Gog Magog Golf Course site) (Wellings & Cooper 1983) show matric potentials above -50 cm H_2O less frequently. For example, at Bridget's Farm in Hampshire (Upper Chalk) (Gardner *et al.* 1990), matrix conductivity was calculated to be as high as 8 mm day^{-1}. As a consequence of this and the buffering effect of the overlying soil and weathered Chalk, fractures rarely conducted water and matric potential rarely exceeded -2 m H_2O. This implies less frequent, and a lower proportion of, by-pass flow than at the Fleam Dyke site.

At sites where by-pass flow rarely occurs, diffusive interchange of solutes between water in fractures and that in the matrix may not be necessary to account for solute concentration profiles. At other sites, such as Fleam Dyke, by-pass flow accounts for a significant proportion of the total flow. However. despite the significance of by-pass flow in terms of quantity of recharging water, the actual proportion of the fracture volume saturated when it is active, is probably relatively small, and it follows that the surface area of block exposed to solute-rich water is also small. Foster (1975) and Barker & Foster (1981) carried out simulations of water movement by assuming that the fracture system is full, and that the total surface area of the Chalk blocks is exposed to solutes. They also showed that the rate of equilibration between fracture- and pore-water is very sensitive to assumptions about block size and effective diffusion coefficient. Such evidence may need to be reassessed in the light of these findings.

Conclusions

Recharge events cannot be related solely to rainfall, but occur when sustained rainfall occurs and antecedent matric potential throughout the profile is greater than -50 cm H_2O. Cooper *et al.* (1990) interpreted this as being due to the fracture system transporting an increasing proportion of total flow once the matric potential rose above -50 cm H_2O. If input to the system is sufficient, matric potentials remain high and by-pass flow may continue for several days (as shown in Fig. 2). Statistical analysis of a 22-month period indicated that by-pass flow at the base of the lysimeter (5 m depth) typically occurred in response to rainfall four days previously.

In June 1985, the rainfall during one day was sufficient to overcome the estimated SMD and matrix suction, and the response was seen as a drainage peak by the following day. This rapid response to rainfall indicates that, under conditions of high-intensity rainfall, the response can be much more rapid than four days.

Comparison of MORECS HEP and lysimeter drainage showed that annual HEP was less than observed drainage, the maximum difference being approximately 30%. This highlights inaccuracies in the regional estimation of evapotranspiration and HEP. Cooper *et al.* (1990) suggested that the magnitude of the difference was related to the timing of the onset of the SMD period due to over-estimation of potential EVT in the spring.

A note of caution is required. The number of Chalk sites at which matric potential measurements have been made is very limited, and all have been situated in the Upper or Middle Chalk of the Southern Chalk Province. There may be considerable differences in Chalk unsaturated zone properties between different units of the Chalk, and in different geomorphological settings, and simple extrapolation of these results may not be appropriate.

This paper is published by permission of the Director, British Geological Survey.

References

BARKER, J. A. & FOSTER, S. S. D. 1981. A diffusion exchange model for solute movement through fissured rock. *Quarterly Journal of Engineering Geology*, **14**, 17–24.

BARRACLOUGH, D., GARDNER, C. M. K., WELLINGS, S. R. & COOPER, J. D. 1994. A tracer investigation into the importance of fissure flow in the unsaturated zone of the British Upper Chalk. *Journal of Hydrology*, **156**, 459–469.

COOPER, J. D. 1981. Hydraulic properties of the English chalk measured in situ. *EOS Transactions of the American Geophysics Union (Washington DC)*, **62**, 210.

——, GARDNER, C. M. K. & MACKENZIE, N. 1990. Soil water controls on recharge to aquifers. *Journal of Soil Science*, **41**, 613–630.

DOWNING, R. A., SMITH, D. N. & WARREN, S. C. 1978. Seasonal variations of tritium and other constituents in groundwater in the Chalk near Brighton, England. *Journal of the Institution of Water Engineers and Scientists*, **32**, 123–136.

FOSTER, S. S. D. 1975. The Chalk groundwater tritium anomaly – a possible explanation. *Journal of Hydrology*, **25**, 159–165.

GARDNER, C. M. G., BELL, J. P., COOPER, J. D., DARLING, W. G. & REEVE, C. E. 1991. Groundwater recharge and water movement in the unsaturated zone. *In*: DOWNING, R. A. & WILKINSON, W. B. (eds) *Applied Groundwater Hydrology*. Clarendon, Oxford, 54–76.

——, COOPER, J. D., WELLINGS, S. R., BELL, J. P., HODNETT, M. G., BOYLE, S. A. & HOWARD, M. J. 1990. Hydrology of the unsaturated zone of the chalk of south-east England. *In*: BURLAND, J. B., CORBETT, B. O., JONES, D. L., MORTIMORE, R. N. & ROBERTS, T. S. (eds) *Chalk*. Thomas Telford, London, 611–618.

HEADWORTH, H. G. 1972. The analysis of natural groundwater level fluctuations in the Chalk of Hampshire. *Journal of the Institution of Water Engineers and Scientists*, **24**, 2, 107–124.

KITCHING, R. & SHEARER, T. R. 1982. Construction and operation of a large undisturbed Lysimeter to measure recharge to the Chalk aquifer, England. *Journal of Hydrology*, **58**, 267–277.

KUTILEK, M. & NIELSEN, D. R. 1994. *Soil Hydrology*. Catena, Cremlingen-Destedt, Germany.

PENMAN, H. L. 1948. Natural evaporation from open water, bare soil and grass. *Proceedings of the Royal Society, London, Series A*, **193**, 120–145.

REEVES, M. J. 1979. Recharge and pollution of the English Chalk some possible mechanisms. *Quarterly Journal of Engineering Geology*, **14**, 231–240.

SMITH, D. B., WEARN, P. L., RICHARDS, H. J. & ROWE, P. C. 1970. Water movement in the unsaturated zone of high and low permeability strata using natural tritium. *1970 Symposium on uses of Isotopes in Geology*, Vienna, 73–87.

WELLINGS, S. R. 1984*a*. Recharge of the Upper Chalk aquifer at a site in Hampshire, England, 1, Water balance and unsaturated flow. *Journal of Hydrology*, **69**, 259–73.

—— 1984*b*. Recharge of the Upper Chalk aquifer at a site in Hampshire, England, 2, Solute Movement. *Journal of Hydrology*, **69**, 275–85

—— & BELL, J. P. 1980. Movement of water and nitrate in the unsaturated zone of the Upper Chalk near Winchester, Hants., England. *Journal of Hydrology*, **48**, 119–36.

—— & COOPER, J. D. 1983. The variability of recharge of the English Chalk aquifer. *Agricultural Water Management*, **6**, 243–253.

Recharge through the drift: a study of contrasting Chalk catchments near Redgrave Fen, UK

R. W. N. SOLEY & J. A. HEATHCOTE

Entec (UK) Ltd, 160–162 Abbey Foregate, Shrewsbury, Shropshire SY2 6AL, UK

Abstract: Redgrave and Lopham Fens lie at the headwaters of the Rivers Waveney and Little Ouse in East Anglia, UK. The Fens are supported by groundwater discharge from the drift-covered Chalk underlying the area. This discharge and the ecological value of the Fens have been derogated by public water supply abstractions from a nearby borehole.

Aspects of the investigations undertaken to construct a groundwater model to assess options for relocating the borehole away from the Fens are described. Particular attention is given to the value of using basic river flow data analysis to identify contrasts in the hydrological characteristics of the two catchments and constrain estimates of recharge.

The Waveney catchment is flashy and dominated by run-off whereas the Little Ouse receives a more significant groundwater baseflow contribution. These catchment contrasts reflect subtle variations in boulder clay thicknesses and topography. They also reflect the contrasting effectiveness of the two rivers as drains on the groundwater system.

Average recharge to the Chalk through the boulder clay-covered interfluves is estimated at around 36 mm a^{-1} (which is similar to other values quoted in the literature). Recharge may locally fall to less than 10 mm a^{-1} where the boulder clay is thick and the Chalk is strongly confined.

Background

Redgrave and Lopham Fens, straddling the Norfolk/Suffolk border near Diss, in East Anglia (Fig. 1), have long been identified as having an unusual flora and fauna (Bellamy & Rose 1960). The international importance of the site has been recognised under the Ramsar Convention. Deleterious ecological changes as a result of reduced discharge of calcareous groundwaters to the Fen have been documented by Harding (1993). The Redgrave borehole is located immediately adjacent to Redgrave Fen (Fig. 2), and is used to produce around 2.8 Ml day^{-1} (licensed capacity 3.6 Ml day^{-1}) of water for public supply, from the Chalk which underlies the whole region.

Aspinwall and Co (1992) produced evidence that the operation of the Redgrave borehole is contributing to the reduced discharge of calcareous water described above. Channel improvement works in the River Waveney which drains the Fens are also thought to have contributed to the problem.

As a result of discussions between Essex and Suffolk Water plc (the borehole operator), the Environment Agency (the licensing authority), and Suffolk Wildlife Trust (representing the ecological interest) a decision was made to relocate the borehole abstraction to an alternative site, to reduce the environmental impacts on Redgrave Fen. The new site was to be selected to avoid similar impacts on other wetlands or abstractors.

An extensive assessment, including long-term pumping tests and the construction of a numerical groundwater model, was undertaken to determine which of five candidate sites would be the best replacement for the Redgrave borehole (Fig. 2).

Topography and geology of the area

The upper catchments of both the Waveney and Little Ouse Rivers have a fairly subdued topography with ground elevations in the range 20 to 60 m AOD. Both catchments are underlain by the Chalk which is blanketed by heterogeneous glacial drift comprising sand and gravel and silty clay till (or 'boulder clay') deposits up to 60 m thick.

In the southern part of the Waveney catchment and beneath the neighbouring Dove catchment, thick deposits of Red Crag (Palaeogene shelly sands) lie between the Chalk and the drift in the deep Stradbroke Depression. The other major feature in the Chalk surface is a drift-filled sub-glacial tunnel valley up to 60m deep which underlies, and is largely congruent with, the present-day course of the River Waveney and the uppermost reaches of the Little Ouse.

SOLEY, R. W. N. & HEATHCOTE, J. A.. 1998. Recharge through the drift: a study of contrasting Chalk catchments near Redgrave Fen, UK. *In*: ROBINS, N. S. (ed.) *Groundwater Pollution, Aquifer Recharge and Vulnerability*. Geological Society, London, Special Publications, **130**, 129–141.

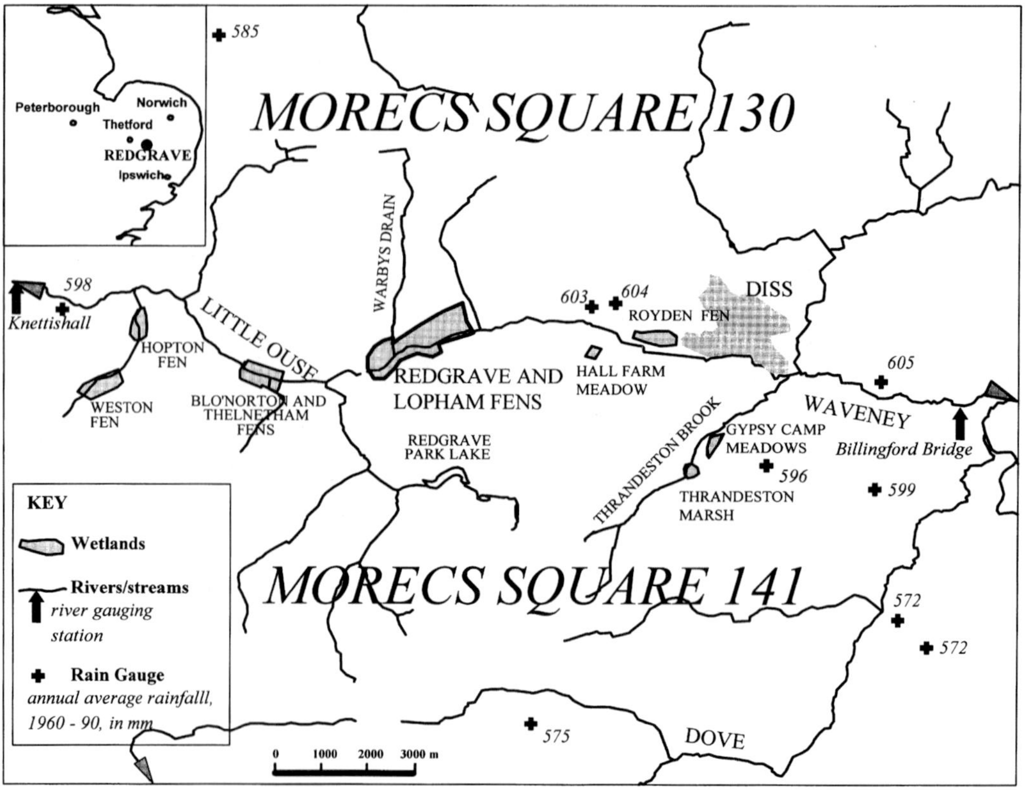

Fig. 1. Study area location, river, wetland and hydrometric data locations.

The groundwater model

The borehole is very close to the surface watershed between the Rivers Waveney and Little Ouse, so the groundwater model covered the upper groundwater catchments of both, down to the first river-flow gauge (Fig. 2). The groundwater and surface water divides between the catchments are not coincident and the borehole draws water from both.

A two-layer distributed groundwater model was constructed using the USGS Modflow code. The lower layer represented the Chalk and any permeable drift or Crag directly overlying it and extended across the whole model to no-flow boundaries delineated by groundwater flowpaths towards the gauge (based on observed head distributions). The upper layer represented drift sand and gravel aquifers in the valley bottoms which are hydraulically well-connected with the Fens and the rivers but may be isolated from the Chalk by variable thicknesses of silty clay aquitards.

The piezometric contours shown in Fig. 2 suggest convergent groundwater flow to both rivers. The existence of a relatively steep vertical gradient observed in many places between the Chalk and the River Waveney particularly also reflects the resistance to discharge provided by the valley bottom silty clay drift and alluvium.

Recharge to the Chalk was originally distributed across the model as a proportion of hydraulically effective rainfall (HER) varying according to the thickness and permeability of the drift (based on borehole logs). The model calculates groundwater flows within boundary constraints using transmissivity and storage parameters. HER not added to the model as recharge was assumed to run-off to the river gauge instantaneously and was combined with groundwater discharge (i.e. 'baseflow') from the model to give a total modelled flow on a monthly time-step.

The model was calibrated by comparing modelled and observed river flows and groundwater head variations over a 20-year period from 1974 to 1994.

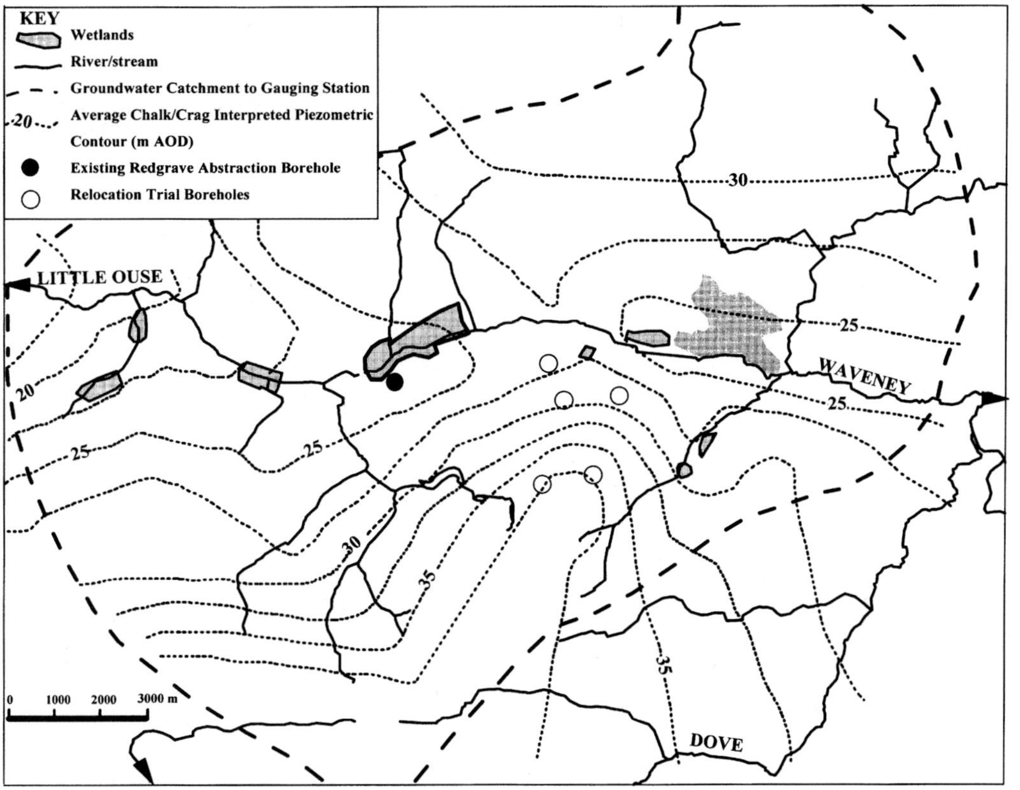

Fig. 2. Borehole relocation options, Chalk piezometry and combined groundwater catchment.

River flow contrasts and early modelling problems

During calibration of the model, it became clear that the two rivers behave in very different ways. The Little Ouse River gauged at Knettishall has persistent summer baseflow, characterized by a baseflow index (BFI) of 0.70, while the River Waveney gauged at Billingford Bridge has very small summer flows and a lower BFI of 0.43. The normalized total discharge ('mean annual run-off' in millimetres) from the Little Ouse is slightly less than the Waveney, probably reflecting groundwater flow across surface catchment divides and through the Chalk beneath the Knettishall gauge (Table 1).

Table 1. *River gauging station statistics*

	Waveney (Billingford Bridge)	Little Ouse (Knettishall)
Surface catchment area (km^2)	149.4	101.0
Period of record	1968–90	1980–90
Mean annual run-off (mm)	166	149
Mean flow (Ml day^{-1})	68.3	41.5
10 percentile (Ml day^{-1})	155.5	77.8
95 percentile (Ml day^{-1})	6.0	11.2
Base flow index (BFI)	0.7	0.43

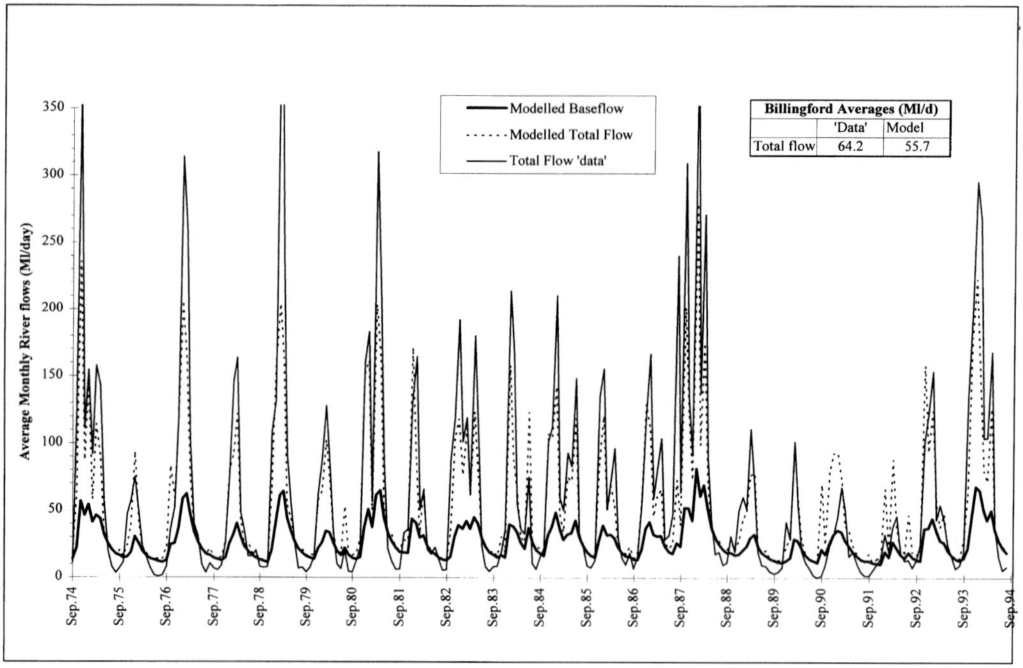

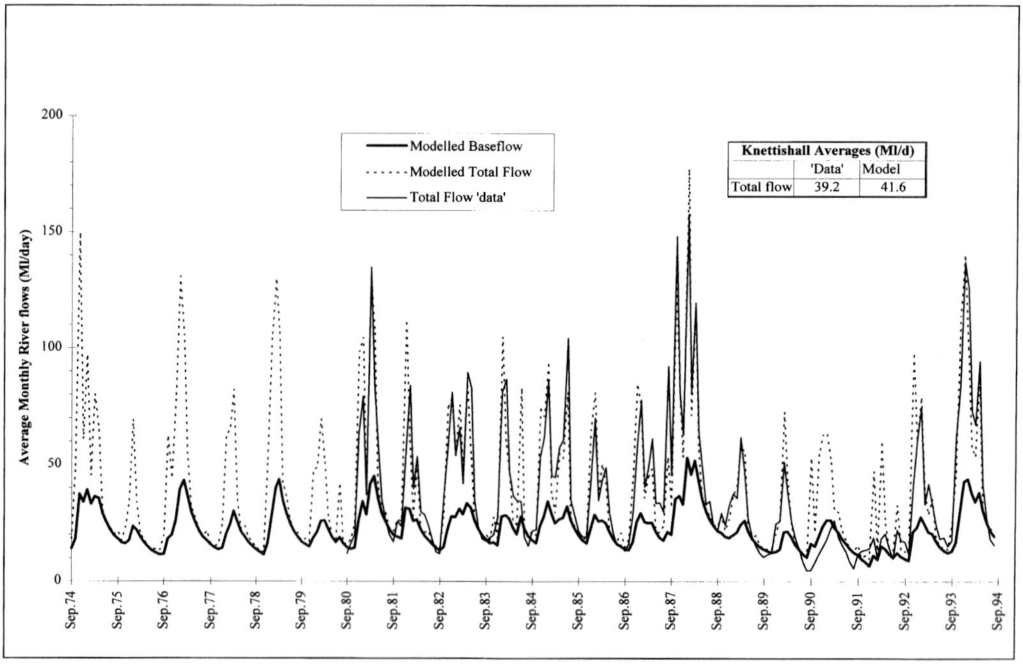

Fig. 3. Modelled and measured river flows (recharge based on Institute of Hydrology baseflow separation): (**a**) Waveney; (**b**) Little Ouse.

Early versions of the groundwater model gave a poor fit to the flows of the River Waveney, failing to reproduce the small summer flows or the high winter flows, although the flows of the Little Ouse were reproduced quite well (Fig. 3).

The geographical proximity of the two river catchments, combined with the subdued relief of the area, rules out the differences in behaviour being caused by meteorological effects. The two catchments also are very similar geologically, with the same basic succession of Chalk covered by variable thicknesses (up to 60 m) of variable glacial drift. This geological succession covers several thousand square kilometres of eastern England. However, there are subtle differences in the distribution of the boulder clay till that covers the area studied, and this seemed to be the only possible explanation for the pronounced differences in behaviour. This led to a more detailed assessment of the hydrology of the two catchments, to support numerical experiments to determine what properties could be so different between the two river catchments.

This paper discusses the methodologies adopted and the insights into controls on groundwater recharge and the consequent effect on modelled river flows that resulted.

Catchment hydrology

Data sources

Three kinds of data are available to support an investigation of the hydrology of the Waveney and Little Ouse catchments:

- daily total river flows from gauges at Knettishall on the Little Ouse, and Billingford Bridge on the Waveney
- daily rainfall from 16 rain gauges spaced irregularly over and around the catchments
- weekly potential evapotranspiration data for MORECS squares 130 and 141.

The locations of some of these measurements are shown in Fig. 1.

Long-term annual average rainfall for the period 1961–1990 shows relatively little variation across the area, from 572 mm a^{-1} to 605 mm a^{-1} with no systematic pattern. On average, February is the driest month and November the wettest. Daily rainfall data were assembled, correlating between gauges where necessary to fill gaps, for the period 1970–1994.

Potential evaporation data from MORECS for the period 1970–1994 give an annual average of 566 mm a^{-1}. By comparison with the rainfall data, it can be seen that this is a dry area. The rainfall and potential evaporation data were combined using a Penman–Grindley type soil moisture balance model, run at daily time-steps, to calculate hydrologically effective rainfall (HER) based on the land use of the area studied. The model incorporated a deficit by-pass mechanism for summer rainfall events exceeding 20 mm day^{-1}, which adds up to 2% extra HER. This gave an average HER of 182 mm a^{-1} for the period 1980–1994.

Statistics for the two river gauges are summarized in Table 1. It should be noted that the Billingford Bridge gauge may be less accurate during summer low flows, because of weed growth. Before analysing the flow data further, they were corrected for surface water abstractions and discharges. The Little Ouse is supported by groundwater abstractions during some summers; this scheme was assumed to produce a net yield of 40% (Seccombe 1992). Measured flows in the River Waveney include treated sewage effluent from Diss.

Overall water balance

In general terms, any inconsistencies in the basic measured or calculated inflows and outflows discussed above will frustrate any attempt to investigate subtle behaviour using a numerical groundwater model. For this reason the data were tested by calculating an overall water balance: can it be shown that the HER needed to account for the known discharges from a specified area is the same as that which is calculated as having occurred? The largest discharges are the two river flows, although further allowance must be made for import and export of public supply water. A correction is necessary because groundwater divides and surface water divides are not coincident in this region (Fig. 2); it is estimated that the recharge into 15 km^2 is lost from the combined river discharges because of this.

Figure 4 presents a plot of the cumulative difference between estimated inputs and outputs. It is clear that there is a calculated net input to the system over the 14-year period illustrated, which groundwater levels suggest has not occurred in reality. The assumption was made that any actual imbalance over this period was negligible, and the measured rainfall was factored by 0.93 to balance the water budget.

The decision to factor the rainfall is arbitrary and needs defending. The water balance error, which is a real error, is the accumulation of measurement errors in rainfall and river flow, errors in the estimation of potential evapotran-

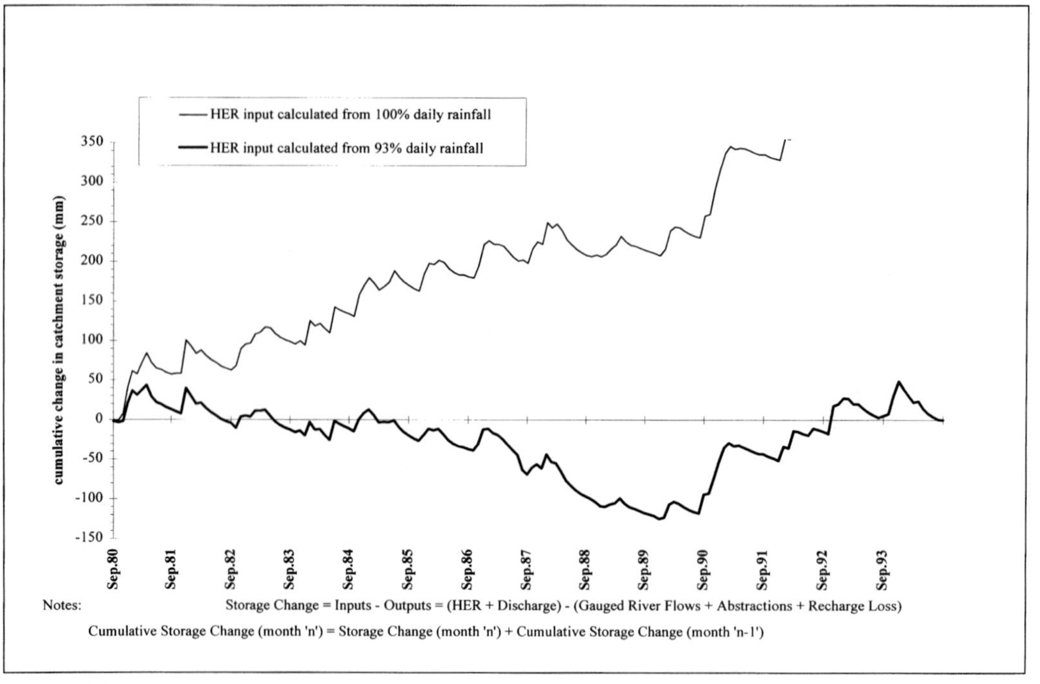

Fig. 4. Cumulative catchment storage calculated by lumped total water balance with effective rainfall (HER) series derived from 100% and 93% of measured rainfall series.

spiration and its conversion to actual evapotranspiration, errors in the estimation of actual abstractions and discharges from licensed quantities, and errors in the correction for non-coincident catchments. In a water resources study, it is more conservative to reduce inputs than to increase outputs, and the rainfall is a large and easily reduced input.

To put the adjustment into perspective, Institute of Hydrology work at Plynlimon (Kirby *et al.* 1991) suggests that in order to measure upland catchment rainfall inputs to within 10% of the actual value over any one month, a rain gauge density of around one gauge for every 3.2 km^2 is required. For the low relief topography of East Anglia this requirement may be slightly relaxed but it is still expected to be considerably in excess of the actual gauge density in the area studied (around one gauge per 36 km^2), and therefore an error of 7% in rainfall input is plausible. The size of the error gives an idea of the accuracy to which total available resource can be estimated using routinely acquired data, in an area where geography and hydrology are relatively simple.

River flow hydrograph analysis

Gauged river flows are one of the best available datasets for understanding recharge to a catchment. Flows reflect the response of the catchment to rainfall which may be 'flashy' or more sustained depending on the balance between rapid run-off and slower groundwater interaction.

Groundwater discharge to rivers is often considered the same as river baseflow (Shaw 1983; Domenico and Schwarz 1990), but estimating baseflow from total flow is not straightforward. The Institute of Hydrology (1989) has derived a widely applied method, based on *n*-day turning points, where *n* is usually five days. This effectively assumes that flow peaks of less than five - days' duration are predominantly run-off and interflow. The five days is arbitrary but produces sensible results for most English rivers and is readily amenable to bulk processing of routine data.

The result of applying this method to data from the Little Ouse and the Waveney is shown in Fig. 5. For the Little Ouse, the resulting baseflow hydrograph looks reasonable and is a fairly

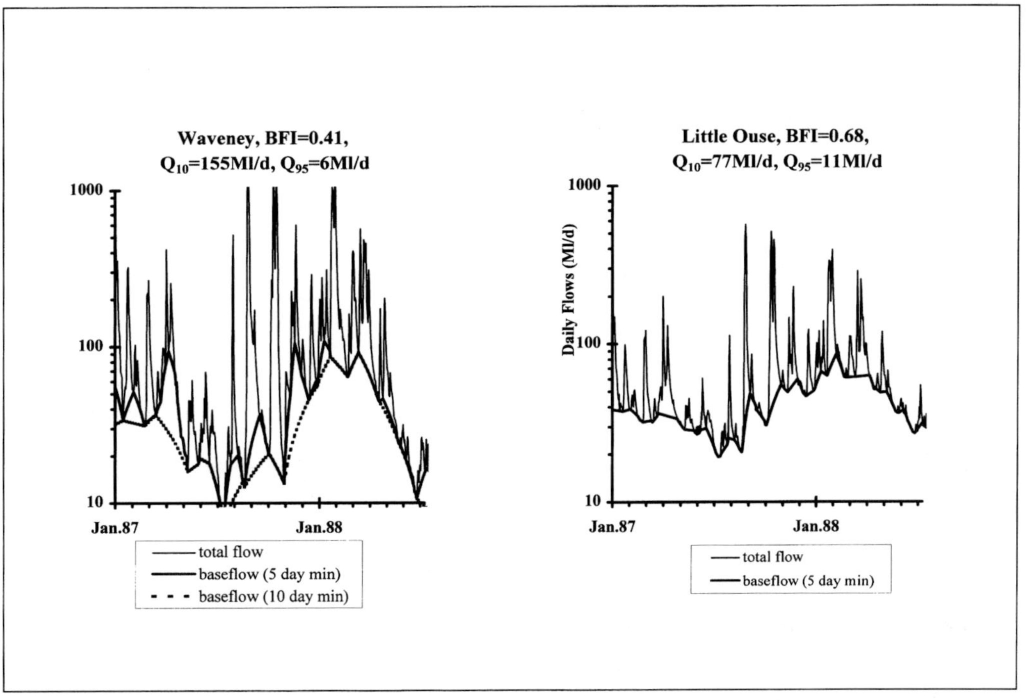

Fig. 5. Baseflow separation for both river hydrographs based on the Institute of Hydrology (five-day minima) technique.

smooth curve, but for the Waveney the 'five-day' baseflow has large peaks. A more reasonable, less 'peaky' baseflow hydrograph would appear to result from $n=10$ days (Fig. 5). Even in this case, however, there is still an important contrast between the two rivers: the winter baseflows of the Little Ouse are around two to three times the summer baseflows, but for the Waveney the ratio is around eight to ten. This is reflected in the ratio of high flows (Q_{10}) to low flows (Q_{95}) which are also routinely reported by the Institute of Hydrology (Table 1): for the Little Ouse $Q_{10}/Q_{95}=7$ whereas for the Waveney $Q_{10}/Q_{95}=26$.

As an alternative to this arbitrary approach, baseflow can be estimated as the slower of two parallel first-order processes (Young & Beven 1991; Jakeman *et al.* 1993). Effective rainfall to the catchment is assumed to produce a rapid (run-off) river flow response and a slower ('groundwater discharge') response, each of which is characterized by an initial amplitude (or 'gain') and an exponential decay term. River flow at any time is the sum of these two components generated dynamically by the preceding series of effective rainfall events. The 'gain' and 'decay' parameters of both processes can be combined into a single transfer function which derives a modelled river flow series from an effective rainfall series.

This is a purely objective approach, relying on a statistical optimization of the transfer function to result in a best fit between modelled and observed river flow. The gain and decay parameters can also be expressed in terms which can be considered analogous to a percentage split between fast and slow flow and a 'residence time' in the catchment for each route. The transfer function analysis was carried out using microCaptain software (Young and Benner 1991). The fits to observed flows were fairly poor, and therefore the results (in terms of the flow split and 'residence times' shown in Fig. 6) are perhaps best regarded as illustrative rather than definitive.

Distinct differences in behaviour are evident from Fig. 6. For both rivers, the 'fast' pathway has a residence time of around two days, compatible with the idea of surface run-off for catchments of this size. However, the 'slow' pathway for the Waveney takes a smaller proportion of the total flow than it does for the Little Ouse,

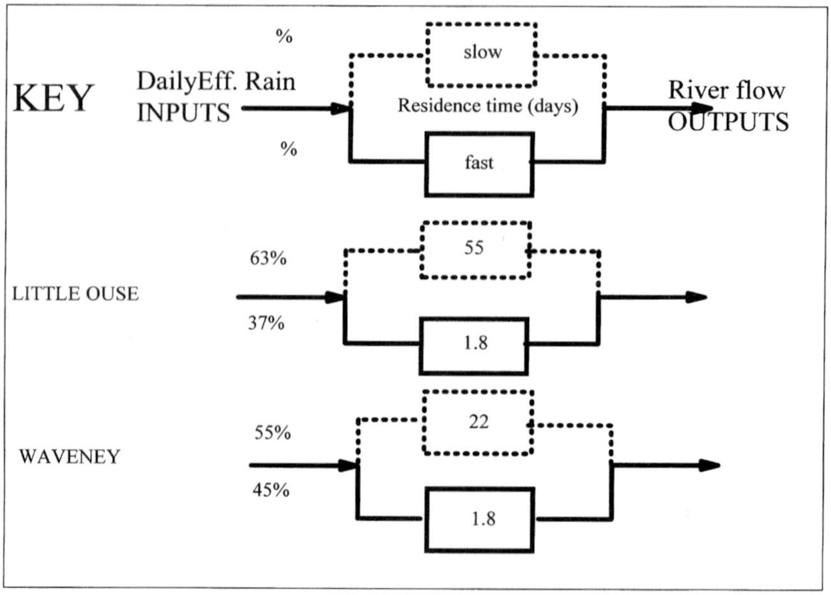

Fig. 6 Effective rainfall – river flow transfer function modelling results.

and the residence time is much shorter, less than one month for the Waveney and nearly two months for the Little Ouse.

As a further illustration of the significance of this contrast (using data aggregated over a longer time period), Fig. 7 shows monthly average actual river flows, and monthly average estimated river flows, assuming that all HER becomes river flow in the month that the precipitation falls. This simple picture is, as expected, a very poor explanation of the flows of the Little Ouse, but is a surprisingly good description of the Waveney flows. The implication of this is that very little of the HER falling on the Waveney catchment is stored for more than one month. In other words, recharge to the Chalk in the Waveney catchment is much less than in the Little Ouse catchment.

Groundwater recharge

Overall recharge estimation

Groundwater recharge cannot be measured directly on a regional scale, and in this case was derived during model calibration. River flows are directly sensitive to recharge amounts and are therefore a useful guide in calibrating recharge.

Initial attempts were based on groundwater discharge derived from river baseflow using the Institute of Hydrology method as discussed above, which gave a regionally averaged recharge of 93 mm a^{-1}. However, all models based on this gave very poor fits for the River Waveney total hydrograph as described in the Introduction.

From Darcy's law, groundwater discharge is proportional to the product of a groundwater gradient and a transmissivity, and therefore a variability in groundwater discharge to the Waveney of eight to ten times, as documented above, requires either nearly an order of magnitude change in groundwater gradient, or nearly an order of magnitude change in transmissivity, or a significant change in both. Examination of available data shows that regional groundwater gradients in the Redgrave area do not change substantially over the year, neither do gradients close to discharge zones. Although dramatic changes in Chalk transmissivity with water level are well-known, the water-level fluctuation in the Chalk at the interfluves in the Redgrave area is only of the order of 1 to 2 m compared with an average interfluve to valley head difference of 10 m, and borehole geophysical logs show that fissuring typically extends to depths of at least 15 m below the top of the Chalk. Thus no combination of change of gradient or change in transmissivity in the Chalk seems likely to be able to combine to produce the large seasonal variations apparent in the Waveney 'baseflow'. It was concluded that the picture of the Chalk aquifer supplying all the baseflow

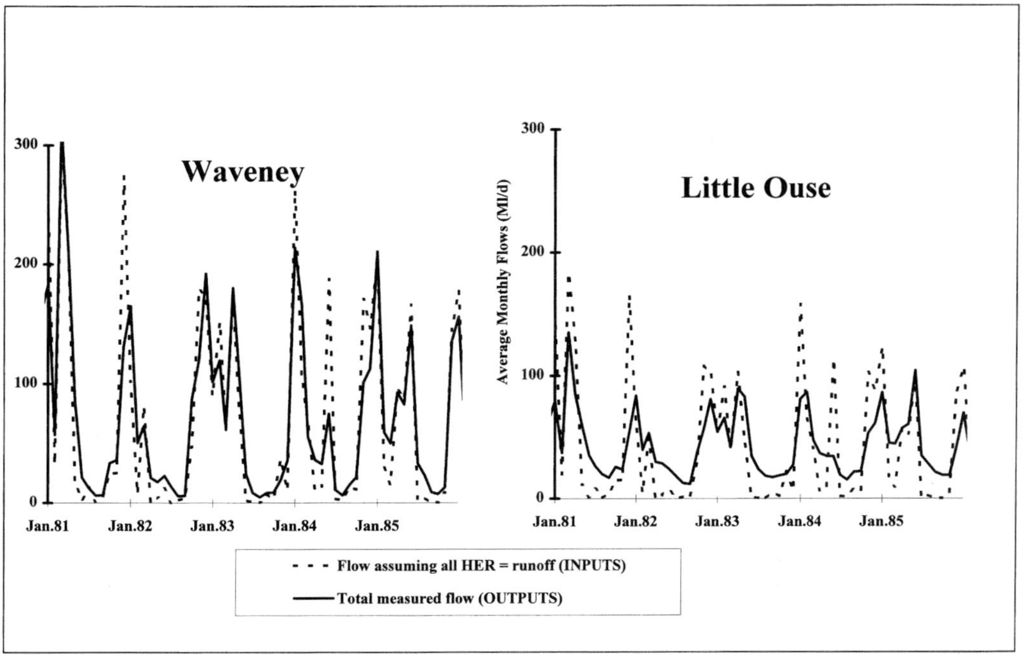

Fig. 7. Comparison of measured flows with flows simply modelled by assuming all effective rainfall (HER) inputs become river flow in the same month.

separated by the minimum turning point technique for the Waveney was too simple.

The transfer function approach to baseflow separation described above further highlighted the contrasting behaviour of the two catchments, but fits of modelled to observed river flows were too poor to provide a reliable recharge series directly.

The problem was then approached from a different direction. On the basis of observed seasonal regional hydraulic gradient variations it was assumed that twice the minimum observed summer flows, after several months with no HER, might represent average Chalk groundwater discharge which, in the longterm, must be balanced by recharge. The proportion of HER represented by this long-term average discharge was assumed to be constant with time and was applied to the transient HER dataset to provide a transient recharge input series for the groundwater model. Taking this approach, estimated average annual recharge quantities for the Little Ouse and Waveney catchments were 71 mm a^{-1} and 39 mm a^{-1} respectively.

The resulting groundwater model discharge, when combined with the balance of the HER (assumed to be run-off), then produced a much better fit to the total flows of both rivers (compare Fig. 8 with Fig. 3).

It is inferred that in the Waveney catchment a significant amount of the HER does not produce Chalk recharge. This water may enter the ground rather than forming direct run-off, but drains to the rivers via the dense networks of ditches and tile drains which make the heavy soils of the area workable, without entering the Chalk. Large variations in hydraulic gradient and transmissivity in this thin, partially saturated layer of near-surface field drains are quite feasible, such that flows from it simply reflect the availability of recent HER.

Recharge distribution: modelling catchment contrasts

It has been deduced from river hydrographs that the average annual recharge into the two adjacent catchments is different, although the HER input to each is effectively the same. This difference must therefore result from the subtle differences in the topography and geology of the two catchments.

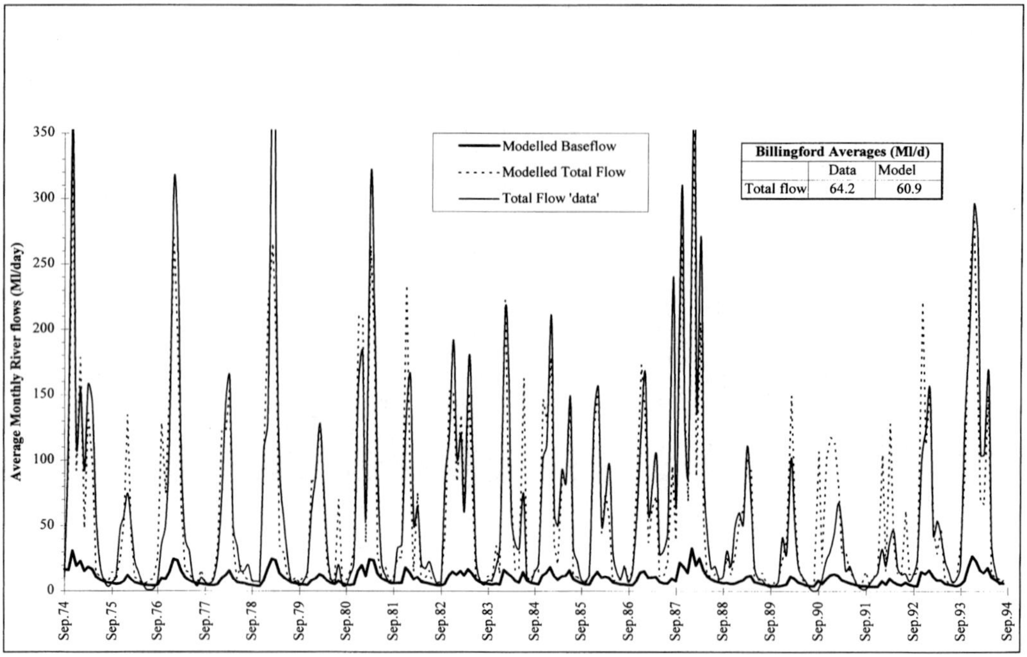

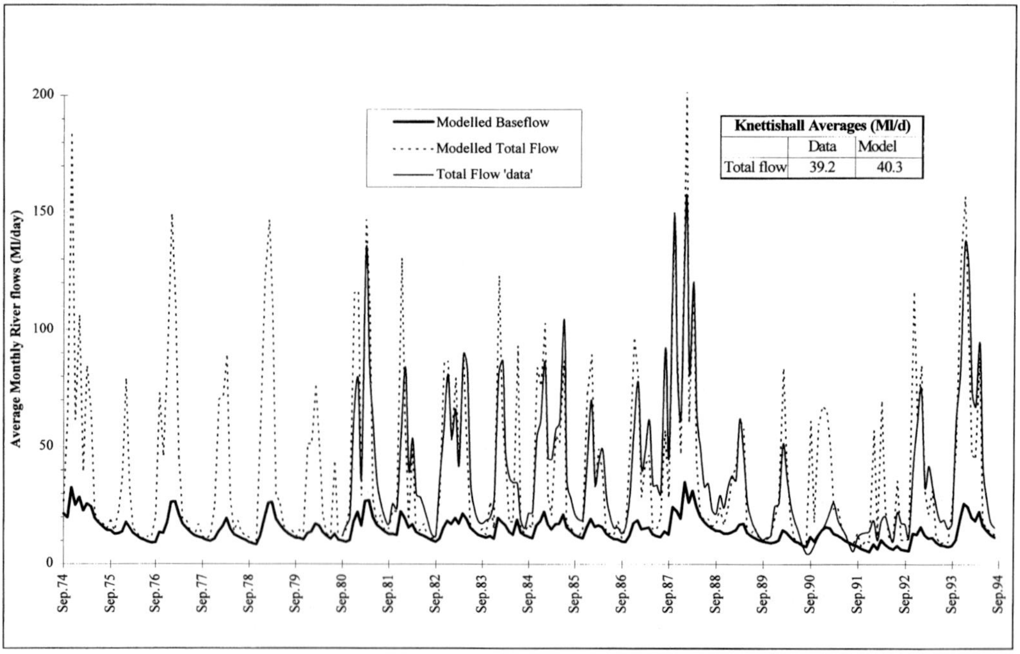

Fig. 8. Modelled and measured river flows (recharge reduced after hydrograph reanalysis): (**a**) Waveney; (**b**) Little Ouse.

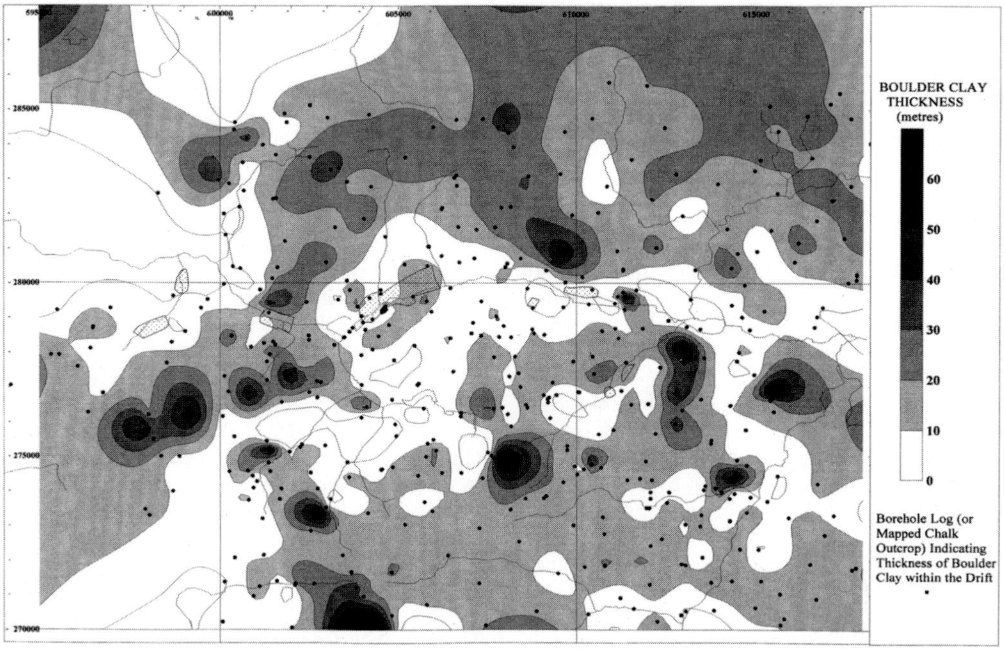

Fig. 9. Contoured total boulder clay thickness based on borehole logs and mapped Chalk outcrop data.

The presence of low-permeability boulder clay within the drift is shown in Fig. 9 based on borehole logs. The boulder clay covers a greater proportion of the Waveney catchment than the Little Ouse catchment and the average thickness is also greater in the Waveney catchment. Furthermore, the Chalk piezometric surface is confined by the boulder clay, reducing the vertical hydraulic gradient available to drive recharge through it, over a greater proportion of the Waveney catchment.

This difference in extent of confinement is probably a function of differences in topography and the effectiveness of the river as a drain in the Chalk groundwater system. The Little Ouse catchment is steeper (river stage gradient 0.001) than the Waveney catchment (river stage gradient 0.0003) and the Waveney runs for most of its length over thick drift (deposited in a sub-glacial tunnel valley). Both of these factors appear to result in the Waveney being less well hydraulically 'connected' to the Chalk than the Little Ouse, which is a more effective groundwater drain.

In the final model runs, the proportion of HER resulting in recharge to the Chalk was distributed according to the product of the estimated vertical hydraulic conductivity of the drift (assigning different values to boulder clay, sand, and gravel based on borehole logs) and the available vertical gradient. By calibration, it was possible to arrive at a distribution that reproduced the different target values of annual average recharge for the two catchments (as deduced in the previous section). Within this distribution an allowance was also made for downslope recharge of run-off from boulder clay-covered areas, as described by Jackson & Rushton (1987). Simpler distributions ascribing recharge factors on the basis of drift type alone (Rushton 1975), were insufficient to produce the differences in recharge between the two catchments.

Inspection of the final modelled recharge distribution suggests that average recharge to the Chalk over boulder clay-covered areas is around 36 mm a^{-1}, with values locally as low as 10 mm a^{-1} on the interfluves where the clay is thick and confines the Chalk. This compares favourably with experimental work by Klinck *et al.* (1996) for Garboldisham (20–40 mm a^{-1}), only a few kilometres outside the study area, and a value of 24 mm a^{-1} estimated from model calibration by Jackson and Rushton (1987) for the Gipping catchment, which adjoins the south side of the present study area.

In parts of the Little Ouse catchment however, drift cover is thin or non-existent and the Chalk is better drained by the river. The modelled distributions suggest that locally such conditions may result in Chalk recharge up to 160 mm a^{-1}, approaching total HER, and may even exceed HER where boulder clay run-off is routed as recharge into unsaturated Chalk.

Conclusions

This work has demonstrated the value of careful examination of basic hydrological data (particularly river flow hydrographs) in the understanding of recharge in an area covered with thick clayey drift.

Comparison of estimates of water input and output is a basic validation of key data for any numerical modelling exercise, and also gives a realistic idea of the certainty to which water resources may be known in any area. In dry areas such as the present study area, where there is pressure to use 100% of the available resource, an appreciation of resource uncertainty is important.

The baseflow index, published by the Institute of Hydrology for all routinely gauged sites in England, is a useful first indicator of the possible contribution of groundwater discharge to river flows. However, the value is determined by a simple algorithm based on five-day minima which may present an over-simplified view of system behaviour. Other published statistics, for example Q_{10} and Q_{95}, provide additional useful information on the variability of river flow and therefore on the possible contribution of groundwater.

The direct graphical comparison of HER with river flow (averaged on a monthly basis in this study) provides a clear indication of the possible significance of storage mechanisms, which may be supported more rigorously by transfer function analysis of daily time series data.

In the end, baseflow is not a rigorously defined concept, and will to some extent depend on its intended use. In this example, the dominant 'baseflow' for the upper Waveney to the Billingford gauge remains in ground storage for less than a month, and is effectively 'run-off', when used in a numerical model with one month time-steps. Some judgement may be necessary to determine an appropriate baseflow for a specific modelling application, as part of calibration, but it should be possible to demonstrate the validity of this judgement by comparison of whole-flow hydrographs, which are not subject to the same uncertainty.

Through model calibration, a value of average recharge through boulder clay has been derived (36 mm a^{-1}), which is similar to others for this same area of England. Some insight into possible variability, and simple controls that can produce this variability, has also been gained during the present work. Recharge through boulder clay in the present study area is considered to vary from 10 mm a^{-1}, where the clay is thick and the available vertical gradient is reduced by a high Chalk piezometric surface, to 160 mm a^{-1} where the clay is thinner and the vertical gradient reaches unity.

Finally, on a catchment-wide basis the study also suggests that the overall recharge getting into an aquifer depends on the river's effectiveness at draining water out of it. The Little Ouse River down to Knettishall is a steeper and better hydraulically connected drain in the Chalk groundwater system than the River Waveney to Billingford. Consequently the relatively poorly drained Waveney Chalk remains 'fuller' and more confined, further reducing the potential for recharge.

The work reported here was carried out under contract to Essex and Suffolk Water plc, whose permission to publish this paper is gratefully acknowledged. The authors accept full responsibility for the work. The detail of the numerical modelling was performed by Alastair Hoare. The work benefited from discussions with many people, especially John Elliott of Essex and Suffolk Water, David Seccombe of the Environment Agency, Mike Harding of Suffolk Wildlife Trust, and Professor Ken Rushton.

References

Aspinwall and Company. 1992. *Redgrave Stage II Study: Data Collection and Analysis*. Report for Suffolk Water Company plc.

Bellamy, D. & Rose, F. 1960. The Waveney-Ouse Valley Fens of the Suffolk-Norfolk Border. *Transactions of the Suffolk Naturalists Society,* **11**, 368–385.

Domenico, P. A. & Schwarz, W. 1990. *Physical and Chemical Hydrology*. Wiley, Chichester.

Harding, M. 1993. Redgrave and Lopham Fens, East Anglia, England: A Case Study of Changes in Flora and Fauna due to Groundwater Abstraction. *Biological Conservation*, **66**, 35–45.

Institute of Hydrology. 1989. *Flow Regimes from Experimental and Network Data (FREND)*. NERC.

Jackson, D. & Rushton, K. R. 1987. Assessment of recharge components for a Chalk aquifer unit. *Journal of Hydrology*, **92**, 1–5.

Jakeman, A. J., Littlewood, I. G. & Whitehead, P. G. 1993. An assessment of the dynamic response characteriztics of streamflow in the Balqhuidder catchments. *Journal of Hydrology*, **145**, 337–355.

KIRBY, C., NEWSON, M. D. & GILMAN, K. (eds) 1991. *Plynlimon Research: The First Two Decades*. Institute of Hydrology, Report **109**.

KLINCK, B. A., BARKER, J. A., NOY, D. J. & WEATHALL, G. P. 1996. *Mechanisms and rates of recharge through glacial till: Experimental and modelling studies from a Norfolk site*. British Geological Survey Technical Report **WE/96/1**.

RUSHTON, K. R. 1975. Aquifer analysis of the Lincolnshire Limestone using mathematical models. *Journal of the Institution of Water Engineers and Scientists*, **29**, 373–388.

SECCOMBE, D. 1992. *Relocation of Suffolk Water Company Borehole*. NRA Anglian Region report.

SHAW, E. M. 1983. *Hydrology in Practice*. Van Nostrand Reinhold.

YOUNG, P. C. & BENNER, S. 1991. *Micro*CAPTAIN 2.0 *Handbook*. Centre for Research on Environmental Systems, Lancaster.

—— & BEVEN, K. J. 1991. Computation of instantaneous unit hydrograph and identifiable component flows with application to two small upland catchments – comment. *Journal of Hydrology*, **129**, 389–396.

Estimation of groundwater recharge on Jersey

J. R. BLACKIE, H. A. HOUGHTON-CARR, M. P. McCARTNEY & J. P. MOORES

Institute of Hydrology, Maclean Building, Crowmarsh Gifford, Wallingford, Oxfordshire OX10 8BB, UK

Abstract: There has been growing concern that groundwater abstraction on the island of Jersey may exceed recharge, with the consequent depletion of the island's finite water resources. A three-year study to determine groundwater recharge through hydrological modelling is nearing completion. A daily model has been developed which predicts evapotranspiration, soil moisture, recharge and streamflow from inputs of rainfall and potential evapotranspiration. Preliminary calibration of the model has used data collected from a small, intensively instrumented catchment around Trinity, Jersey, between April 1993 and December 1994. Subsequent validation, for the pre-study period January 1990 to December 1992, has shown estimated net recharge for the catchment as a whole to be 50 mm, with an initial depletion of groundwater reserves in 1990 followed by two years of positive recharge. Net recharge under a grass site was estimated to be slightly more than under arable land. Model calibration and validation will be reworked following collation of the 1995/96 dataset. The model will then be run on 25 years of catchment rainfall and potential evapotranspiration data to estimate recharge patterns over the period 1968–92, and to assess the impact of different abstraction rates.

Jersey is the largest of the Channel Islands, with an area of 117 km^2 (Fig. 1). It is essentially a Precambrian rock plateau with a maximum altitude of 138 m a.s.l. in the north, draining towards the south by a series of incised valleys. The island is predominantly rural, consisting mainly of intensively worked arable land, which commonly supports two or even three crops in one year, and grassland used for dairy farming. The island has a resident population of 82000, although this can almost double during the peak tourist season. The south coast towns of St Helier and St Brelade are the most significant areas of urbanization, though there are also numerous small villages scattered around the island, both inland and on the coast. Hence, water demands, for both domestic supply and agricultural use, are already high and steadily increasing. Water use for irrigation alone increased by over 100% in the five years to 1991 (Agriculture and Fisheries Committee 1992).

Mains water supply on the island is the responsibility of the Jersey New Waterworks Company (JNWWC). Water is supplied to about 80% of the population from six impounding reservoirs (James 1989), the largest of which is the recently constructed Queen's Valley Reservoir with a capacity of 1.14 Mm^3. Farmers and other rural dwellers rely on private boreholes and shallow wells for both domestic water supply and agricultural needs. According to Norman law, landowners own the water on or under their land, and can therefore dam a stream or sink a borehole, irrespective of the downstream consequences or the impact on neighbouring water sources. There were last reported to be at least 5400 boreholes into the shallow fractured basement aquifer beneath Jersey, which also provides baseflow to the surface reservoirs (Robins *et al.* 1993).

Robins & Smedley (1994) suggest that the island's annual average net groundwater recharge is approximately equivalent to present groundwater abstraction levels, at around 30 mm. Although they acknowledge some uncertainty in their estimate, due to the difficulties of identifying water table monitoring sites free from the effects of local pumping, it does correspond well with that of James (1989), who proposed an annual average recharge figure of 31 mm. Hence, there is growing, and understandable, concern that groundwater abstraction on the island may already, or will soon, exceed recharge. In April 1993, Jersey Public Services Department (PSD) commissioned a three-year study by the Institute of Hydrology to determine groundwater recharge through hydrological modelling. The ensuing model will ultimately be used to estimate long-term recharge patterns and assess the impact of different abstraction rates.

Study catchment and monitoring network

The study catchment selected for the project was the Trinity Valley catchment (Figs 1 and 2). The

Blackie, J. R., Houghton-Carr, H. A., McCartney, M. P. & Moores, J. P. 1998. Estimation of groundwater recharge on Jersey. *In*: Robins, N. S. (ed.) *Groundwater Pollution, Aquifer Recharge and Vulnerability*. Geological Society, London, Special Publications, **130**, 143–152.

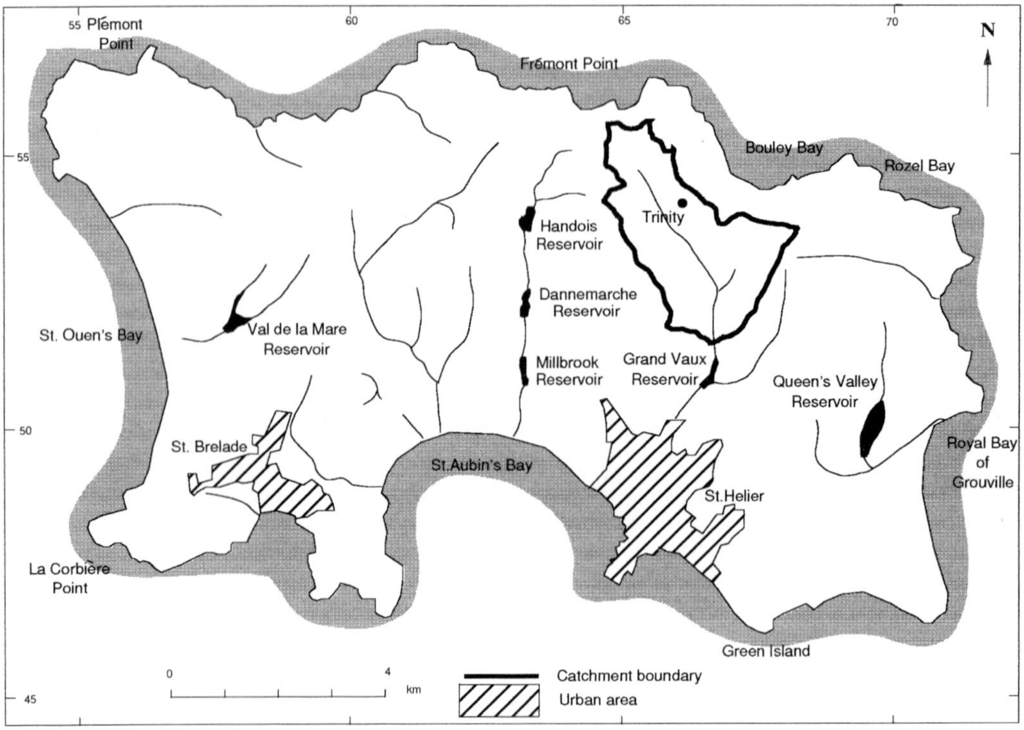

Fig. 1. Catchment location.

catchment has an area of 6.34 km^2, thereby representing just over 5% of the island. It is underlain by volcanic rock (andesite and rhyolite), and overlain by clay-loam soils derived from loess. Like all major catchments on Jersey, it is oriented north–south, draining from the higher ground in the north; in fact, the catchment encompasses the highest point on the island at an elevation of 138m a.s.l. The catchment outfall, just upstream of the Grand Vaux Reservoir, is at 46m a.s.l. Prevailing winds are westerly and southwesterly from the Atlantic Ocean. The catchment is predominantly rural, with only two villages of any real size: Trinity Village in the north and Victoria Village in the south. Table 1 shows a breakdown of agricultural land cover within the catchment over an 18-month period, illustrating the common practice of changing land use several times a year. The catchment is assumed to be 'water-tight', with any artificial inputs and outputs (i.e. mains water supply and mains drainage) balancing each other.

The catchment was intensively instrumented to provide adequate data to calibrate and validate the hydrological model to be developed. The hydrological instrument network was installed in March 1993 (Fig. 2), and was designed to monitor precipitation, streamflow, potential evaporation and soil moisture storage changes under representative vegetation types over a three-year period. Rainfall is measured at six sites, chosen to sample both the areal and altitudinal ranges of the catchment. Storage gauges at ground level are read once a week, whilst three event recorders are used to time-distribute the weekly totals. Streamflow is measured, using a sensitive recorder and logger programmed to produce mean 15–min flows, at a compound sharp-crested weir structure built by JNWWC in June 1988. Meteorological data, required for computation of Penman potential evapotranspiration (Penman 1948), are obtained from an automatic weather station installed at Howard Davis Farm, close to the centre of the catchment at an elevation of 101m a.s.l. Soil moisture is measured approximately weekly at four locations using a down-tube capacitance probe. The four sites have been chosen to represent different vegetation types within the catchment. Since June 1994, an automatic water sampler installed at the catchment outlet has been used to collect daily composite samples

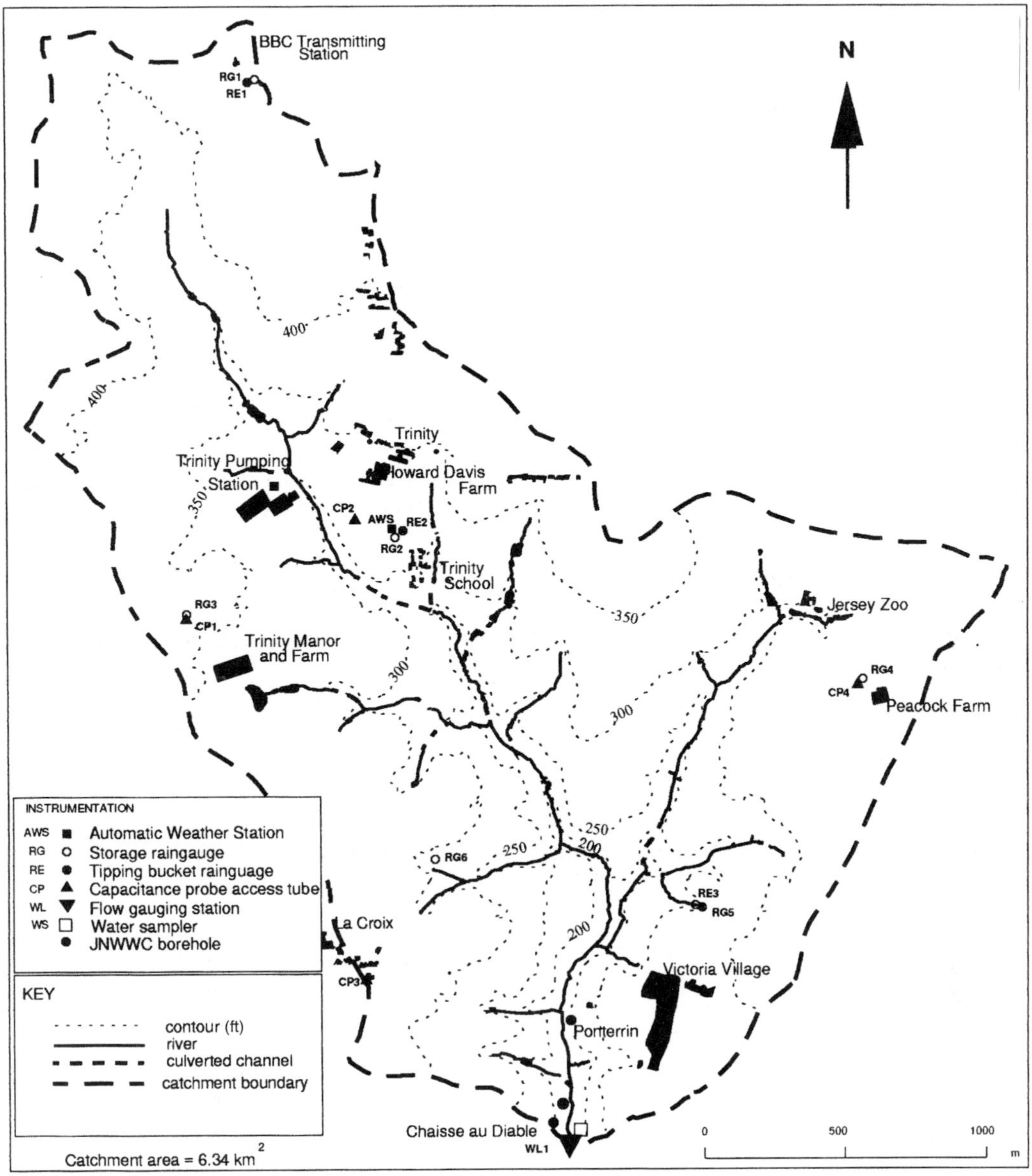

Fig. 2. Map showing catchment topography and instrumentation.

for chemical analyses. Although the data collection phase of the study ended in March 1996, PSD will continue to maintain the monitoring network, and collect and archive the data.

In recent years, JNWWC have supplemented the flows into the Grand Vaux reservoir by adding groundwater to the Trinity stream. This water is pumped from three boreholes within the catchment, located just upstream of the Grand Vaux weir. In absolute terms, the total volume added is small compared to the natural flow of the stream. Nevertheless, it significantly affects the summer flow regime. Isotopic analysis of the borehole water indicates that it is old water (i.e. predating the atmospheric detonation of atomic bombs) originating from a deep aquifer. Consequently, as this water is unlikely to have entered the stream naturally as baseflow, the

Table 1. *Changes in agricultural land cover from a sample of 146 fields*

Land cover	Percentage of sample area		
	May 1994	Aug 1995	Nov 1995
Grass	21.7	55.6	66.8
Bare soil or emerging crops	35.4	25.8	14.1
Vegetables	28.9	9.9	19.1
Grain crops	14.0	8.7	0.0
Total	100.0	100.0	100.0

streamflow record was naturalized by deducting the added borehole water.

Modelling strategy

The hydrological modelling strategy involves a number of stages, initially building upon the foundations provided by the catchment monitoring programme, and eventually culminating in the estimation of groundwater recharge for Jersey as a whole. Modelling, using standard spreadsheet software, commenced in Year 2 of the catchment study, with development of a simple, conceptual model, operating at a daily time-step, to predict changes in soil moisture from inputs of daily rainfall and potential evapotranspiration. No attempt was made to model streamflow, and recharge was estimated simply by subtracting observed streamflow from total drainage through the soil. The model was optimized through parameters representing crop water use and soil drainage rate. This model underwent substantial development during Year 3 of the study, and now has inputs of daily rainfall and potential evapotranspiration, and outputs of evapotranspiration, soil moisture, streamflow and recharge.

The model will eventually be calibrated for the two-year period April 1993 to March 1995 using three sets of data: a catchment average dataset to determine model parameters for the observed mixed land use in the catchment, and then datasets for each of a notional grassed catchment and a notional arable catchment. The model will then be validated on the one-year period April 1995 to March 1996. The model will then be run on 25 years of catchment rainfall and potential evapotranspiration data to estimate recharge patterns over the period 1968–92, and to assess the impact of different abstraction rates.

To date, preliminary calibrations of the model have been done using data collected from the Trinity catchment between April 1993 and December 1994, and a tentative validation based on these parameter sets has been performed for the period January 1990 to December 1992. This validation uses rainfall from a gauge at Howard Davis Farm and potential evapotranspiration values computed from a meteorological site outside the catchment as inputs, and predicted streamflow is compared with streamflow measured at an existing JNWWC gauge. Since no soil moisture data were available before April 1993, validation of this output was not possible. These data used for the validation cannot be expected to be of the same quality as the data collected specifically for the catchment study, but nevertheless allow a first check on the quality of the model fit.

Model description

The model is a simple, conceptual water balance model. It consists of four stores, with 12 model parameters controlling the store sizes and rates of flow into and out of the stores. Figure 3 shows the structure of the model. Table 2 lists the 12 parameters, together with the ranges associated with each parameter. The ranges are governed by mathematical constraints (e.g. to ensure that the model is dealing with positive terms), as well as hydrological constraints.

Several of the parameters listed in Table 2 can be estimated from field observations or taken from documented studies. For example, many studies have determined crop factors *CF*, root constants *RootC* and permanent wilting points *PWP* for a range of plant types. The rate of capillary rise *CRM* in various soil types has also been studied; in a silt soil, the rate of capillary rise can be up to 28 mm a day, with a lower rate with increasing particle size (Todd 1980).

In the model, rainfall onto the catchment enters a surface store, representing interception of the rainfall by vegetation and other surfaces. Water may be lost from this store by evapora-

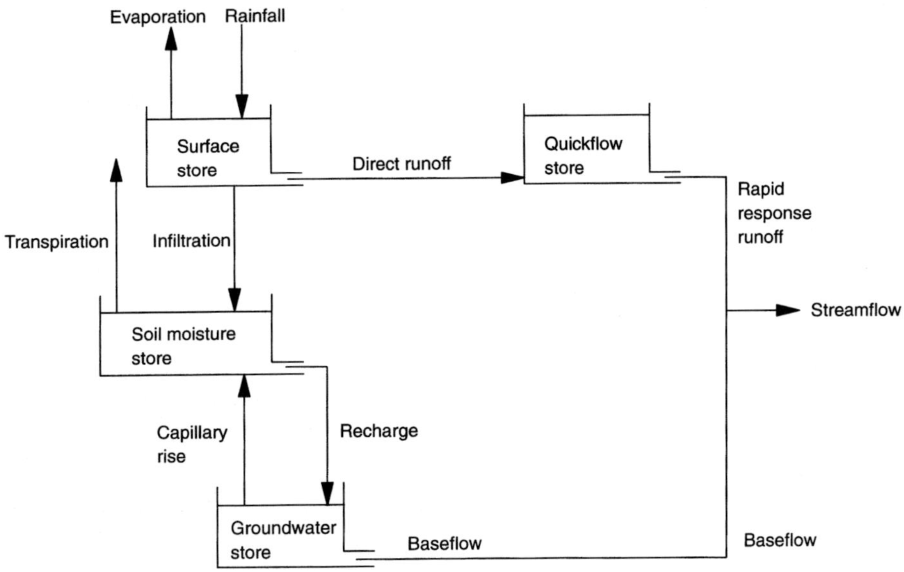

Fig. 3. Jersey rainfall–recharge–run-off model.

Table 2. *Parameter value ranges*

Parameter	Parameter name	Range
SS	Surface store (interception) capacity (mm)	>0
RC	Run-off coefficient	0–1
RK	Quickflow store factor	0–1
RX	Quickflow store index	>1
CF	Crop factor	0–1
RootC	Root constant (mm)	50–150
PWP	Permanent wilting point (mm)	75–200
SDR	Soil moisture store drainage rate	0–1
GSU	Groundwater store factor	>30
GSP	Groundwater store index	>1
GST	Groundwater store threshold (mm)	>0
CRM	Capillary rise maximum rate (mm)	>0

tion. If the rainfall exceeds the capacity of this store *SS*, the excess is transferred to two other stores: a quickflow store representing the direct run-off process, and a soil moisture store representing the infiltration process. The proportion of excess rainfall going to each store is controlled by the run-off coefficient *RC*; the higher the value of *RC*, the more excess rainfall enters the quickflow store.

Water in the soil moisture store may be depleted by transpiration from vegetation or by drainage to a groundwater store, or by a combination of both depending on the range of controlling factors, namely the crop factor *CF*, the root constant *RootC*, the permanent wilting point *PWP* and the content of the store relative to field capacity (i.e. zero soil moisture deficit). When the soil moisture store content is at or above field capacity, surplus water drains to the groundwater store at a drainage rate *SDR*; increasing the value of *SDR* increases the drainage rate from the soil. When the soil moisture store content is less than field capacity, drainage to the groundwater store ceases. During particularly wet periods, the water-table may rise to be close to the soil surface. This process is represented by the soil moisture and groundwater stores combining, with transpiration loss occurring directly from the groundwater store. During very dry spells, when the water-table is below the root range, water may also move upwards from the groundwater store back to the soil moisture store by the capillary rise process. The transfer of water back to the soil moisture store via capillary rise is governed by the groundwater store threshold *GST*, and the capillary rise maximum rate *CRM*.

Rapid-response run-off is released from the quickflow store at a rate controlled by the store factor *RK* and store index *RX*. Baseflow is released from the groundwater store at a rate controlled by the store factor *GSU* and store index *GSP*. The rapid-response run-off and baseflow combine to become streamflow in the river channel.

Over the modelling period, the quantity of water in each of the stores fluctuates with the input of rainfall, loss through evapotranspiration and run-off, and transfers to other stores. The fluctuation of water in the groundwater store represents the rise and fall of the water-table. Net recharge is simply the difference between the depth of water in the store at the beginning and end of the modelling period. A more detailed description of the model is given in Blackie *et al.* (1996).

Model results

To date, the model parameters have been determined for a calibration period from April 1993 to December 1994. Parameters have been estimated for three datasets: a dataset for each of a notional grassed catchment (based on the Trinity Manor site) and a notional arable catchment (based on the La Croix site), and then a catchment average dataset to determine model parameters for the observed mixed land use in the catchment. The best-fit model parameters for each dataset were then used in applying the model to a validation dataset, from January 1990 to December 1992.

The classic approach to fitting a conceptual model to observed data to obtain an optimum parameter set involves minimizing an objective function, generally by automatic optimization techniques (e.g. Rosenbrock 1960). However, it is not generally possible to define hydrological model performance solely on the basis of objective criteria, as subjective measures, such as quality of the simulation of the daily flow regime, must also be considered. Therefore, model performance during calibration was based on both quantitative and qualitative measures. Objective functions, indicating the average difference between the observed values of streamflow or soil moisture and those predicted by the model, were calculated by the formula:

$$OJ_X = \sqrt{\frac{\sum(X_o - X_e)^2}{n_X}}$$

where X_o = observed value, X_e = estimated value, n_X = number of observed values.

Nash–Sutcliffe (1970) efficiency criteria, which are essentially normalized objective functions, were also calculated by the formula:

$$Eff = 1.0 - \frac{\sum_{i=1}^{N}\left(X_{o_i} - X_{e_i}\right)^2}{\sum_{i=1}^{N}\left(X_{o_i} - X_m\right)^2}$$

where X_m = the mean of the observed values over the n-value period. A perfect agreement between the observed and simulated values yields an efficiency of 1.0, whilst a negative efficiency represents a lack of agreement that is worse than if the simulated values were replaced with the mean of the observed values.

The ability of the model to predict the total run-off volume over the calibration period was also examined, as a significant error would upset the overall water balance and the estimate of recharge. Finally, the graphical output was inspected subjectively, to assess both how well the different data series were reproduced and the extent to which behaviour was consistent over the modelling period. Table 3 shows the model parameters, together with the values of each objective function and efficiency, the error in the total predicted run-off, and the estimate of recharge over the calibration and validation periods.

In Table 3, several of the model parameter values are identical for each of the datasets; these are mainly parameters related to the modelling of run-off and, after these parameters were initially determined for the mixed catchment case, they were not varied as the release of run-off was considered to be independent of the land cover of the catchment. The differences between mixed, grass and arable catchments were considered to be much more significant in terms of variation in water use and soil drainage. An indication of the proportional division between the rapid-response run-off and baseflow components of the flow hydrograph was derived from hydrograph analysis during model calibration.

The model calibration produced a good fit between the observed and simulated data for each of the three datasets. Streamflow efficiencies were all around 80%, with the arable catchment giving the highest value, and the grass catchment the lowest. Soil moisture efficiencies were lower, ranging from 61% for the arable catchment to 80% for the grass catchment. The errors in

Table 3. *Parameter values and results of model calibration (1993–94) and validation (1990–92)*

	Mixed catchment	Grass catchment	Arable catchment
SS	5	0.3	7
RC	0.18	0.18	0.18
RK	0.07	0.07	0.07
RX	1.55	1.55	1.55
CF	0.8	0.9	0.76
RootC	75	65	85
PWP	125	100	140
SDR	0.1	0.06	0.2
GSU	159	159	159
GSP	6	6	6
GST	125	125	125
CRM	1	1	1
Cal. OJ_Q	0.580	0.630	0.502
Cal. Eff_Q	0.82	0.79	0.86
Cal. OJ_S	18.085	22.549	18.439
Cal. Eff_S	0.75	0.80	0.61
Cal. ErrorQ (%)	3.7	−2.5	1.5
Cal. Net Rech (mm)	60.9	59.0	66.8
Cal. Net Rech (% rain)	2.95	2.96	3.24
Val. OJ_Q	0.530	0.601	0.478
Val. Eff_Q	0.83	0.76	0.82
Val. ErrorQ (%)	−8.3	5.5	−9.7
Val. Net Rech (mm)	49.7	52.5	42.0
Val. Net Rech (% rain)	1.97	2.08	1.67

Cal. = calibration period, Val. = validation period, OJ_Q = streamflow objective function, Eff_Q = streamflow efficiency, OJ_S = soil moisture objective function, Eff_S = soil moisture efficiency, ErrorQ (%) = error in total run-off volume, Net Rech (mm) = net recharge over period, Net Rech (% rain) = net recharge over period as percentage rain

total run-off volume were small, ranging from a 3.7% over-prediction for the mixed catchment to a 2.5% under-prediction for the grass catchment. In all three cases, the predicted net recharge over the 21-month period was about 3% of the total rainfall, being greatest for the arable catchment. Figure 4 shows the calibration fit for the mixed catchment case.

For the model validation, the values of the efficiencies remained at the same magnitudes as in the calibration (only a streamflow function was calculated), being slightly worse for the grass and arable catchments, but improving slightly for the mixed catchment. The errors in total run-off volume increased, now ranging from a 5.5% over-prediction on the grass catchment to a 9.7% under-prediction on the arable catchment. Recharge over the three-year period was estimated to be only around 2% of the total rainfall in each case, this time being greatest for the grass catchment. Net recharge for the catchment as a whole was estimated to be 50 mm. Figure 5 shows the validation fit for the mixed catchment case.

Table 4 shows the estimated water balance for the mixed catchment during each of the three years in the validation period. The years 1990 and 1991 were both relatively dry, with rainfall below the long-term average. During 1990 predicted net recharge was negative. Although rainfall was lower in 1991, the distribution of the rainfall was such that the model predicted there would be net recharge over the 12 months. In contrast, 1992 was fairly wet, with predicted net recharge increasing accordingly. Figure 6 depicts the mixed catchment water balance over the validation period.

Discussion

Several strategies were considered to allow estimation of groundwater recharge for Jersey, given the frequent changes in land use across the island. An approach which estimated recharge under some 'average' pattern of land use was adopted. The monitoring network established within the Trinity catchment has provided the data to develop a simple, conceptual hydrolo-

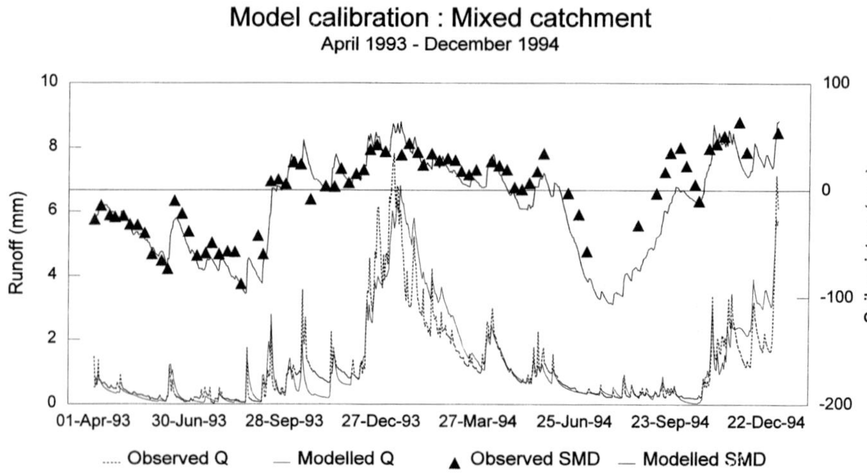

Fig. 4. Results of model calibration on mixed catchment.

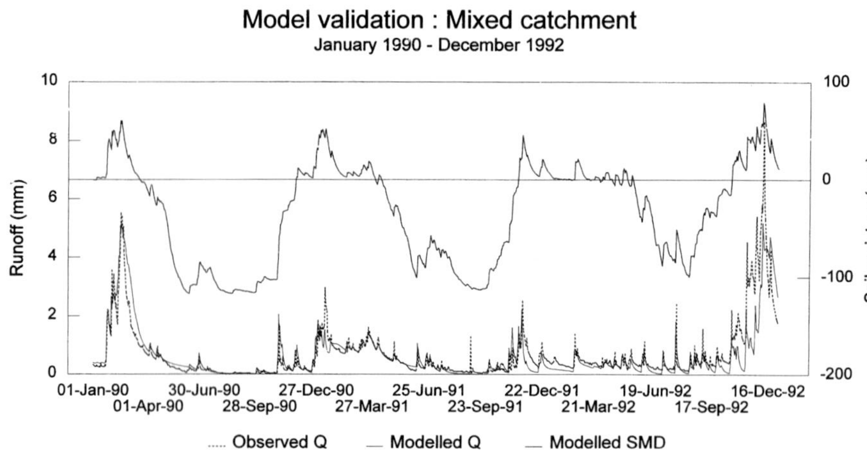

Fig. 5. Results of model validation on mixed catchment.

gical model, operating on a daily time-step, and used to simulate evapotranspiration, soil moisture, recharge and streamflow, given inputs of daily rainfall and evapotranspiration. The model has been calibrated for the Trinity catchment with its 'average' pattern of land use. The predicted net recharge to groundwater for the mixed catchment case over the calibration period April 1993 to December 1994 was 61 mm, and that over the preliminary validation period January 1990 to December 1992 was 50 mm. Whilst these values are similar to existing estimates, the relatively short periods involved preclude any long term conclusions being drawn from them.

When finalized and proven on the complete three-year data series, the model will be run on 25 years of catchment rainfall and evapotranspiration data (1968–92) to identify fluctuations in groundwater and streamflow in the catchment through a range of wet and dry years and to estimate the impacts of different groundwater abstraction rates. The land cover proportions, rainfall and evapotranspiration for Jersey as a whole will then be compared with those of the Trinity catchment and the model, adjusted to take account of these differences, will be rerun to provide estimates of recharge patterns over the 25-year period for the island.

Table 4. *Estimated water balance for the mixed catchment over the validation period*

Year		Water balance
1990	Rainfall	793.9
	Actual evapotranspiration	538.4
	Soil moisture change	+43.3
	Run-off	264.9
	Net recharge	−63.1
1991	Rainfall	725.5
	Actual evapotranspiration	547.9
	Soil moisture change	−37.3
	Run-off	184.5
	Net recharge	+39.5
1992	Rainfall	1002.6
	Actual evapotranspiration	681.6
	Soil moisture change	+5.1
	Run-off	242.7
	Net recharge	+73.3

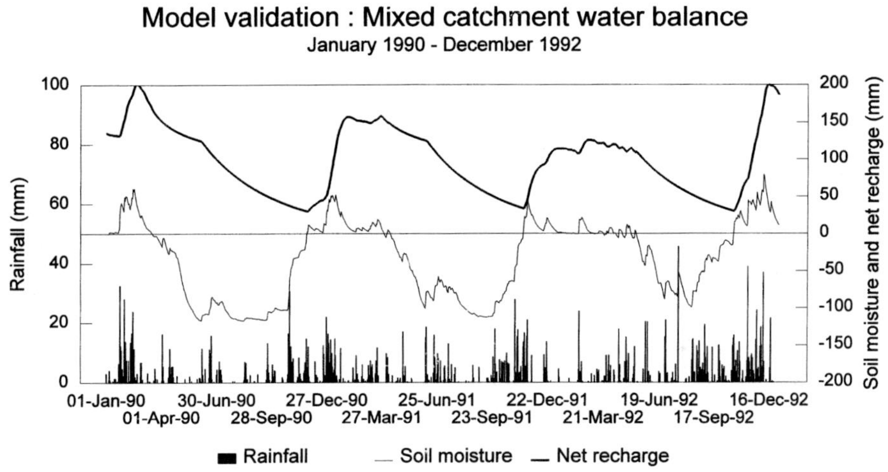

Fig. 6. Water balance over validation period on mixed catchment.

The work described in this paper was funded by Jersey Public Services Department over a three-year period. The cooperation of the various organizations and individuals who collected data and provided other relevant information is gratefully acknowledged; these include Jersey Public Services Department, Howard Davis Farm, Jersey New Waterworks Company and Jersey Meteorological Department. The authors wish to point out that the views expressed in the paper are their own, and not necessarily those of any of the above organizations.

References

Agriculture and Fisheries Committee. 1992. *Agricultural statistics 1991.* Annual report.

Blackie, J. R., Houghton-Carr, H. A., McCartney, M. P., Moores, J. P. & Davies, H. N. 1996. *The Trinity Catchment Study: Final Report.* Institute of Hydrology Technical Report.

James, I. 1989. *Water supply management on the island of Jersey.* Commonwealth Engineer's Council

Seminar on Water Management for Small Island States.

Nash, J. E. & Sutcliffe, J. V. 1970. River flow forecasting through conceptual models: 1. A discussion of principles. *Journal of Hydrology*, **10**, 282–290.

Penman, H. L. 1948. Natural evaporation from open water, bare soil and grass. *Proceedings of the Royal Society, London, Series A*, **193**, 120–145.

Robins, N. S. & Smedley, P. L. 1994. Hydrogeology and hydrochemistry of a small, hard- rock island: The heavily stressed aquifer of Jersey. *Journal of Hydrology*, **163**, 249–269.

——, P. L., Chilton, P. J. & Stuart, M. E. 1993. *Jersey Groundwater Year 3: Further observations and potential sources of pollution*. British Geological Survey Technical Report, WD/93/28.

Rosenbrock, H. H. 1960. An automatic method of finding the greatest or least value of a function. *Computer Journal*, **3**, 175–184.

Todd, D. K. 1980. *Groundwater Hydrology*, (2nd edn). Wiley, New York.

Groundwater recharge and pollutant transport beneath wastewater irrigation: the case of León, Mexico

P. J. CHILTON[1], M. E. STUART[1], O. ESCOLERO[2], R. J. MARKS[1], A. GONZÁLEZ[2] & C. J. MILNE[1]

[1]*British Geological Survey, Wallingford, OX10 8BB, United Kingdom*
[2]*Comisión Nacional del Agua, Av. Insurgentes Sur 1960, Mexico DF*

Abstract: Agricultural irrigation with wastewater is widely established in Mexico. The city of León is the centre of the Mexican tanning industry, and produces wastewater with a significant component of industrial effluent. The city is one of the fastest growing in Mexico and is highly dependent on groundwater for public supply. Much of this is drawn from aquifers around the city, including those beneath the wastewater area. The sources of recharge to these aquifers and the types of pollutants are reviewed. Regionally, groundwater levels are falling as a result of excessive groundwater abstraction but, within the area of wastewater irrigation, groundwater levels are maintained close to the ground surface. Groundwater sampling from existing boreholes, surface geophysics, soil sampling and the drilling of cored investigation boreholes have been able to characterize a body of shallow, polluted groundwater beneath the wastewater area, and contrast this with the regional groundwater quality. The most immediate threat to public supply from pollutants originating in the wastewater is posed by salinity. The predicted consequences of various management options for reducing the impact of wastewater on the quality of deeper groundwater are briefly discussed.

The very rapid urban growth of the last few decades has produced increasing demands for potable water. As a result of this growth and the accompanying industrialization, surface water resources are now either fully exploited or of poor quality, and groundwater resources have become increasingly important. Although often lagging behind provision of water supplies, improving coverage of water-borne sewerage systems in large cities produces enormous volumes of wastewater for disposal. As a result, expanding demands on groundwater and greater problems of wastewater disposal often occur together.

Water is an extremely scarce resource in the semi-arid parts of central and northern Mexico. The need for greater food production is readily apparent, and can only be met by irrigation. Increasing recognition is therefore being given to the value of wastewater as an important resource, and strategies for reuse are likely to become more widely adopted. The use of wastewater for irrigation is already widely practised in Mexico. Where this occurs over important aquifers, there could be major impacts on groundwater. In some situations, the substantial volumes of additional recharge may completely alter the local hydrogeology. Perched aquifers and new groundwater flow regimes and discharge points may be created. Impacts may be simultaneously positive in terms of water resources conservation and negative in relation to groundwater quality. Improper disposal of untreated wastewater directly into aquifers used for water supply can cause serious pollution problems. On the other hand, controlled disposal of untreated or partially treated wastewater at the ground surface for irrigation can provide significant additional resources of tolerable quality water, and can take advantage of the available nutrients for improved crop growth.

The overall objective of the study described in this paper was to determine the effects of wastewater reuse on groundwater resources. Both the positive impacts in the form of increased recharge and the negative impacts from deterioration in groundwater quality were examined, and the results have been used to recommend strategies for improving the management of wastewater irrigation to maintain groundwater quality. The present paper summarizes the overall water resource and groundwater quality impacts of the wastewater irrigation system in León. Further papers are in preparation which present more detailed interpretations of the results in terms of hydrochemical processes and pollutant behaviour.

The wastewater irrigation system

The city of León is situated at about 1780 m above sea level in a wide upland valley in the western part of the state of Guanajuato, some 500 km northwest of Mexico City. León was chosen for

Chilton, P. J., Stuart, M. E., Escolero, O., González, A. & Milne, C. J.. 1998. Groundwater recharge and pollutant transport beneath wastewater irrigation: the case of León, Mexico. *In*: Robins, N. S. (ed.) 1998. *Groundwater Pollution, Aquifer Recharge and Vulnerability*. Geological Society, London, Special Publications, **130**, 153–168

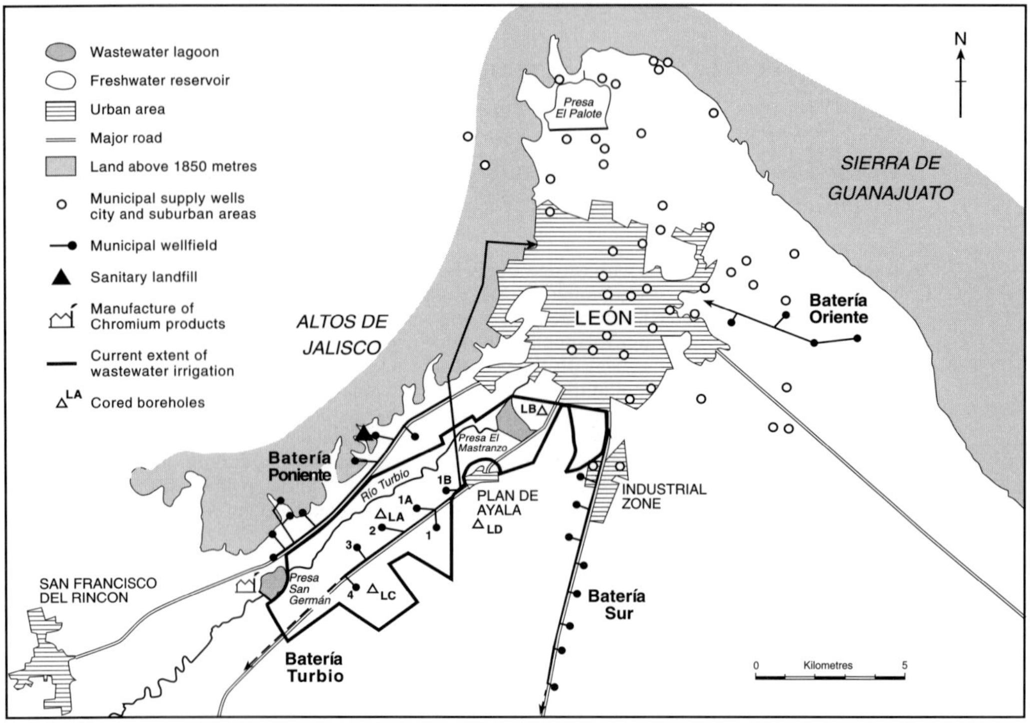

Fig. 1. Location map showing area of wastewater irrigation and municipal wellfields.

the present study because it is one of the fastest growing urban areas in Mexico and is highly dependent on groundwater for public supply. Groundwater is drawn from aquifers beneath and down-gradient of the city and some of the wellfields are situated within and close to the area of wastewater irrigation (Fig. 1). Irrigation with wastewater commenced about 40 years ago. The first area to be irrigated was located below El Mastranzo lagoon close to the village of Plan de Ayala (Fig. 1). The irrigated zone has grown and spread southwards with the expansion of the urban area and the consequent increase in availability of wastewater. The land between Plan de Ayala and the Presa San Germán storage reservoir (Fig. 1) has been irrigated for more than 20 years, and the most recently incorporated land in the south has been irrigated for some 10 years. The current (1995) area licensed by the National Water Commission (CNA) for irrigation with wastewater is 3286 ha of which, from study data, some 3030 ha were actually under irrigation.

León is extensively but not comprehensively sewered. Mains sewerage coverage was stated by the municipal water utility, SAPAL, to be 84% of the urban population of 1.1 million in 1994, and is planned to rise to 91% by the year 2000. Sewage is collected from the city into six main open wastewater canals which supply the irrigated area. Collectors 3 to 5 carry water to El Mastranzo storage lagoon (Fig. 2). Collector 1, the largest, carries water from the southeast of the city directly to the irrigated land. Collector 6 drains from an industrial area to the southwest of the city. At the time of the last detailed survey in 1991, the system was producing some 82×10^6 $m^3 a^{-1}$, equivalent to $2600\, l\, s^{-1}$. More recent estimates by SAPAL suggest a figure of 91×10^6 $m^3 a^{-1}$.

León generates highly polluted wastewater because the city is one of the most prominent leather-processing centres in Latin America. There are said to be 530 separate establishments in León involved in the leather industry. Estimates of the industrial component of the wastewater, obtained from several sources during the study, vary from 9×10^6 to 47×10^6 $m^3 a^{-1}$, or 10 to 40% of the total. The upper figure seems high, and the lower may represent the volume of more heavily contaminated effluent from the leather-processing industry. Leather manufac-

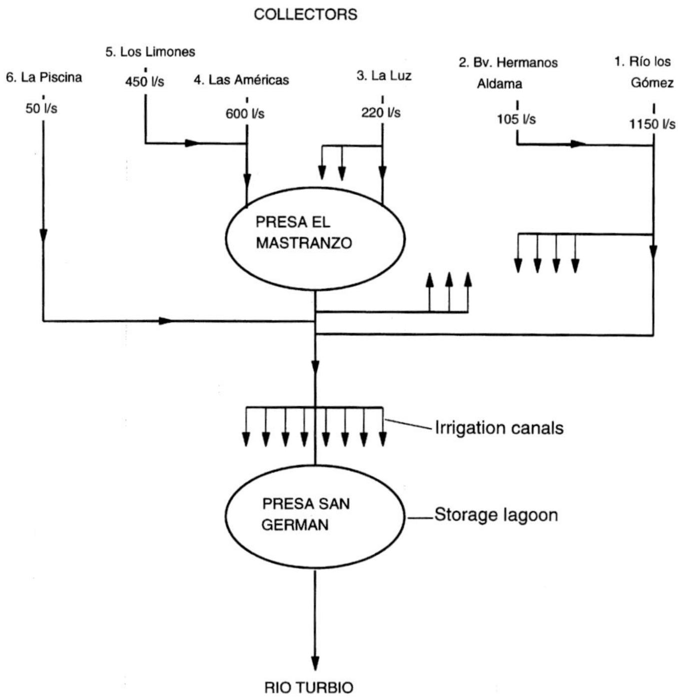

Fig. 2. Schematic layout of wastewater distribution system.

turing comprises a number of distinct processes, all of which give rise to highly polluting effluents. Hides are received salt-cured and initial processing involves soaking in water to remove salt and dirt and in lime solution to remove hair. The resultant effluent contains high loadings of organic carbon and nitrogen and of sodium chloride. Initial processing is followed by one of two alternative further stages which produce effluents of somewhat different qualities (Table 1). The curing of fine leather with chromium compounds produces an effluent containing high concentrations of chromium, sodium, sulphate and chloride. The tanning of thicker leathers produces effluents with a high content of organic carbon and nitrogen.

Such highly concentrated effluents receive significant dilution within the large volumes of domestic wastewater. Average quality of wastewater in the six main collectors from a 1991 study is shown in Table 2. At the relatively dry time of the year when the study was carried out, discharges were much lower than the averages given in Fig. 2.

Table 1. *Quality of tannery effluents from different processes*

Tannery process (no. of sites assessed)	Average flow ($m^3 a^{-1}$)	pH	Total Cr ($mg l^{-1}$)	Total N ($mg l^{-1}$)	Conductivity ($\mu S cm^{-1}$)	Fats and oils ($mg l^{-1}$)
Chromium (25)	32 000	8.5	195	400	19 000	500
Tannin (3)	14 000	5.8	7.2	1000	40 000	400
Mixed (3)	7600	6.8	51	250	35 000	200

Table 2. *Quality of wastewater in collectors, February 1991*

	Los Gomez	La Luz	Hermanos Aldama	Americas	Los Limones	La Piscina
Flow ($l\,s^{-1}$)	649	100	20	352	232	27
pH	8.2	7.4	7.5	8.6	7.7	9.0
Fats and oils	266	99	124	144	103	218
BOD	527	273	411	527	214	1000
COD	1191	666	919	1118	421	2847
NH_4-N	61	42	57	54	28	145
Org-N	11	8	6	14	10	93
NO_3-N	0.21	0.20	0.22	0.16	0.22	0.84
PO_4	14	12	17	9	9	9
Detergents	2.0	1.4	1.3	1.9	1.9	0.9
Hardness (as $CaCO_3$)	393	435	399	399	243	785
SO_4	308	169	132	401	101	1083
Cl	517	165	186	594	42	2437
Alkalinity (as $CaCO_3$)	582	492	554	494	415	825
Conductivity ($\mu S\,cm^{-1}$)	3216	1729	1805	3401	923	8939
Fe	4.02	14.8	10.5	3.44	3.21	4.4
Zn	0.66	0.40	0.76	0.42	0.24	0.41
Sb	0.84	0.79	0.60	1017	1.00	0.91
Mn	0.38	0.53	0.76	0.36	0.36	1.08
Cr	17.2	2.78	4.79	21.7	0.49	38.1
Pb	-	-	0.02	0.02	0.01	0.03
Co	-	-	-	-	-	0.01
S	4.88	2.60	1.84	6.27	0.11	24.9
Phenols	0.18	0.11	0.29	0.16	0.07	0.48
Tannin	28.0	28.5	18.6	27.6	8.78	147

Units are $mg\,l^{-1}$ unless otherwise stated. BOD, biological oxygen demand; COD, chemical oxygen demand.
Source: CYDSA (1991)

The usual pattern of cultivation in the area irrigated with wastewater is one of rotation of sorghum and maize in the spring and summer, and wheat and oats in the autumn and winter. Some farmers grow the fodder crop alfalfa continuously. Thus, in 1995, the winter cropping was estimated at 1975 ha of wheat and 650 ha of oats, and the summer at 2630 ha of wheat and sorghum, together with some 400 ha of perennial alfalfa. In the area of wastewater irrigation, cultivation of vegetables is forbidden, and the restrictions have been more stringently imposed since the 'clean water' programme of 1991, established following concern over the possible spread of cholera. Vegetables are widely grown under groundwater irrigation on the land to the south and east of the zone of wastewater irrigation (Fig. 1). Wastewater is applied to the fields using both flood and furrow methods. Application rates are up to 2600 $mm\,a^{-1}$, but there is a considerable range depending on rainfall, crop type and wastewater availability.

Hydrogeological setting

The climate is semi-arid. The average rainfall of only 600 $mm\,a^{-1}$, the majority of which falls in summer, and which is unreliable and cannot be depended upon for cultivation. The area of the city was originally crossed by numerous southward-flowing streams which joined to form the River Turbio. This was dammed in several places below the city and the system has been used for more than 30 years to collect, transport and distribute wastewater.

In the study area, Cretaceous intrusive igneous rocks of dioritic composition forming the core of the Sierra de Guanajuato are overlain by extensive Middle and Upper Palaeogene volcanic formations. Within this sequence, rhyolitic rocks form the Cuatralba Ignimbrite, a complex series of ignimbrites and ashes deposited in lacustrine conditions, interbedded with alluvial and lacustrine sands, clays, gravels and conglomerates. The alluvial plain forming the valley floor is covered by up to 3 m of black humus soil overlying Palaeogene sediments made up of volcanic ashes deposited in a lacustrine environment, interbedded with alluvial and lacustrine sands, gravels and clays, themselves largely formed of reworked volcanic ash. More comprehensive descriptions of the geology are given by Hernández-Laloth (1991) and BGS *et al.* (1996).

As is commonly the case for intermontane basins such as this, the highly variable geology produces complex hydrogeological conditions. The Cuatralba Ignimbrite forms the main deep aquifer in the valley. It is well exposed on the flanks of the valley, providing scope for direct recharge from rainfall, and occurs at variable depths of greater than 250 m beneath the valley floor, as a result of extensive faulting. Low to moderate intergranular permeability is increased by fracturing, which is seen clearly at outcrop. Transmissivity values from test pumping range from 50 to 1700 $m^2 day^{-1}$, averaging around 600 $m^2 day^{-1}$. The overlying valley fill provides a thick aquifer of variable permeability which sustains much of the agricultural abstraction.

Recent estimates suggest that there are some 820 boreholes and 80 dug wells drawing groundwater for potable supply, agricultural irrigation and industrial uses. The municipal water utility, SAPAL, is heavily dependent on groundwater and operates some 105 production boreholes, some within the urban area and the rest grouped into six peripheral wellfields. The total licensed abstraction is some 3500 $l s^{-1}$, which could provide up to 144×10^6 $m^3 a^{-1}$ if all of the boreholes operated continuously. This is obviously not the case, and actual abstraction figures suggest a current total provision of some 85×10^6 $m^3 a^{-1}$. Of this, the boreholes of the northern part of the Turbio, the Poniente and the Sur wellfields within or close to the area of wastewater irrigation (Fig. 1) contribute 21%. Loss of these wellfields as a result of quality deterioration would pose considerable supply problems for the rapidly growing city. The value of these sources is correspondingly great, as indicated by the fact that SAPAL has given preliminary consideration to bringing surface water to supply León from 160 km away.

The principal natural sources of recharge are regional inflow from the valley flanks, together with local infiltration within the valley of excess rainfall and surface runoff. To these sources must be added the infiltration from beneath the land irrigated with wastewater and groundwater, and infiltration from mains and sewer leakage and unsewered sanitation within the urban area. The different sources could provide recharge of very different qualities, and hence greatly influence the evolution of groundwater quality in the valley. Characterizing and quantifying these recharge sources was, therefore, a major task in the study.

Methods of study

The first phase of the study consisted of collation and review of existing data, which were widely dispersed in consultants' reports for the CNA, in the archives of federal, regional and state offices of the CNA and in University reports and theses. It became apparent very soon that there had been no regular monitoring of groundwater quality and a programme of regular sampling of a representative number of wells and boreholes in the study area was established. These were selected to give:

- an indication of trends by making use of the infrequent data from previous studies;
- good coverage of potentially polluted and unpolluted zones, and urban and rural areas;
- a range of depths (although few construction details were available).

This programme was subsequently extended to lagoons and canals of the wastewater system and to private boreholes close to a factory producing chromium products for the tanning industry (Fig. 1), at which a chromium plume in groundwater had been identified previously (UNAM 1991; Rodríguez-Castillo & Armienta 1994). Unstable chemical parameters (dissolved oxygen, redox, pH, temperature, alkalinity), as well as electrical conductivity, were determined in the field at the well head. Analyses for major ions were carried out by both BGS and the León Regional Public Health Laboratory. Coliforms and hexavalent chromium were determined by the León Laboratory, and minor ions, trace metals and ^2H and $^{18}0$ stable isotopes by BGS. Full details of the methods used are given in BGS *et al.* (1994, 1996). A six-monthly programme of water-level monitoring was also implemented, collecting rest water levels wherever possible from boreholes and wells for which some information already existed.

Surface geophysics was used to assist in delineating the area affected by wastewater infiltration and in defining the geometry and form of any perched aquifers (Ogilvy 1994). Four boreholes were drilled in the study area in 1995, three within the area of wastewater irrigation where wastewater application history was known (LA, LB, LC, Fig. 1) and the fourth in the groundwater-irrigated area to provide a control site (LD, Fig. 1). Core samples were obtained from selected depths, using a Denison core barrel, to provide detailed lithological information and for the extraction of pore-water. The pore-waters were analysed by BGS to give profiles of groundwater quality with depth to indicate the extent and degree of contamination of the aquifer. The boreholes were geophysically logged before being completed as multilevel piezometers.

Soils were sampled from 12 sites to assess the degree of accumulation of toxic and trace metals, particularly chromium, in soils irrigated with wastewater. Sites included former wastewater lagoons, fields irrigated with water from different collector canals, fields with different periods of wastewater applications and control sites where wastewater had never been applied. Sub-samples at each site were taken by hand-augering at nine points about 50 m apart in a large W-formation. The sub-samples were mixed on site on a large plastic sheet before coning and quartering to give a representative sample. Soils were analysed by BGS for total organic matter content and for nitric acid-extractable metals and phosphate.

A simple analytical modelling approach was developed to simulate the build-up of pollutants. A four-cell mixing model was developed in which migration of water and pollutants through both the unsaturated zone and the aquitard between the upper and lower layers occurs without dispersion. The model was used to predict changes in water quality in the upper and lower aquifers for various scenarios representing the possible management options discussed below. Full details of the analytical model are given in BGS *et al.* (1996).

Results and discussion

Groundwater resources

The study confirms the findings of previous work (Ariel 1982; UNAM 1991) with respect to the overall groundwater resource situation in the León valley, but illustrates the difficulty of estimating recharge in such complex situations, as described by Lerner (1986) and Lerner *et al.* 1993). Groundwater balances for the most important representative areas of the valley were estimated separately. These are summarized in Table 3, and given in detail in BGS *et al.* (1996). The lateral inflow to the valley provided by infiltration over the extensive outcrop of the ignimbrites was estimated by application of Darcy's law, although considerable uncertainty remains in respect of both local hydraulic gradients and transmissivity.

The current total volume of water supplied by SAPAL to the city is about 85×10^6 to 90×10^6 $m^3 a^{-1}$, some two-thirds of which is estimated by

Table 3. *Elements of groundwater balance*

	Urban area (2600 ha)	Wastewater area (3026 ha)	Groundwater irrigated area (9011 ha)
		$\times 10^6$ $m^3 a^{-1}$	
Recharge			
Lateral inflow	17	10	10
Rainfall	5	18	included in irrigation
Mains leakage	32	–	–
Sanitation	15	–	–
Wastewater irrigation	–	35	–
Groundwater irrigation	–	small	22–32
Total	**70**	**63**	**32–42**
Abstraction			
SAPAL	15.1	17.4	4.5
Industrial and private	40	small	small
Agricultural	–	small	176
Groundwater discharge to south	–	7	–
Total	**55**	**25**	**180**
Groundwater balance	**15**	**38**	**–140**

the most recent surveys to actually reach the consumers. Confirmation that this is of the right order comes from estimates of domestic water consumption given in the current CNA licence document issued to SAPAL. This represents a loss by mains leakage of some 32%. Some of the water leaking from the mains may be intercepted by the sewer system, as the sewer pipes are likely to be buried more deeply than the water mains (Lerner *et al.* 1993), but this clearly represents a significant potential source of recharge of good quality water (Table 3). The city occupies an area of approximately 2600 ha, so this is equivalent to an infiltration of some 1200 mm a^{-1} below the city, even higher than comparable estimates for the city of Merida in Yucatan, Mexico (Morris *et al.* 1994).

Recharge contributions from unsewered sanitation and sewer leakage are more difficult to estimate. Sewers may gain and leak in different parts of the system, but the net losses are likely to be relatively small although of much poorer quality, having elevated concentrations of chloride and nitrate. Infiltration from septic tanks has been estimated from the per capita consumption of the 16% of the urban population remaining without sewered sanitation, and assuming most of this water returns to the ground. The volume of sewer leakage is unknown, and a nominal figure has been used in Table 3. Most rainfall occurs in heavy summer storms and will run off rapidly into the wastewater collection system in the urban area, leaving little scope for groundwater infiltration (Table 3).

Abstraction of groundwater within the urban area from SAPAL boreholes is well-documented (Table 3), but the volume of private abstraction for potable and industrial supply is not known. These were together estimated at 7.4 x 10^6 $m^3 a^{-1}$ in 1982 (Ariel 1982), but both have increased significantly since then. The recent audit by SAPAL referred to above suggested that all but 10% of the 55×10^6 to 60×10^6 $m^3 a^{-1}$ delivered to consumers was returned to the sewer system. If so, this makes up just over half of the estimated 90×10^6 $m^3 a^{-1}$ total of wastewater collected. The deficit must be made up of discharges to the sewer system from premises with private supplies, supplemented by interception of mains leakage and the collection of storm run-off. Current estimates suggest there are some 100 private boreholes within the city and, with average discharges of 10 to 15 l s^{-1}, they could be producing up to 40 x 10^6 $m^3 a^{-1}$ (Table 3).

The discussion of the preceding paragraphs illustrates the difficulty of estimating water balances within urban areas, and the potential for double accounting of water volumes. Nevertheless, overall there may be a small excess of recharge over abstraction, providing a positive balance in Table 3, and this may allow for limited movement of reasonably good quality groundwater southwards towards the area of heavy abstraction.

In the area of wastewater irrigation, the volume of water delivered by the collector system has changed little in recent years and can be estimated with some confidence at 85×10^6 to 90×10^6 $m^3 a^{-1}$. Of this, only 2×10^6 to 3×10^6 $m^3 a^{-1}$ could be lost by evaporation from the lagoons. Consumptive use by the crops listed above is likely to account for a further 51×10^6 $m^3 a^{-1}$, leaving some 35×10^6 $m^3 a^{-1}$ available for recharge (Table 3). Rainfall is equivalent to 18.2×10^6 $m^3 a^{-1}$, all of which is assumed to contribute to infiltration from below the irrigated fields, although this may be an over-estimate. Groundwater abstraction in the wastewater-irrigated area is largely from SAPAL public supply boreholes, with some local potable supplies for small communities and limited irrigation. Groundwater outflow at the south west of the area has again been estimated from Darcy's law (Table 3).

As expected, there appears to be a considerable excess of recharge over abstraction within the area of wastewater irrigation, giving a strong positive balance in Table 3. The limited shallow abstraction, due both to the poor quality of groundwater and to the high cost to farmers of groundwater abstraction as against free provision of wastewater, contributes to this difference. Confirmation of this volume of recharge is given by the chloride balance. Taking an average chloride concentration of the wastewater of 300 mg l^{-1} and assuming that some 2700 mm a^{-1} are applied, that there is 600 mm a^{-1} rainfall with negligible chloride and that the chloride concentration of pore-waters in the upper part of the aquifer is 800 mg l^{-1}, then it can be calculated that the recharge in the wastewater irrigated area must be some 1000 mm a^{-1}. This is equivalent to 30×10^6 $m^3 a^{-1}$, which is very close to the recharge figure given in Table 3.

The main groundwater-irrigated area lies to the east of the wastewater irrigation area (Fig. 1). Groundwater abstraction for agriculture was estimated as 176×10^6 $m^3 a^{-1}$ (Ariel 1982), having risen from 105×10^6 $m^3 a^{-1}$ in 1970. The 1982 figure was obtained from a valley-wide inventory of some 713 irrigation boreholes and, as it uses the diameter of the discharge pipe as a crude indicator of yield, and makes general and uniform assumptions about pumping regime, must be considered a very uncertain figure. In addition, it is not clear to what proportion of the valley these figures apply. The general validity of this figure and the scale of groundwater abstraction for irrigation can be ascertained from the irrigated area and average application rates. Thus, if some 1200 mm a^{-1} is applied to the 9011 ha involved, then the annual abstraction would be the somewhat lower figure of 108×10^6 m^3.

As for the wastewater area, it is necessary to estimate the volume of groundwater recharge. It is likely that irrigation with groundwater is significantly more efficient than that with wastewater because of the high cost of pumping. Taking account of the consumptive use of the crops and the suggested efficiencies of the sprinkler and furrow systems employed, perhaps only 22×10^6 to 32×10^6 $m^3 a^{-1}$ might be available for recharge from both irrigation water and rainfall (Table 3). Again, this can be compared with the chloride balance. The profile obtained from the cored borehole LD in the groundwater-irrigated land suggests that the chloride concentration in recharge is about 50 mg l^{-1}. Assuming that 1200 mm of irrigation water and 600 mm of rain are received annually, at 5 and 2 mg l^{-1} respectively, then total recharge would be some 13.5×10^6 $m^3 a^{-1}$, a significantly lower figure and in poorer agreement than the chloride balance for the wastewater area.

There are clearly major difficulties in estimating both abstraction and recharge for the area which is irrigated with groundwater, but equally clearly there must be a considerable groundwater deficit, of the order of 100×10^6 to 140×10^6 $m^3 a^{-1}$ (Table 3). As a result of the rapid growth of groundwater usage, abstraction for the whole valley is now approximately double the annual recharge. The need for improved management of these resources is clearly demonstrated.

The earliest groundwater-level data available (for 1959) indicate that even then a shallow cone of depression had begun to develop, associated with concentrated agricultural abstraction immediately to the south of the city. Over the years the cone of depression has deepened and extended both eastwards and southwards. Current depths to groundwater range from 10–20 m below ground level in the west of the valley close to the flanks, 5 m or less within the area of wastewater irrigation, to more than 100 m to the south and southeast of the city. The picture of stable groundwater levels within the area of wastewater irrigation and declining levels further east is confirmed by data collected during the present project. Fig. 3a shows hydrographs of deep and declining groundwater levels at five boreholes within the area of heaviest abstraction. Water levels have fallen by 20 to 60 m since 1982, and by a total of perhaps 90 m in the centre of the cone of depression between 1959 and 1995, and are still falling at rates of 1 to 3 m a^{-1} (Fig. 3a). In contrast, hydrographs from within the wastewater irrigation area indicate water levels which are much closer to the ground surface and which are relatively stable (Fig. 3b).

A conceptual model of the evolution of groundwater conditions is shown in Fig. 4. Figure 4a represents the initial conditions, with a relatively shallow regional water-table, recharge on the valley floor and from the mountains, and modest groundwater abstraction. With the increasing population in the valley, groundwater abstraction for potable supply and crop irrigation has grown, and the wastewater collected from León has been used for irrigation. Infiltration and recharge has resulted from excess irrigation water. Figure 4b represents conditions around 15 to 20 years ago, with the onset of declining piezometric levels in the area to the south of the city. More recently, increased abstraction has produced the more extensive cone of depression shown in Fig. 4c. Beneath the area of wastewater irrigation, strong downward hydraulic gradients would be introduced by the deep abstraction, and poor quality water could be drawn downwards. Groundwater flow directions in the centre of the valley would be modified by the heavy pumping, and the steepening hydraulic gradients could introduce lateral movement of the poorer quality water eastwards. The relative importance of lateral and vertical movement of pollutants would depend partly on the vertical distribution of permeability, which is known to be complex, and the degree of hydraulic separation between the polluted shallow water and the deeper aquifer.

Groundwater quality

Regional groundwater quality in the sequence of volcanic and lacustrine aquifers is generally very good. The water is of mixed calcium/sodium bicarbonate type with exceptionally low chloride and sulphate concentrations. Average electrical conductivities are in the range 350 to 450 $\mu S\,cm^{-1}$. This low level of background mineralization enables groundwater affected by infiltrating wastewater to be readily distinguished from unpolluted waters by the elevated concentrations of most major ions, including calcium, sodium, bicarbonate, chloride, sulphate and nitrate, and by low dissolved oxygen. Comparison of groundwater chloride concentrations in the vicinity of the River Turbio in 1984 and 1994 (Fig. 5) indi-

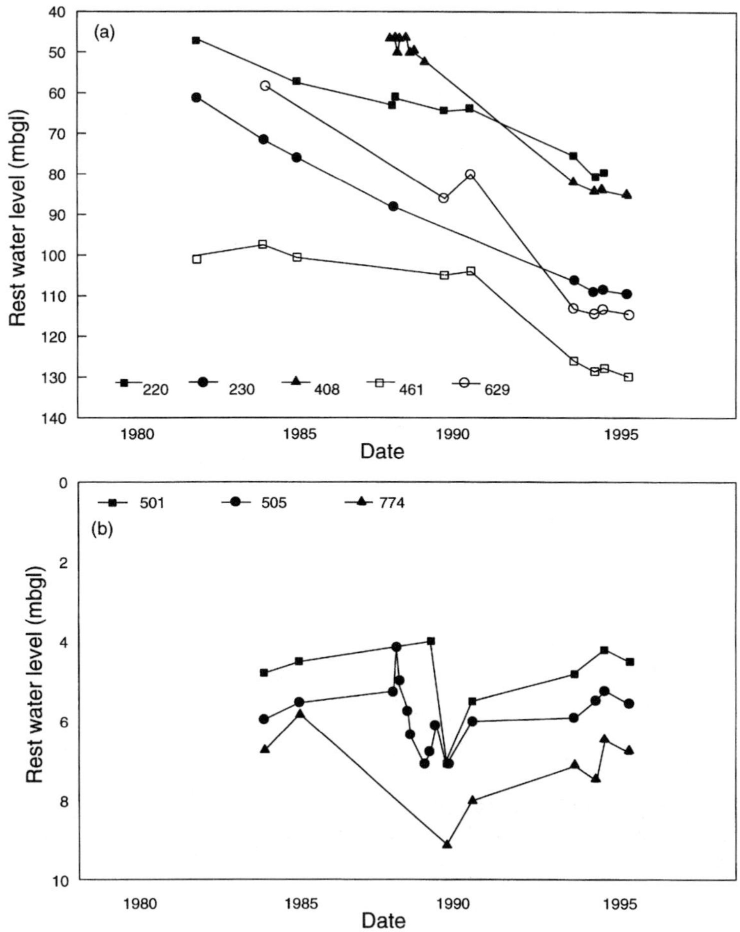

Fig. 3. Groundwater level hydrographs from the study area: (a) within the area of heaviest groundwater abstraction; (b) within the area of wastewater irrigation.

cates the rapid growth of the area of shallow groundwater affected as the wastewater irrigation has extended southwestwards along the course of the river.

Establishing the vertical dimension of hydraulic behaviour and contaminant fate and transport is critical to understanding the timescales of groundwater quality evolution and for assessing the risk of pollution of deep groundwater supplies. That poor-quality water had penetrated to significant depths in the aquifer sequence was suggested by the initial sampling of groundwater from deep boreholes, and confirmed by the geophysical survey. Interpretation and modelling of the transient electromagnetic (TEM) traverses across the area of wastewater irrigation suggested penetration of poor-quality water of salinity high enough to provide adequate resistivity contrasts to at least 40 m, with a broad transitional zone below (Ogilvy 1994). The deepest and most consistent shallow zones of poor-quality water were encountered in the traverses across the oldest-established wastewater irrigation and shallower and less persistent conductivity contrasts were observed beneath the more recently incorporated land further to the southwest (BGS *et al.* 1996).

The results of the core drilling confirmed the presence of poor-quality water to depths of 50 to 100 m (LB and LC, Fig. 6). Chloride concentrations of 800 to 1000 $mg\,l^{-1}$ in the upper parts of the three study boreholes within the wastewater area (Fig. 1) contrasted with concentrations of 100 $mg\,l^{-1}$ and less at the groundwater-

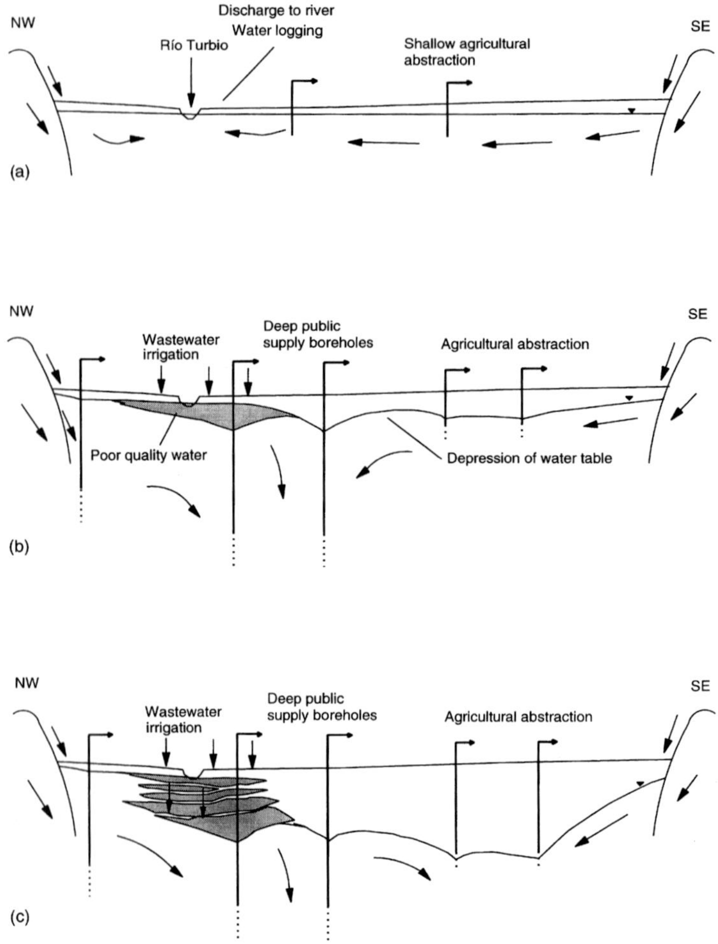

Fig. 4. Conceptual model of the evolution of the groundwater system in the Turbio Valley: (**a**) original conditions; (**b**) development of shallow aquifer and regional falling water levels; (**c**) present situation.

irrigated site (LD in Fig. 6). The three boreholes LA to LC were located on land which had been irrigated with wastewater for different periods, but a simple relationship between history of irrigation and depth of poor-quality water, from which the rate of downward movement of a 'pollution front' could be estimated, was not observed. The highly variable nature of the valley-floor deposits, which was confirmed by the core samples, and local impact on hydraulic conditions of high-discharge pumping prevented this relationship from being preserved.

The overall changes occurring to key pollutants in the wastewater during reuse and infiltration to the sub-surface are summarized in Table 4. Samples taken from the exits of the major storage lagoons, from the River Turbio which connects the two lagoons (Fig. 1), and from a distribution canal had broadly similar chemistry. The incidental role of surface impoundment in the improvement of water quality during distribution can clearly be seen, particularly for chromium (Table 4). The nitrogen content is reduced, but nitrogen remains as reduced species, such as ammonia. Nitrogen reduction may be due to degradation of organic nitrogen directly to ammonia, which can be volatilized or by denitrification of nitrate to nitrogen gas in the anaerobic bottom sediments of lagoons and canals.

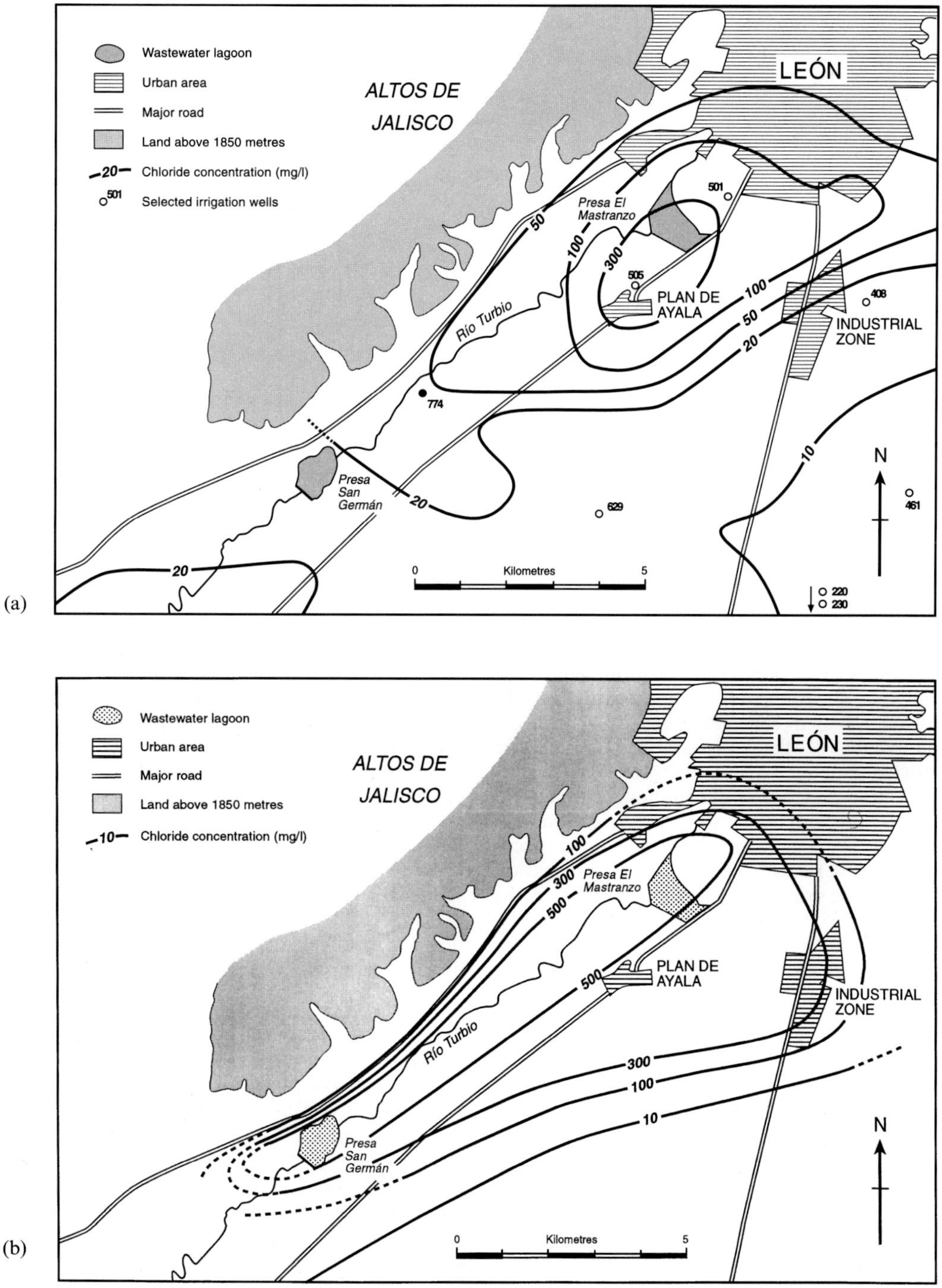

Fig. 5. Groundwater chloride concentrations in the Turbio Valley: (a) 1984; (b) 1994.

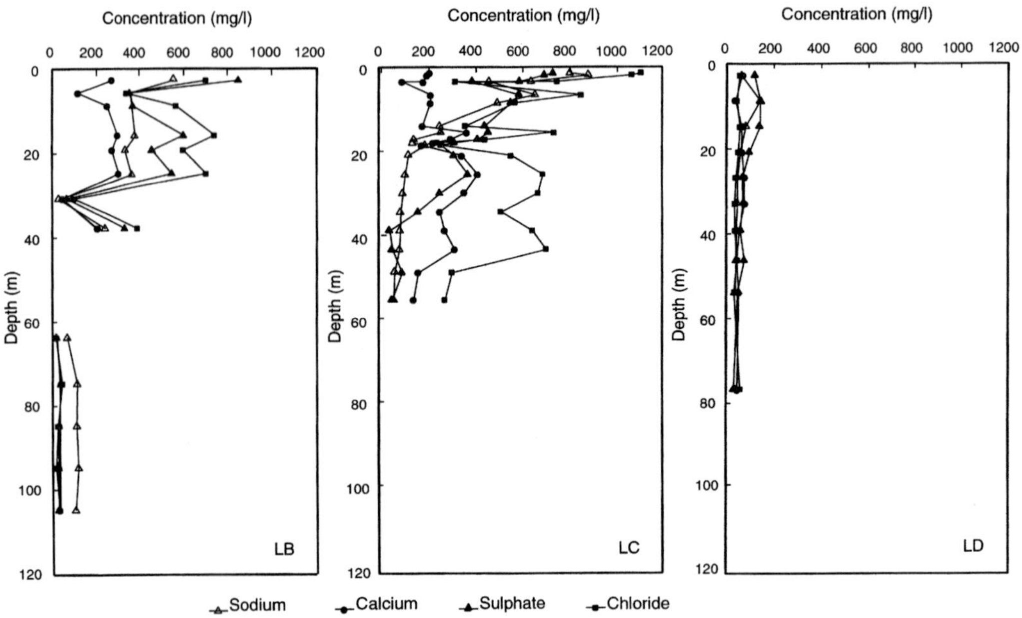

Fig. 6. Vertical profiles of pore-water chemistry for boreholes LB, LC and LD.

Chromium is rapidly lost by precipitation or sorption of Cr^{III} species in the bottom sediments. Analysis of sediments from disused lagoons showed elevated concentrations of chromium. The results of the soil sampling, which are reported in detail by BGS *et al.* (1996), also provided clear evidence of chromium accumulation in soils affected by wastewater irrigation and there was some indication that the amount accumulated was related to the period of irrigation (Fig. 7). The role of the settling lagoons in the attenuation of chromium in irrigation water is also illustrated; the site irrigated with water directly from the lagoon has accumulated little chromium. As a consequence, virtually none of the chromium originally present in the tanning effluent reaches the water-table, and chromium concentrations in the shallow, polluted groundwater beneath the area of wastewater irrigation are not elevated above regional background concentrations (Table 4).

The SAPAL boreholes of the Turbio wellfield within the area of wastewater irrigation (Fig. 1) draw water from the lower part of the aquifer sequence, with screened sections from 200 to 400 m depth. These are clearly affected by the downward penetration of poorer-quality water, and chloride concentrations have risen steeply

Table 4. *Summary of water quality in the distribution system*

Water type	Total Cr ($mg\,l^{-1}$)	Total N ($mg\,l^{-1}$)	Chemical oxygen demand ($mg\,l^{-1}$)	Conductivity ($\mu S\,cm^{-1}$)	pH
Tannery discharge	200	400	–	20 000	8.0
Industrial collector (No. 6)	40	240	5000	9000	9.0
Domestic collector (No. 1)	20	100	1000	3500	8.4
Rio Turbio	4	50	250	1800	8.3
Lagoon exits	0.1–0.4	60	170–350	2000–2300	8.3
Irrigation canal	0.06	40	200	1200	7.9
Shallow polluted groundwater	0.002–0.005	2–14 (as NO_3)	10–60	2700	6.7
Regional background quality	0.002–0.005	0–2 (as NO_3)	0–4	480	7.2

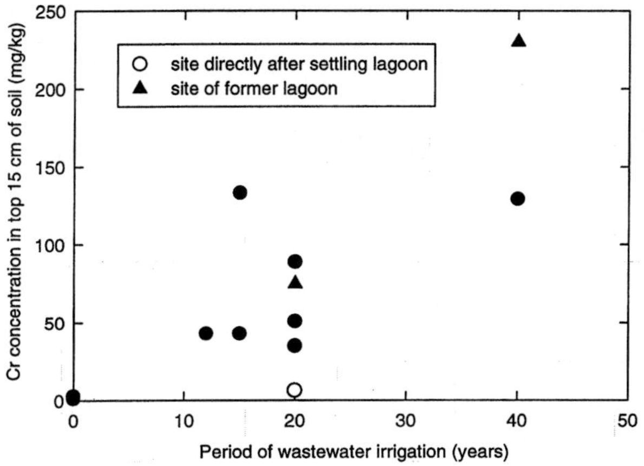

Fig. 7. Variation of chromium concentrations in agricultural soils with the number of years of wastewater application.

during the study. At the most affected borehole, Turbio 2, the chloride concentration was 15 $mg\,l^{-1}$ at the time of construction in 1972, and rose from 105 to 230 $mg\,l^{-1}$ in the two years of the present study (Fig. 8). While this may partly reflect a locally accentuated downward hydraulic gradient in response to a very high discharge rate, the other nearby SAPAL boreholes in this wellfield show consistent, albeit less rapid, increases in chloride concentration (Fig. 8). Boreholes of the Poniente wellfield show minor and inconsistent rises and boreholes of the Sur wellfield appear to be as yet unaffected.

Security of potable supplies

The most pressing issue for SAPAL in relation to the future security of potable water supply is the trend in salinity in the wellfields closest to the wastewater area, which currently supply some 15% of the city's water. A longer-term issue is the security of the Sur wellfield to the east, which supplies a further 6%. This would be threatened if the wastewater area were extended towards them as a result of increased water availability or management measures. A further concern is the overall fate of metals, organic carbon

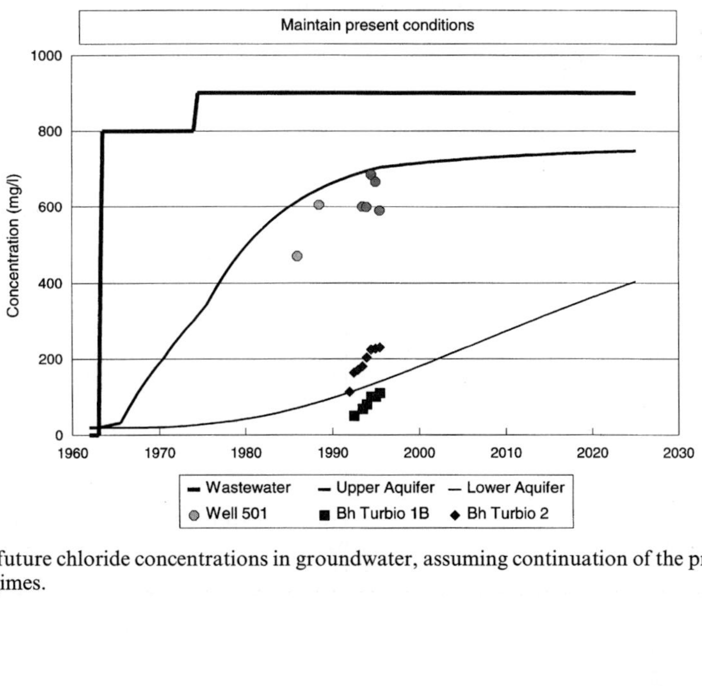

Fig. 8. Predicted future chloride concentrations in groundwater, assuming continuation of the present abstraction and irrigation regimes.

and pathogens originating in the wastewater. High concentrations in the shallow part of the aquifer could penetrate damaged lining tubes or poorly grouted supply boreholes. As expected, the entire wastewater distribution system contains high concentrations of coliform bacteria. These were regularly detected in shallow groundwater within the wastewater area, and occasionally at SAPAL supply boreholes within the city, which are generally shallower and older than those of the wellfields. Isolated positives were seen at individual boreholes in the Sur and Turbio wellfields, usually associated with heavy rainfall. These may result from poor sanitary sealing of the individual boreholes.

The simple mixing-cell model suggests that continuation of the present regime could result in further deterioration of groundwater quality in the upper aquifer until salinity approaches that of the infiltrating groundwater (Fig. 8). Salinity would also continue to increase steadily in potable supplies drawn from the lower aquifer unit, rising to about 400 $mg\,l^{-1}$ by the year 2025 (Fig. 8). The observed chloride concentrations at shallow well 501 (Fig. 5) drawing from the upper part of the aquifer and at boreholes Turbio IB and 2 (Fig. 1) are plotted for comparison.

A number of management options are available for ameliorating this risk to potable supply and these are summarized in Table 5. The first is effectively an extension of the current situation. The most obvious option is to establish a programme of taking polluted boreholes out of supply as they exceed a predetermined threshold. This should also have the benefit of slowing the downward movement of poor-quality water, but might encourage lateral movement instead. The effects of this option on the more distant wellfields cannot be tested with the current model.

Conventional wastewater treatment would remove pathogens and reduce the loading of nutrients and organic carbon, but would not reduce the salinity of the effluents significantly. Separate collection and disposal of the most polluting industrial effluents would reduce the salinity of the applied irrigation water, but this would be difficult and costly (Table 5). The amount of the reduction would depend on the proportion of the effluent so treated, and an illustrative reduction from 900 to 600 $mg\,l^{-1}$ has been modelled. This would produce a rapid improvement in the quality of the shallow aquifer, but the chloride concentration in the lower aquifer would continue to rise for many years.

Partial replacement of wastewater for irrigation by increased abstraction from the shallow aquifer (option 7 in Table 5) would intercept poorer-quality water on its way down to the deeper aquifer. Deeper groundwater quality would be more strongly influenced by better-quality lateral inflow and quality deterioration would be significantly reduced. This would, however, have implications for both additional expense to the farmer and the disposal of the excess wastewater further down the valley. A combination of these two options is modelled in Figure 9, taking an additional shallow abstraction of 12×10^6 $m^3\,a^{-1}$, sufficient to irrigate about 1000 ha for a double cropping regime of 1200 $mm\,a^{-1}$. This approach appears to have a

Table 5. *Management options for maintaining aquifer quality*

Management option	Maintenance of potable supply quality	Effectiveness in aquifer protection	Disadvantages
1. Blending with production of other wellfields	Short term	None	Continued deterioration of aquifer quality
2. Programmed shutdown of polluted wells	Short term	Less drawdown of poor-quality water	Alternative supplies costly; saline water may move laterally
3. Separate disposal of saline industrial efffluents	Removes Cl	Very effective for Cl	Difficult to organize; requires factory relocation
4. Wastewater treatment	Removes NO_3, Cr and organics	Limited: Cl not removed	Also requires 3 for maximum benefit
5. Groundwater trading for wastewater	–	–	Only possible if 3 and 4 undertaken but should be energy cost-saving
6. Improved wastewater irrigation efficiency	–	–	Only possible if 3 and 4 undertaken
7. Exploiting contaminated shallow groundwater for agriculture	–	Protects deeper water	Incentive needed; crop delection may be constrained

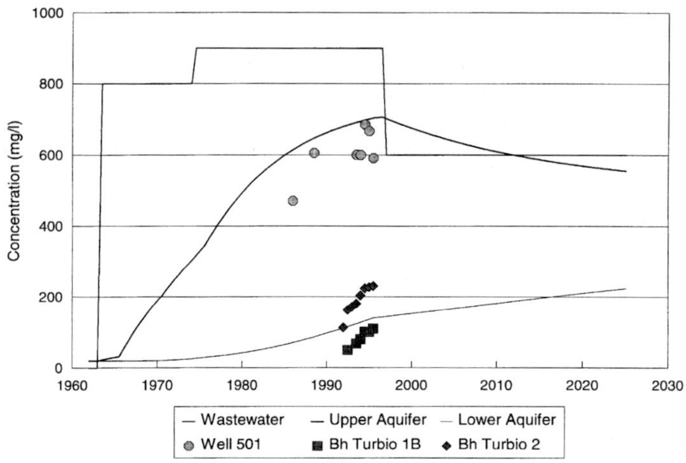

Fig. 9. Predicted future water quality, represented by chloride concentrations, assuming both additional pumping from the upper aquifer and treatment of industrial effluents.

significant quality benefit for the lower aquifer, suggesting the mean chloride concentration might not reach 250 $mg\,l^{-1}$ by 2025.

Conclusions

As a result of the rapid growth of groundwater usage for agriculture, for potable supply and for industrial uses, abstraction of groundwater significantly exceeds recharge. Although uncertainties in the assessment of recharge remain, the study confirms the results of previous work in suggesting that abstraction is now approximately double the recharge, and may have been so for up to 20 years. Regional groundwater levels have been falling by 1 to 3 $m\,a^{-1}$ and this decline is continuing. In the area of most concentrated abstraction, water levels have declined by about 90 m between 1959 and 1995. In contrast, within the area of wastewater irrigation, groundwater levels are maintained within 5 to 10 m of the surface by the infiltration of excess wastewater which sustains a shallow groundwater body.

Regional groundwater quality is very good, and the low background levels of mineralization allow the impact of wastewater to be readily distinguished from unpolluted waters by elevated concentrations of most major ions and by low dissolved oxygen. Comparison of groundwater chloride in 1984 and 1994 indicates rapid growth in the area of shallow groundwater affected by wastewater infiltration.

Suggestions of significant vertical penetration of poorer-quality water from the initial groundwater sampling and TEM geophysical survey were supported by the vertical profiles of porewater quality obtained from the cored boreholes. These confirmed the presence of poor-quality water to depths of 50 to 100 m. Chloride concentrations of 800 to 1000 $mg\,l^{-1}$ in the upper portions of the three profiles beneath the wastewater area contrast with concentrations of 100 $mg\,l^{-1}$ or less at the site irrigated with groundwater. Hydrochemical changes and attenuation processes affecting the pollutants contained in the wastewater mean that high concentrations of residual chloride pose the most serious threat to potable water supplies abstracted within the wastewater area. Municipal supply boreholes within the area of wastewater irrigation are now beginning to be significantly impacted and, at the worst affected, groundwater chloride concentration has risen from 100 to 230 $mg\,l^{-1}$ during the two years of the study.

Simple analytical modelling indicates that, under the present regime, the salinity of groundwater in the deeper part of the aquifer will continue to increase steadily. Average chloride concentrations would, under the present scenario, exceed 250 $mg\,l^{-1}$ by the year 2009. A number of water management options have been considered and their effectiveness modelled. Separate collection, treatment and disposal of industrial effluents would reduce the chloride concentration of the applied wastewater. This would produce rapid improvement in shallow groundwater quality, but chloride concentrations in the lower part of the sequence would continue to rise for many years. Increased groundwater abstraction from the upper part of the aquifer for irrigation would intercept poorer-quality water on its way down through the sequence, and quality in the lower part would be more strongly influenced by lateral movement of less saline groundwater. This approach appears to have significant quality benefits for the deeper part of the aquifer, suggesting average chloride concentrations would not reach 250 $mg\,l^{-1}$ until after 2025. This would, however, displace wastewater irrigation further down the Turbio Valley, and short- to medium-term benefits in one location might be offset by longer-term and less predictable disadvantages elsewhere. In addition, for overall resource management, increased groundwater abstraction in an already heavily exploited aquifer might have to be offset by comparable reductions elsewhere.

The study has demonstrated the complexity of the sources and mechanisms of recharge in areas of wastewater irrigation and the neighbouring urban areas from which the wastewater is collected. These sources are difficult to quantify, but as they can be of significantly different quality and their scales define the regional hydrogeological environment, estimates of their past, present and future magnitude are required so that the impact of wastewater irrigation can be determined.

Wastewater irrigation can significantly modify the quality of the underlying groundwater, as the applied water contains a range of potential pollutants. However, attenuation processes within the distribution system and the soils of the irrigated land help to prevent pollutants such as organic carbon, nutrients, heavy metals and pathogens from reaching the groundwater. The most mobile and persistent pollutant, in this case chloride, will have the quickest and most severe impact on quality, and deep public supply boreholes are beginning to be affected.

Adequate knowledge of the hydrogeology, the recharge processes and the movement and natural attenuation of pollutants is required for effective management of wastewater irrigation. The study has demonstrated that a combination of groundwater and soil sampling, water level measurement, surface geophysics and investigation drilling can provide the field data from which the impact of wastewater irrigation can be assessed.

This paper is published by permission of the Director of the British Geological Survey (NERC). The study was jointly funded by the UK Overseas Development Administration under contract 93/11661A, the European Commission under research contract CI1*-CT92-0043, and the National Water Commission of Mexico. Considerable support was also receive from SAPAL, the municipal water undertaking of León, particularly in respect of the costs of drilling, transport for sampling and for provision of laboratory facilities. The study was a joint activity between the Hydrogeology Group of the British Geological Survey, the state, regional and federal offices of the National Water Commission, and the Autonomous University of Chihuahua. The authors are grateful to the many staff of these organizations, too numerous to list here, who have contributed to this work and without whom none of this would have been possible

References

ARIEL CONSTRUCCIONES S. A. 1982. *Estudio geohidrológico cuantitavo de la Cuesca Alta del Rió Turbio. Informe elaborado para la Secretaría de Agricultural y Recursos Hidráulicos, Contrato GZA-8244–EG, Tomos l-II.*

BGS, CNA & UACH. 1994. *Effects of wastewater reuse on groundwater resources, León, Mexico: Phase I Report.* British Geological Survey, Technical Report **WD/94/25**.

——, ——, SAPAL & UACH. 1996. *Effects of Wastewater reuse on urban groundwater resources of León, Mexico.* British Geological Survey, Technical Report **WD/95/4**.

CYDSA. 1991. *Reporte de aforo, caracterización y la primera etapa de tratabilidad de las aguas residuales de la Ciudad de León, Guanajuato.*

HERNÁNDEZ-LALOTH, N. 1991. *Modelo conceptual de funcionamiento hidrodinámico del sistema acuífero del Valle de León, Guanajuato.* Tesis Profesional, Licentiatura en Ingenería Geológica, Facultad de Ingenería, UNAM, Mexico.

LERNER, D. N. 1986. Leaking pipes recharge groundwater. *Groundwater*, **24**(5), 654–662.

——, BURSTON, M. W. & BISHOP, P. K. 1993. Hydrogeology of the Coventry region (UK): an urbanised, multi-layer dual porosity aquifer system. *Journal of Hydrology*, **149**, 111–135.

MORRIS, B. L., LAWRENCE, A. R. & STUART, M. E. 1994. *The impact of urbanization on groundwater quality (Project Summary Report).* British Geological Survey, Technical Report **WC/94/56**.

OGILVY, R. D. 1994. *Use of transient electromagnetics for mapping contaminated groundwater and hydrostratigraphy, León valley, Guanajuato State, Mexico.* British Geological Survey, Technical Report **WN/94/33R**.

RODRÍGUEZ-CASTILLO, R. & ARMIENTA, M. A. 1994. Chromium aquifer pollution: evaluation of environmental impact. *In*: *Future Groundwater Resources at Risk, Proceedings of the Helsinki Conference*, June 1994, IAHS Publ. No. **222**.

UNAM. 1991. *Estudio hidrogeoquímico y modelación matemática del acuífero del Río Turbio para definir las acciones encaminadas a proteger de contaminantes a la fuente de abastdecimiento de la cuidad León, Guanajuato.* Informe elaborado para la Secretaría de Agricultural y Recursos Hidráulicos, Comisión Nacional del Agua, Contrato SARH CC-88-306-D.

Recharge components in a semi-arid area: the Sana'a Basin, Yemen

A. M. ALDERWISH* & J. DOTTRIDGE

Hydrogeology Group, Department of Geological Sciences, University College London, Gower Street, London WC1E 6BT, UK

Abstract: Credible estimates of recharge are essential for assessment of sustainable water resources in semi-arid areas, which rely on groundwater for irrigation and domestic supply. Investigation of the mechanisms and rates of recharge to the aquifers of the Sana'a Basin, Yemen, showed that the main components of recharge are infiltration of surface flows, irrigation return and urban leakage. Limited data and inaccessible, mountainous terrain required the use of robust, low-cost methods of quantifying the recharge components. Infiltration of surface flows from 12 wadis averages 39 $Mm^3\,a^{-1}$, with high interannual variation. The recharge was calculated using a combination of a rainfall–run-off model and computation of the water balance for the wadi. Irrigation return flows of 30 $Mm^3\,a^{-1}$, 20 to 40% of the water applied, were derived from measurements of soil moisture and matric potential for the three most extensive crops. Urban recharge from mains leakage and seepage through septic tanks was estimated from a well inventory, water supply records and usage rates, as 59% of the water supplied. The total recharge to the aquifers of the Sana'a Basin in 1993 was estimated to be 102 Mm^3, significantly less than the total groundwater abstraction, of about 184 Mm^3.

Evaluation, sustainable development and management of groundwater in arid and semi-arid areas requires a good understanding and accurate quantification of the main components of recharge. Uncertainty about recharge rates frequently leads to a reluctance to implement aquifer management strategies and consequently may result in over-exploitation of the groundwater resource. In many areas, lack of data results in estimates of recharge rates which often vary widely, are based on a variety of assumptions and approximate calculations, and are inadequately supported by measurements. This was the case for the Sana'a Basin, Yemen, in 1992, at the start of a comprehensive investigation of recharge rates and processes (Alderwish 1995).

The objectives of this study were to:

- understand the mechanisms of recharge;
- quantify the components of recharge and their variation with time;
- define the sustainable groundwater resource.

To compensate for the lack of data and to ensure significant improvement on the previous estimates, it was considered essential that the recharge rates should be based as closely as possible on values measured during fieldwork (Alderwish 1995).

The location of the study area and main features are shown in Fig. 1. The Sana'a Basin is a high-altitude (over 2200 m above sea level) intermontane basin, surrounded by steep mountains with deeply incised wadis. The annual rainfall of 200 to 300 mm is concentrated in two rainy seasons, April–May and August–September, when there are occasional, short-lived surface flows. Although natural vegetation is sparse, traditional terracing and water conservation methods allow rain-fed cultivation of cereals. However, irrigation is required for the main cash crops – qat (a bushy plant, chewed as a mild stimulant), grapes, fruit, vegetables and alfalfa – and this accounts for 80% of the total groundwater abstraction. Demand for water has increased steeply over the last 10 years and continues to rise, to meet both water and food requirements of the rapidly growing population.

The main aquifer is the Tawilah Sandstone, a thick, permeable, Cretaceous sandstone, which dips steeply southwards from outcrops in the central part of the basin (Fig. 2). In the southern half of the basin, the sandstone aquifer is confined by extensive Palaeogene basalts, which act as an aquitard except where vertical faults provide a connection to the sandstone. The northwestern part of the Tawilah sandstone is overlain by Quaternary volcanics, which are more permeable, forming an unconfined, highly variable aquifer and transmitting recharge to the underlying aquifer. Unconsolidated alluvium covers most of the irrigated areas in the central plain and lower reaches of the wadis, and provides storage for water from recharge. With a general downward head gradient, the groundwater leaks slowly into the deeper aquifers.

*Present address: Department of Geology, University of Sana'a, PO Box 1247, Sana'a, Yemen

ALDERWISH, A. M. & DOTTRIDGE, J. D. 1998. Recharge components in a semi-arid area: the Sana'a Basin, Yemen. *In*: ROBINS, N. S. (ed.) *Groundwater Pollution, Aquifer Recharge and Vulnerability*. Geological Society, London, Special Publications, **130**, 169–177.

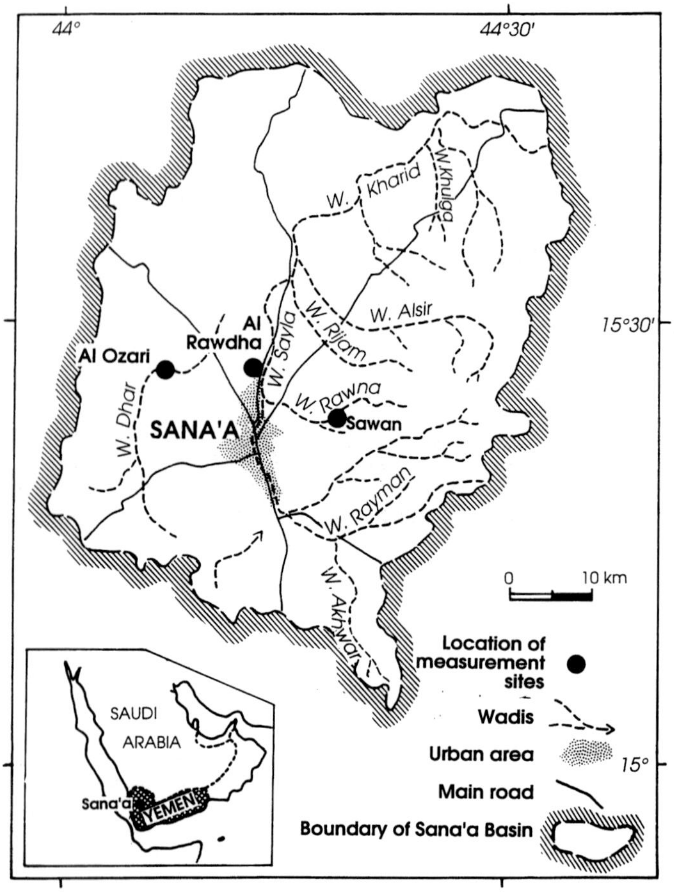

Fig. 1 The Sana'a Basin: location, main features and location of fieldwork sites.

Components of recharge

In common with many semi-arid areas (Lerner *et al.* 1990), the main components of recharge to the aquifers of the Sana'a Basin are:

- infiltration of surface flows
- irrigation return
- urban recharge from water and sewerage.

The objectives of this paper are to outline the methods used to estimate recharge components, to discuss the relative contribution of each component of recharge, and to highlight the implications of these results for future management of the aquifer.

Infiltration of surface flows

Methods. Infiltration of surface flows in ephemeral wadis is the most important and least predictable component of recharge. The problems are aggravated by the difficulties in measuring wadi flows, due to their flashy nature, short duration and inaccessibility during floods. Although a few spot measurements were made during a previous study (Mosgiprovodkhoz 1986), there is no routine gauging of surface flows in the basin. Adoption of an indirect method for defining wadi flows was therefore essential, and the US Soil Conservation Service (SCS) curve number method was used (SCS 1964, 1972). This method is appropriate for the study area because it was developed for similar climatic conditions, is widely used, and requires data only on geomorphological characteristics and daily rainfall. An important feature of the application was careful calibration of the model, which is described in more detail in Alderwish & Dottridge (1995*b*). Appropriate curve numbers under local conditions were

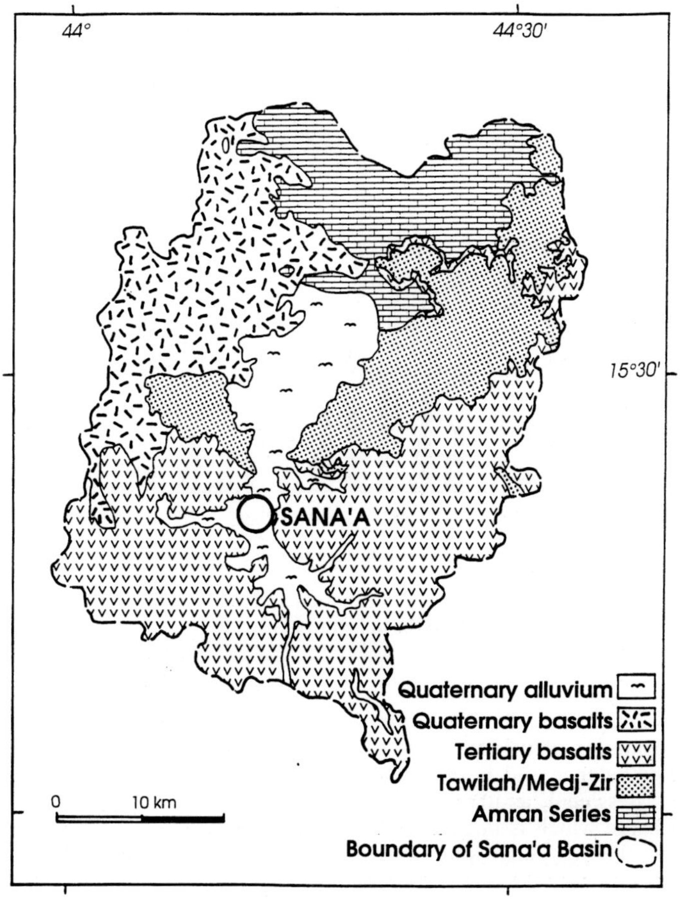

Fig. 2 Simplified geology of the Sana'a Basin.

derived for both run-off-producing and run-off-absorbing zones, through simulation of measured floods in a series of small, homogeneous catchments. The model was verified using data collected in 1993 from two wadis with larger and more varied catchments.

During the field season in 1993, detailed observations were made of two streams, Wadi Dhar and Wadi Alsir, including measurement of flood duration, extent and maximum stage, groundwater levels in observation wells and estimation of flows. These data were used to calibrate the rainfall–run-off model and to derive infiltration rates by two independent approaches: a daily water balance calculation for the channel, and a groundwater model for the alluvium along the same reach. Although the limitations of the data inevitably required a large number of assumptions, consistent results were obtained from the two methods. This generated increased confidence in their extrapolation to other years and wadis, through the development of a regression equation relating recharge to run-off from the upland and total catchment areas. During this process, the sensitivity of the results to uncertainties in the data was tested, and the critical parameter was found to be duration of the flood (Alderwish & Dottridge 1995*a*). Fortunately, flood duration is the easiest parameter to measure as it can be reliably observed and recorded by the local residents.

Results. The total annual recharge from all 12 major wadis in the Sana'a Basin varies widely, from 3 to 129 Mm^3 as shown in Fig. 3, with an average of 39 $Mm^3\ a^{-1}$, equivalent to 4.5% of the average annual rainfall. However, wadi recharge is not related to annual rainfall, but depends on the distribution, intensity and duration of individual storms, which determine the

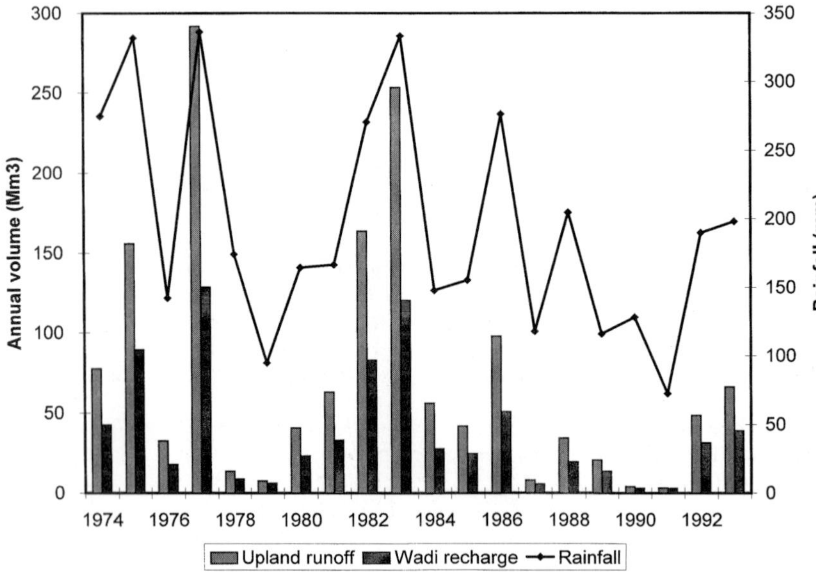

Fig. 3 Wadi recharge estimated using the regression equation and total run-off from the mountains derived using the SCS curve number method, for the Sana'a Basin from 1974 to 1993.

run-off from the mountains (Fig. 3). This component makes a substantial contribution to the overall groundwater balance and is the critical factor in resource assessment, as it remains the only source of recharge which is not a form of recycling. However, the very high inter-annual variation makes this an unpredictable component of recharge on a short-term basis.

Irrigation return

Methods. Although irrigated agriculture uses 80% of the groundwater abstracted from the Sana'a Basin, no reliable estimates of crop use and return flows were available at the start of the study in 1992. Improved values were derived by combining field measurements at representative sites with analysis of soil moisture characteristics and modelling of flow in the unsaturated zone.

Three contrasting sites (Fig. 1) were selected for detailed investigation, each with a different crop and soil type, and characteristic of a part of the study area (Table 1). Values of residual moisture content and saturated hydraulic conductivity in Table 1 were based on existing data from a soil survey (Mosgiprovodkhoz 1986). From May to August 1993, field measurements were made every 10 days of soil moisture content, using a neutron probe calibrated for each site, and suction pressure using tensiometers, to a

Table 1. *Soil Properties at the field sites*

Location	Crop	Soil type	Saturated moisture content θ_s	Residual moisture content θ_r	Saturated hydraulic conductivity K_s (m day^{-1})
Al Rawdha (central plain)	apples	sandy loam/loam	0.47	0.09	0.115
Sawan (Wadi Rawna)	grapes	medium loam	0.434	0.103	0.048
Al Ozari (Wadi Dhar)	qat	loam/silty loam	0.431	0.142	0.072

depth of 1.5 m. Information on irrigation practices was provided by the farmers.

In the first stage of interpretation, the field data were used to derive the relationship between unsaturated hydraulic conductivity, moisture content and suction pressure by fitting an analytical expression for soil moisture characteristics (Van Genuchten 1980). The fitting was made to the average characteristics for each site using the computer program SOIL (El Kadi 1987). Due to the heterogeneity of the soil, this approach produced a range of values for the moisture characteristics, leading to some uncertainty in the relative conductivity function, which was refined in the second phase of calibration.

The second stage of interpretation used a numerical model, CHEMFLO, which solves the Richards equation (Richards 1931) for vertical flow in the unsaturated zone, to refine the values of soil hydraulic properties, and compute soil moisture and cumulative flux (Nofziger *et al.*1989). To improve the estimated hydraulic properties, the model was applied to a period with a constant evaporation rate; the soil properties were adjusted until the model results agreed with the measured moisture profiles.

The same model was used to simulate unsaturated flow over a period of four months, with actual rates of evapotranspiration, rainfall and irrigation. The results were verified by comparing the full sequence of measured and modelled moisture profiles, as illustrated by the example in Fig. 4. The irrigation return was computed as the flux leaving the lower boundary of the model at a depth of 1.5 m; the results for irrigated qat at Al Ozari are shown in Fig. 5.

Results. The results showed that between 19 and 39% of the water applied by irrigation and rainfall infiltrates below the root zone, averaging 20% for the most extensive crops – qat and grapes. When the contribution from rainfall is neglected, these proportions rise to 25 to 49% of the irrigation water. As expected, the highest proportion of return flow occurred on the coarsest soil, a sandy loam at Al Rawdha, but this was also influenced by the choice of crop – widely spaced apple trees – and frequent application of irrigation water, totalling 290 mm. At the other two sites, moisture remaining from early rainfall in April formed an important component of the deep percolation. Over-irrigation also had a significant effect, with the majority of the return flow from irrigated grapes at Sawan originating from a single application of 140 mm of water. Although the proportion of return flow was found to be similar for grapes and qat, the total quantities are lower for qat, which is a drought-tolerant crop with a low evapotranspiration rate when less water is applied.

The final stage in the calculation of recharge to groundwater from irrigated land was the regionalization of the values, using local rainfall, regional climatic data, soil maps (Mosgiprovodkhoz 1986), estimates of the total area of each crop (Al Eryani *et al.*1992) and calculation of changes with time. Extrapolation from the measured data was acceptable because the selected sites are representative of the range of conditions in the study area and irrigation practices are consistent for each crop. For an irrigated area of 22,500 ha, the total return flow in 1993 was estimated as 45 Mm^3. Although this represents the largest component of recharge to the aquifers of the Sana'a Basin, it does not provide an additional source of water, but only reduces the net abstraction for irrigation from 149 Mm^3 to 104 Mm^3. The rising trend in irrigation return shown in Fig 6. reflects the increase in groundwater-fed irrigation, which has accelerated with rapid population growth, particularly since a ban on the import of fruit and vegetables was imposed in 1987. However, the fluctuations in total recharge also demonstrate the influence of rainfall on both the availability of water and the cultivated area.

Urban recharge

Methods. The calculation of recharge in the urban area from leaking pipes, sewers and cesspits was based on data on: population from the census; water use from information collected during the well inventory and detailed monitoring of sample households; water supplied, from an inventory of private wells, their discharges and pumping hours, and government data on wellfield abstraction; and water removed by sewerage from measurement of average daily flows at the treatment works. For each component of water loss, the proportion contributing to recharge was estimated as the difference between water supplied and the probable consumption rates.

A total population of 972000 in the Sana'a urban area was recorded in 1994, reflecting a very high growth rate of 11% over the previous decade. The public water supply, derived from two wellfields in the Tawilah Sandstone on the margins of the city, only serves about 30% of the total population. Their water use is estimated to be between 100 and 120 $l\,day^{-1}$ per person, with about 30% leakage losses from the mains. The remaining 70% of the city dwellers are sup-

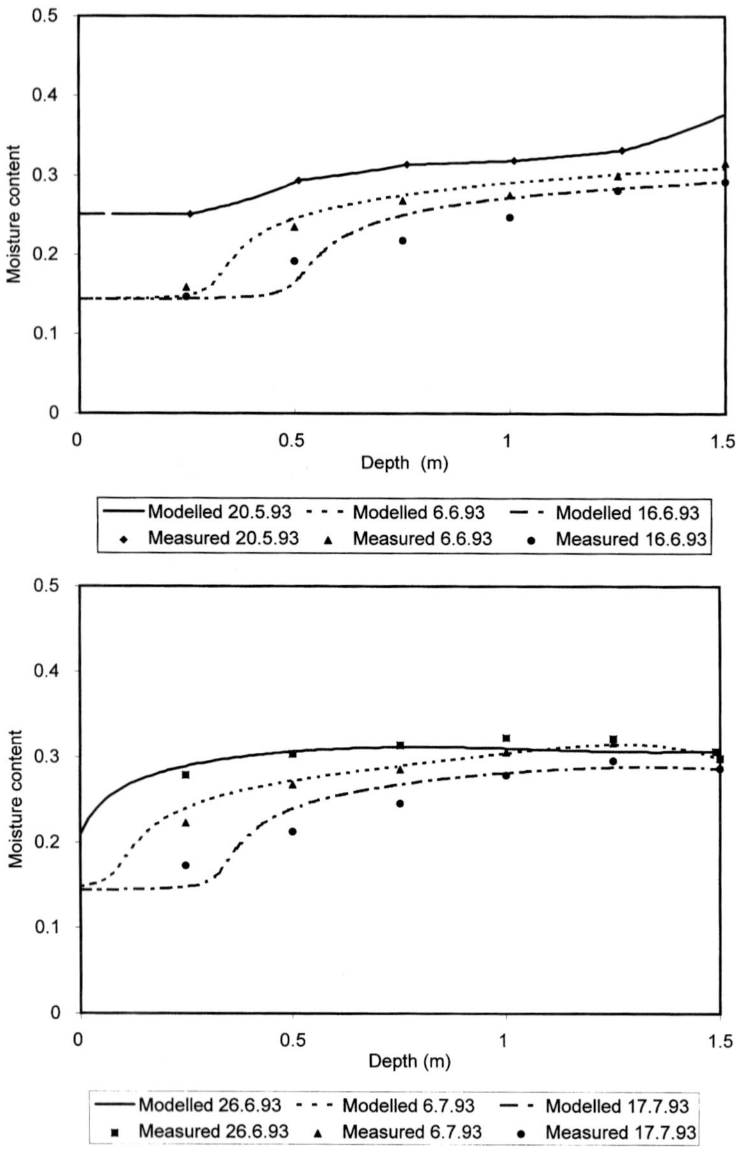

Fig. 4 Results of modelling of unsaturated flow: comparison of measured and observed soil moisture profiles in a loamy soil beneath irrigated qat at Al Ozari, (**a**) at the start of modelling and (**b**) during the following month.

plied privately through pipes or tankers with water pumped from boreholes within or close to the city. The water use per capita is lower, especially where tankers are the main supply, averaging 60 to 80 $\mathrm{l\,day^{-1}}$ and due to the short pipe lengths, losses are also lower, typically 20% of the water supplied. An even smaller proportion of the population is served by main sewerage, only 12%. This wastewater is piped to Al Rawdha to the north of Sana'a, where it undergoes very basic treatment, before being allowed to flow down the main wadi channel, partially infiltrating into the alluvium. The main form of sewage disposal is through cesspits, dug into the alluvium beneath the city.

Results. The results showed that the main component of urban recharge is wastewater derived

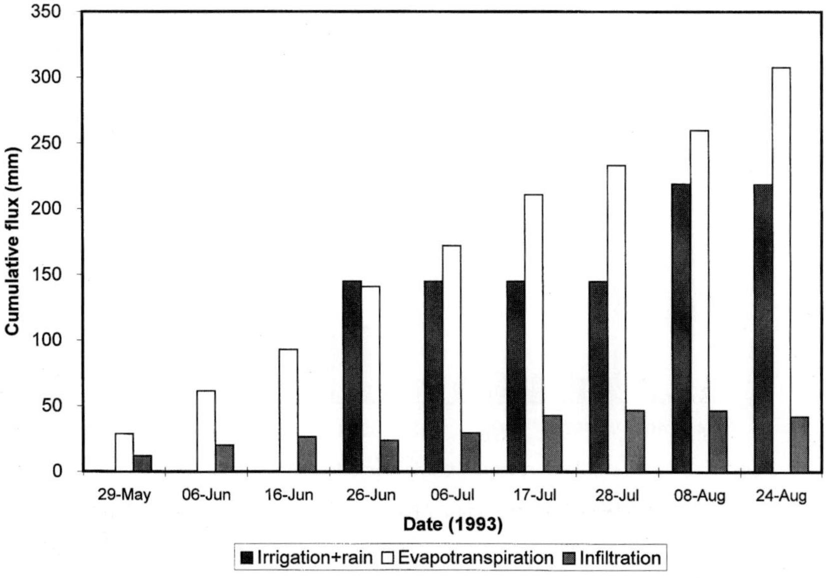

Fig. 5 Vertical flows during irrigation of qat at Al Ozari, showing measured irrigation rate, calculated evapotranspiration and computed infiltration through base of modelled profile, at a depth of 1.5 m.

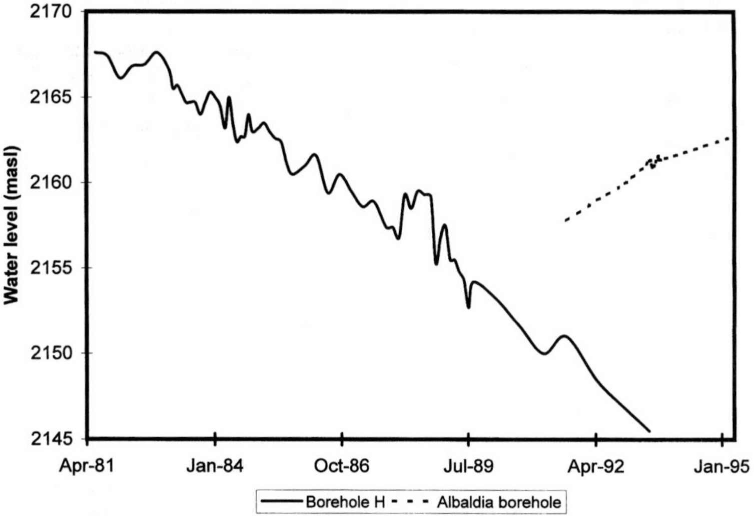

Fig. 6 Hydrographs of observation boreholes in the Tawilah Sandstone giving a comparison of the declining trend in the wellfield area (Borehole H) and rising water levels in the city (Albaldia borehole).

from cesspits, which contributed 12.5 Mm^3 in 1993, over 80% of the total volume of infiltration. Good-quality water from mains leakage accounted for 3.4 Mm^3 of recharge, with an additional 1.0 Mm^3 from other sources, chiefly industrial users and irrigation within the urban area. In addition, infiltration of floods from the main Wadi Sayla contributes to recharge in the city, but the volumes were included in the wadi recharge totals.

The consequences of the location and type of urban recharge are that groundwater levels are observed to be stable or rising, especially in shallow wells tapping the alluvial aquifer. Figure 6

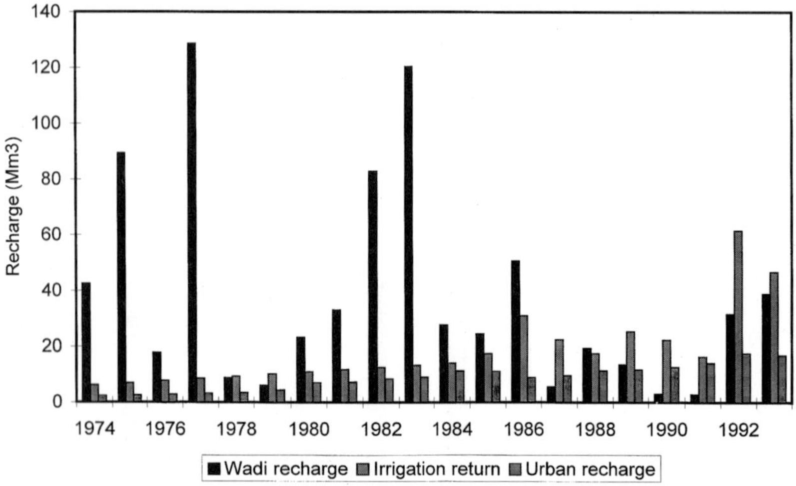

Fig. 7 Calculated recharge components for the Sana'a Basin on an annual basis for 1974–1993.

illustrates the contrasting trends in water levels in the city and the wellfield area. Not surprisingly, the shallow aquifer shows signs of contamination, but fortunately water quality in the deeper boreholes completed in the underlying sandstone remains good. However, due to the inadequate treatment and disposal facilities for the sewered wastewater, cesspits pose a lesser pollution hazard overall.

Total recharge

The estimates of annual recharge to groundwater in the Sana'a Basin from the three main sources are shown in Fig. 7, for a 20-year period from 1974 to 1993. This diagram illustrates the rising trend in the two components of return flow – irrigation and urban recharge – and the wide variation in wadi recharge. Comparison of the three

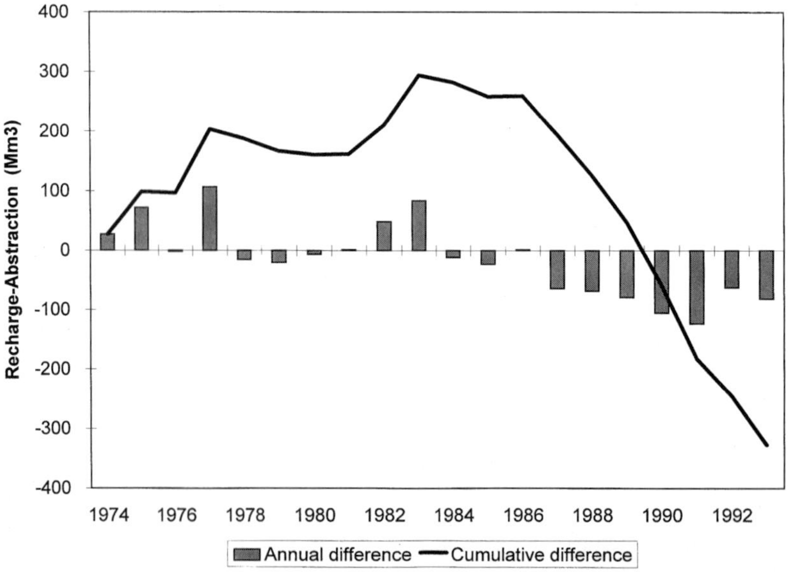

Fig. 8 Annual variation and cumulative total of the differences between recharge and abstraction for the Sana'a Basin. Positive values indicate surplus recharge and negative values show over-abstraction.

components highlights the importance for the total water balance of those years with unusually high wadi flows, such as 1977 and 1983, and emphasizes the potential problems that might result from prolonged series of dry years or decline in rainfall. The effect of a sequence of dry years (1987–1991) combined with increasing abstraction is shown in Fig. 8, where this period coincides with a sharp decline from water surplus to deficit on the basin scale. This accumulating deficit is more alarming when compared with previous estimates of recharge, which were significantly lower than the results presented in Fig. 8. The previous values ranged from 25 to 67 $Mm^3\ a^{-1}$ and were mostly derived as a proportion of rainfall or based on throughflow calculations.

Conclusions

The methods used in this study for quantifying the main components of groundwater recharge proved to be workable, even for an area such as the Sana'a Basin with difficult terrain and very limited existing data. The logistical and financial limitations of the project restricted the number of field sites and period of data collection; extrapolation was essential to produce results for the entire basin over a period of 20 years. However, the revised estimates are based on more systematic measurements than previous figures and therefore are considered to be more reliable (Alderwish 1995), although a high but unquantifiable degree of uncertainty still remains. The accuracy could be improved by increased monitoring of groundwater levels in the alluvial and deeper aquifers, gauging of surface flows, collection of data on the irrigated and urban areas at additional locations and regular time intervals, updating of recharge estimates and, when sufficient data are available, use of improved methods for calculating recharge.

This study concentrated on quantification of recharge to the shallow alluvium, which covers most of the favourable areas for recharge, but a critical issue for management of water resources is to determine the proportion of this recharge which reaches the main Tawilah Sandstone aquifer. Comparison of water levels suggests that the hydraulic gradient is downwards, but there is insufficient information to allow calculation of the rates of leakage to the deeper aquifer. Additional monitoring of shallow wells and deep boreholes is therefore essential to quantify the recharge reaching the sandstone aquifer. Meanwhile, the new information gained from this study of recharge rates should be incorporated into the water planning for Sana'a, through revision of management strategies and development of a groundwater management policy for the Sana'a Basin.

This work forms part of the research by Ahmed Alderwish for a PhD degree, funded by ODA through the British Council, and their support is gratefully acknowledged. Thanks also to Sana'a University for provision of field equipment.

References

ALDERWISH, A. M. 1995. *Estimation of groundwater recharge to the aquifers of the Sana'a Basin, Yemen.* PhD thesis, University of London

—— & DOTTRIDGE, J. 1995*a*. *Modelling infiltration from ephemeral wadi flows in the Sana'a Basin, Yemen*, British Hydrological Society Occasional Paper No. **6**, p 4 – 16.

—— and —— 1995*b*. Estimation of groundwater recharge from wadi flows using sparse data. *In*: *Solutions 95, Proceedings of the IAH Congress*, Edmonton, Canada, June 1995.

AL ERYANI, M. L, BAMTRAF, A. & AL HAKIMI, A. 1992 *Water resource planning models, Issue paper part 2, the Highland region.* High Water Council, Sana'a, Yemen.

EL KADI, A. I. 1987. *SOIL, A computer program to estimate the parameters of soil hydraulic properties.* User's manual. GWMI 84–12, IGWMC, Holcomb Research Institute, Butler University, Indianapolis, Indiana.

LERNER, D. N., ISSAR, A. S & SIMMERS, I. 1990. *Groundwater recharge. A guide to understanding and estimating natural recharge.* International Contributions to Hydrogeology No. 8.

MOSGIPROVODKHOZ, 1986. *Sana'a Basin Water Resources Scheme*, Volume 1 *Hydrology*; Volume 2 *Geology and Hydrogeology*; Volume 3 *Soils*; Volume 4 *Natural and Agricultural Zones.*

NOFZIGER, D. L., RAJENDER, K., NAYUDU, S. K. & SU, P. Y. 1989. *CHEMFLO: One-dimensional water and chemical movement in unsaturated soils.* US EPA, Ada, Oklahoma.

RICHARDS, L. A. 1931. Capillary conduction of liquids through porous media, *Physics*, **1**, 318–333

SOIL CONSERVATION SERVICE (SCS) 1964. *A method of estimating volume and rate of run-off in small watersheds.* US Department of Agriculture, Washington, Technical Publication **149**.

—— 1972 *SCS National Engineering Handbook*, section 4. US Department of Agriculture, Washington.

VAN GENUCHTEN, M. T. 1980. A closed form equation for predicting the hydraulic conductivity of unsaturated soils. *Soil Science Society of America Journal*, **44**, 892–989.

Mapping groundwater vulnerability: the Irish perspective

DONAL DALY & WILLIAM P. WARREN

Geological Survey of Ireland, Beggars Bush, Haddington Road, Dublin 4, Ireland

Abstract: The groundwater protection scheme used in the Republic of Ireland since the 1980s had not encompassed the vulnerability mapping concept. Yet internationally, vulnerability maps were becoming an essential part of groundwater protection schemes and a valuable tool in environmental management. Consequently, following a review of protection schemes world-wide, the scheme used in Ireland was updated and amended to include vulnerability maps as a crucial component of the scheme.

The approach taken to vulnerability assessments and mapping in the Republic of Ireland has been dictated by the following fundamental questions: Vulnerability of what? Vulnerability to what? Which factors determine the degree of vulnerability? What is the appropriate scale for map production? How can limitations and uncertainties be taken into account? How can vulnerability assessments be integrated into environmental and resource management?

The following decisions were made: (i) we should map the vulnerability of groundwater, not aquifers or wells/springs; (ii) the position in the groundwater system specified to be of interest is the water-table (i.e. first groundwater encountered) in either sand/gravel aquifers or in bedrock; (iii) we should map the vulnerability of groundwater to contaminants generated by human activities (natural impacts are a separate issue); (iv) as the main threat to groundwater in Ireland is posed by point sources, we should map the vulnerability of groundwater to contaminants released at 1–2 m below the ground surface; (v) the characteristics of individual contaminants should not be taken into account; (vi) the natural geological and hydrogeological factors that determine vulnerability are the sub-soils above the water-table, the recharge type (whether point or diffuse) and, in sand/gravels, the thickness of the unsaturated zone; (vii) based on these factors, four vulnerability categories are used (extreme, high, moderate and low); (viii) map scales of 1:50 000 and 1:10 000 are preferred; (ix) limitations and uncertainties are indicated by appropriate wording on the maps and a disclaimer; (x) vulnerability maps should be incorporated into groundwater protection schemes, which should be used in decision-making on the location and control of potentially polluting developments.

Vulnerability maps have now been produced for a number of local authority areas. They are an important part of county groundwater protection schemes as they provide a measure of the likelihood of contamination, assist in ensuring that protection schemes are not unnecessarily restrictive of human economic activity, help in the choice of engineering preventative measures, and enable major developments, which have a significant potential to contaminate, to be located in areas of relatively low vulnerability and therefore of relatively low risk, from a groundwater perspective.

Groundwater is a major natural resource in the Republic of Ireland, providing between 20% and 25% of drinking water supplies. In certain counties, the proportion is much greater – over 50%. In rural areas not served by public or group water schemes, groundwater is usually the only source of supply and there are many thousands of wells and springs (at least 10 0000) in use.

Since the mid-1980s, a small number of local authorities have been successfully using groundwater protection schemes. These were based on a relatively basic and simple scheme proposed by the Geological Survey of Ireland (GSI) and were appropriate to the available hydrogeological information and planning needs of that time. With one exception, these schemes did not include vulnerability maps.

Consideration of the vulnerability concept and the use of maps showing groundwater vulnerability to contamination arose in the GSI from the need for a defensible, comprehensive, up-to-date, useful and user-friendly groundwater protection scheme in the Republic of Ireland. In arriving at final views on vulnerability mapping and groundwater protection schemes, the scientific literature was reviewed; visits were made to England, Switzerland, Wisconsin, Iowa and Illinois; discussions were held with academic and consulting hydrogeologists in Ireland; and specific advice and reviews were obtained from the British Geological Survey (BGS) and the Wisconsin Geological and Natural History Survey. Following this, the groundwater protection scheme already in use in the Republic of Ireland was updated and amended, and

Daly, D. & Warren, W. P. 1998. Mapping groundwater vulnerability: the Irish perspective. *In*: Robins, N. S. (ed.) *Groundwater Pollution, Aquifer Recharge and Vulnerability*. Geological Society, London, Special Publications, **130**, 179–190.

vulnerability maps became a crucial component of the scheme.

The approach taken to vulnerability assessments and mapping has been dictated by the the following fundamental questions:

- What is meant by 'vulnerability' and 'vulnerability mapping'?
- What are the objectives of vulnerability maps?
- Vulnerability of what?
- Vulnerability to what?
- Which factors determine the degree of vulnerability?
- What vulnerability categories should be used?
- What is the appropriate scale of map production?
- How can limitations and uncertainties be taken into account?
- How do vulnerability mapping guidelines relate to site assessments of vulnerability?
- How can vulnerability assessments be integrated into groundwater source and resource management and protection?

Answers to these questions are influenced by: (i) the use of the concepts of groundwater contamination risk and risk assessment as a framework for the groundwater protection scheme in use in the Republic of Ireland; (ii) the particular hydrogeology of Ireland, which is significantly different to that of most other European situations; (iii) the main hazards for groundwater in Ireland, which also are different to most other European situations; (iv) the availability of geological and hydrogeological information; and (v) the need to take a pragmatic, practical approach.

This paper describes the framework provided by the risk concept, outlines relevant features of Irish hydrogeology, gives the basis for the definition of vulnerability used by the Geological Survey of Ireland, describes the vulnerability categories and mapping guidelines, and shows how groundwater vulnerability to contamination is incorporated into the groundwater protection scheme.

Groundwater vulnerability – a component of risk and risk assessment

Groundwater protection schemes are usually based on the concepts of groundwater contamination risk and risk management. In the past, these concepts were background, often implicit, sometimes intuitive factors. However, with the language and thought processes associated with risk and risk assessment becoming more common, relating a groundwater protection scheme to these concepts is likely to be beneficial in that it allows consistent application of a protection policy and encourages a rigorous and systematic approach. The GSI is using these concepts, in a qualitative way, as the framework for the groundwater protection scheme.

The **risk** of contamination to groundwater depends on three elements:

(i) the hazard provided by a potentially polluting activity;
(ii) the vulnerability of groundwater to contamination; and
(iii) the potential consequences of a contamination event.

Risk management is based on analysis of these elements followed by a response to the risk. This response includes the assessment and selection of solutions and the implementation of measures to prevent or minimize the consequences and probability of a pollution event.

The **hazard** depends on the potential contaminant loading. The natural **vulnerability** of the groundwater dictates the likelihood of contamination. The **consequences**, if a contamination event occurs, depend on the value of the groundwater, which is normally indicated by the aquifer category ('regionally important', 'locally important' or 'poor') and the proximity to an important groundwater abstraction source (a public supply well, for instance). Preventative measures may include, for instance: control of land-use practices and, in particular, directing developments towards lower risk areas; suitable building codes that take account of the vulnerability and value of the groundwater; lining of landfill sites; installation of monitoring networks; specific operational practices. Consequently, assessing the risk of contamination to groundwater is complex. It encompasses geological and hydrogeological factors and factors that relate to the potentially polluting activity. The geological and hydrogeological factors are (a) the vulnerability to contamination, and (b) the relative importance or value of the groundwater resource. The factors that relate to the potentially polluting activity are (a) the contaminant loading, and (b) the preventative measures.

The groundwater protection scheme used in Ireland integrates these factors and in the process serves to focus attention on the higher risk areas and activities, and provides a logical structure within which contamination control measures can be selected. Vulnerability is clearly a crucial component of the scheme.

Relevant aspects of Irish hydrogeology

In the Republic of Ireland, apart from a small area (20 km^2) of Permo-Triassic rocks, all bedrock types are of Carboniferous age or older (i.e. older than 290 million years) (Geological Survey of Ireland 1962) and all have a fissure permeability only (Wright *et al.* 1982).

While the bedrock permeabilities vary depending on the lithology, tectonic history and, in the case of limestones, the degree of solution, all bedrock types are regarded as having a certain degree of permeability. Consequently, all are capable of yielding at least domestic supplies and all are classed as aquifers by the GSI. In contrast, the only sub-soil type classed as an aquifer is sand/gravel. (This includes glacio-fluvial, glacio-lacustrine, aeolian and alluvial deposits.) While sandy tills have a moderate permeability, they are seldom capable of supplying modern domestic and farm supplies and few wells are located in them. In many of the bedrock units, faulting and folding have resulted in zones with a higher permeability than the surrounding rock. Also, solution of limestones has resulted in a high degree of karstification in some areas. These factors have influenced the classification of Irish aquifers by the GSI, which is given below.

Regionally Important (R) Aquifers

(i) Karstified aquifers (where conduit flow is dominant) (Rk)
(ii) Fractured bedrock aquifers (Rf)
(iii) Extensive sand/gravel (Rg)

Locally Important (L) Aquifers

(i) Sand/gravel (Lg)
(ii) Generally Moderately Productive (Lm)
(iii) Moderately Productive only in Local Zones (Ll)

Poor (P) Aquifers

(i) Generally Unproductive except for Local Zones (Pl)
(ii) Generally Unproductive (Pu)

The main regionally important bedrock aquifers are certain (usually purer) Carboniferous limestones, Carboniferous sandstones and Carboniferous volcanics. Only a small proportion of the readily usable groundwater in these aquifers is confined by low-permeability bedrock. Regionally important aquifers confined by bedrock units are not common.

In view of the fissure permeability that characterizes Irish bedrock, it is assumed that contaminant attenuation in bedrock is minimal in both the unsaturated and saturated zones. (While this is a simplification, experience and investigations consistently validate it.) This contrasts with the situation in Britain where major aquifers, such as the Sherwood Sandstone and the Chalk, are not only far more productive than the regionally important aquifers in the Republic of Ireland but also have a far greater capacity to attenuate contaminants.

Definition of groundwater vulnerability

Background

Vulnerability mapping has been carried out in many countries in recent years. Examination of these maps and descriptions of vulnerability in the scientific literature shows considerable variation in the definition and the usage of the vulnerability concept. (Lindstrom & Scharp (1995) provide a good review of this issue.) Up to now there has been no generally accepted definition or methodology for the construction of vulnerability maps. The variation is highlighted by the following points:.

- The definition may be limited to the intrinsic geological and hydrogeological characteristics of an area or may also include land-use and management practices.
- The geological and hydrogeological parameters considered relevant in vulnerability assessment vary widely fron country to country. For instance, in the USA, Aller *et al.* (1985) proposed a system called DRASTIC, which uses seven hydrogeological parameters: Depth to groundwater, net Recharge, Aquifer media, Soil media, Topography, Impact of vadose zone and hydraulic Conductivity of aquifer. Each parameter is given a rating range and a weight factor. A somewhat simpler method, using fewer parameters, was proposed by Foster (1987). It uses three parameters – Groundwater occurrence, Overall lithology and Depth to groundwater – and is termed GOD. It also uses a scoring system.
- The aspect of groundwater that is considered vulnerable can vary. It can be 'groundwater' itself; the 'groundwater system', 'aquifers', 'groundwater sources' such as karst springs, or 'groundwater resources'.
- Groundwater can be taken to be vulnerable to a variety of impacts such as 'natural impacts', 'human impacts including groundwater abstraction', 'contamination caused by human activities', 'conservative pollutants', 'universal pollutants', 'specific pollutants',

'point-source pollution' and/or 'diffuse pollution'.
- The number of vulnerability categories vary: five categories (extremely high, high, medium, low, very low) are suggested in the IAH guide to hydrogeological maps (Struckmeier & Margat 1995). Zaporozec (1985) suggests three categories (best, moderate and least (sub-surface) attenuation potential) for use in Rock County, Wisconsin, but the terminology is likely to be confusing if used directly in relation to vulnerability as 'best' means 'least vulnerable' and 'least' means 'most vulnerable'.
- There are wide variations in map scales, from < 1:10 000 to > 1:500 000.

These differences are understandable and largely inevitable because vulnerability, as used by most hydrogeologists, is a generalized, relative, qualitative term which can depend on several factors, such as the intended usage of the vulnerability concept, the specific hydrogeology of the mapped area, professional judgement and the availability of data.

The GSI's definition

Having considered the different options, the GSI uses the following definition.

> Vulnerability is a term used to represent the intrinsic geological and hydrogeological characteristics that determine the ease with which groundwater may be contaminated by human activities. Vulnerability mapping is the technique of assessing the geological and hydrogeological factors and displaying it on a map in a manner that is understandable and useful.

Factors determining the vulnerability assessment process in Ireland

Objectives

The overall aim of the vulnerability assessment process is to provide relevant information for land-use planning decisions, so that potentially polluting developments can be located and controlled in an environmentally acceptable way. The objectives, which are interrelated, are as follows:

- to indicate the degree of natural protection against contamination and therefore the likelihood of contamination;
- to provide an essential zoning component for the groundwater protection scheme in use in Ireland;
- to enable rapid and consistent decision-making by environmental managers;
- to help in providing a hierarchy of levels of risk and, in the process, assist in targeting and setting priorities for technical resources and investigations;
- to help comply with the water pollution legislation and with the sustainable development and precautionary principles;
- to provide useful information in designing monitoring networks;
- to assist in creating public awareness of groundwater and the susceptibility of groundwater to human activities.

Vulnerability of what? or reference location within the groundwater system

The concept of vulnerability to contamination is applied to 'groundwater' in bedrock and sand/gravel aquifers and not, for instance, to 'wells' or 'major aquifers'. Vulnerability is assessed on the basis of the (largely) vertical transport of contaminants to the water-table in these aquifers. Consequently, the vulnerability maps show the vulnerability of the first groundwater encountered in either sand/gravel aquifers or in bedrock. This groundwater may not be the main resource beneath a site where there is a deep confined bedrock aquifer.

Vulnerability to what?

Groundwater is taken to be vulnerable to contaminants generated by human activities and not, for instance, to natural impacts or groundwater abstraction. Pollution of groundwater in Ireland tends to be mainly microbiological rather than chemical; wells with levels of chemicals greater than the EC MAC (maximum admissible concentration) and arising from human activities are not common. In certain extremely vulnerable areas, more than 50% of the wells (mainly private wells) are polluted, usually intermittently, by faecal bacteria.

Release point of contaminants into the geological environment

The main threat to groundwater in Ireland is posed by point sources: farmyards, septic tank systems, any pollutants in sinking streams in karst areas, and, to a lesser extent, landfills and leakages and spillages from storage tanks. Diffuse sources, such as inorganic fertilizers and landspreading of organic wastes and pesticides, are currently far less of a threat in Ireland than in England and most other European countries. Consequently, it was decided that, when using the vulnerability concept in the production of vulnerability maps, it should be assumed that the contaminants are being released from point sources below the ground surface at depths of 1–2 m. Therefore, the topsoil is not taken into account in general vulnerability mapping. (In local investigations, the usage of the GSI guidelines on vulnerability mapping can be amended to take account of this, if appropriate.)

Geological and hydrogeological parameters used in vulnerability assessments

The vulnerability of groundwater is assumed to depend on the time of travel of infiltrating water (and contaminants) to the water-table, on the contaminant attenuation capacity of the geological materials through which the water and contaminants infiltrate, and on the relative quantity of contaminants that can reach the groundwater. As all groundwater is hydrologically connected to the land surface, it is the effectiveness of this connection that determines the relative vulnerability to contamination. Groundwater that readily and quickly receives water (and contaminants) from the land surface is considered to be more vulnerable than groundwater that receives water (and contaminants) more slowly and in lower quantities. The travel time, attenuation capacity and quantity of contaminants are viewed by the GSI as a function of the following natural geological and hydrogeological attributes of any area:

1. the sub-soils that overlie the groundwater;
2. the recharge type – whether point or diffuse;
3. in the case of unconfined sand and gravel aquifers, the thickness of the unsaturated zone below the point at which the contaminant is introduced.

The following are the reasons why we restrict ourselves to just these attributes.

(a) As the reference location is the water-table, the possible flow rates and attenuation of contaminants, once they have entered groundwater in either bedrock or sand/gravel, are not taken to be factors in vulnerability mapping because by then the impact has occurred and there is no further capacity to prevent the impact or protect the groundwater. Obviously the degree of impact is another issue, which depends on a variety of factors including the contaminant loading. Also, particularly in the case of sand/gravel aquifers, if the reference point was the supply source, this would not apply.

(b) The bedrock hydrogeology is not taken to be a significant factor in defining vulnerability. It is assumed, firstly, that once contaminants enter the bedrock there is, owing to the fissure permeability that characterizes Irish bedrock, little attenuation other than by (usually relatively limited) dilution when groundwater is reached, and secondly that an unsaturated zone in bedrock is not a significant factor.

(c) With the exception of karstic morphological features that indicate rapid recharge (swallow holes, for instance), the only factor determining the vulnerability of groundwater in bedrock is the nature (type, permeability and thickness) of overlying sub-soils or Quaternary deposits. It is considered that certain aspects of the hydrogeology of different bedrock types can be taken into account when the aquifer map is linked with the vulnerability map in preparing the groundwater protection map and the code of practice. For instance, regionally important bedrock aquifers that are karstified are indicated on the aquifer map.

Role of sub-soils

The sub-soils are regarded as the single most important natural feature in influencing groundwater vulnerability to pollution. They can act as a protecting filtering layer over groundwater, depending on their type, permeability and thickness. So, for instance, the higher the clay content (or the lower the permeability) and the greater the thickness, the better the protection from contaminants. Groundwater is most vulnerable where the sub-soil is absent or very thin.

Vertical preferential flowpaths are assumed not to be important below a depth of 3 m from the ground surface. While they undoubtedly are present in places, we await specific research to indicate their frequency of occurrence in the sub-soils in Ireland.

Role of recharge

The type and rate of recharge are particularly important in karstic limestone areas, where point recharge through swallow holes and sinking streams can occur. Attenuation is minimal, flow velocity is fast, and so groundwater is very vulnerable to pollution in these situations.

Role of unsaturated zone

In sand/gravel aquifers, a deep water-table reduces the likelihood of contamination because contaminants have to travel further and are slower to reach the groundwater. This allows the various beneficial physical, chemical and biological processes, that occur in the unsaturated zone, to attenuate the pollutants. However, as the permeability of sand/gravel is high, vulnerability will always be relatively high, even where the unsaturated zone is thick.

Guidelines for vulnerability mapping

General

In producing the guidelines, account was taken of general Quaternary geology and hydrogeology of Ireland (for instance, sub-soil thicknesses), the release point of contaminants and the experiences of hydrogeologists, particularly with contamination situations. All guidelines depend on the quality and availability of relevant data. Consequently, the GSI has produced guidelines, which are closely interrelated but are varied to suit data availability.

Optimum vulnerability mapping

The basis for the vulnerability guidelines being used by the GSI is given in Table 1. The four ratings described – extreme, high, moderate and low – assume the following:

- that 1–2 m of sub-soil and, in the case of sand and gravel aquifers, a 1–2 m thick unsaturated zone below the point of release of contaminants have a sufficient protecting capacity to allow a rating of 'high';
- that sand and gravel do not have a sufficient protecting capacity, no matter how thick, to merit a 'moderate' rating;
- that a minimum of 8 m of sandy till or 3 m of clayey till or clay, below the point of release of contaminants, enables a 'moderate' rating;
- that sandy till does not give sufficient protection to allow a 'low' vu1nerability rating no matter how thick it is;
- At least 8m of clayey till or clay below the point of release of contaminants are needed to merit a 'low' vulnerability rating.

In order to draw a vulnerability map based on these ratings, the following geological and hydrogeological information must be available on maps:

Table 1. *Vulnerability mapping guidelines*

Vulnerability rating	Hydrogeological requirements				
	Sub-soil permeability (type) and thickness			Unsaturated zone	Recharge type
	high permeability (sand/gravel)	moderate permeability (sandy till)	low permeability (clayey till, clay, peat)	(sand & gravel aquifers *only*)	
Extreme	0–3.0 m	0–3.0 m	0–3.0 m	0–3.0 m	point (< 30 m radius)
High	> 3.0 m	3.0–10.0 m	3.0–5.0 m	> 3.0 m	diffuse
Moderate	N/A	> 10.0 m	5.0–10.0	N/A	diffuse
Low	N/A	N/A	> 10.0 m	N/A	diffuse

N/A = not applicable.
Precise permeability values cannot be given at present.
Release point of contaminants is assumed to be 1–2 m below ground surface.
Adapted from Daly & Warren (1994)

(a) areas where the Quaternary sediment is generally less than 1 m thick (bedrock outcrop or sub-crop);
(b) sand/gravel deposits;
(c) till (boulder clay) deposits with details of basic matrix (textural) characteristics;
(d) peat: both cutover and intact bog;
(e) alluvium;
(f) lake clays;
(g) depth to bedrock map showing contours at 3 m, 5 m (in the case of clayey till and clay), and 10 m (these values are obtained by combining the sub-soil depths in the previous paragraph with the assumed point of release of contaminants from point sources at depths of 1–2 m below ground level);
(h) sand and gravel aquifers; with a differentiation between areas with thin ($\leq$ 3 m) and thicker (> 3 m) unsaturated zones;
(i) karstic features such as sinking streams and collapse features.

Information with this level of detail is available for only a few small areas in Ireland and is never likely to be more widely available, except perhaps around a limited number of major public supply sources. Routine GSI mapping programmes provide adequate data with regard to the sediments, but the paucity of depth to bedrock data is a problem. Consequently, these guidelines are aspirational. However, they do provide a basis for adaptation.

In extensive karst areas, detailed studies by a karst specialist are advisable. Further research on karstification may allow future refinement of the vulnerability ratings in karst areas.

Diffuse pollution sources, such as landspreading of organic wastes, are likely to become more important in the future. With a slight adaptation, the ratings can be used to take account of this by producing a map specifically oriented towards situations where the release point of contaminants is at the ground surface and which applies the 'extreme' rating to areas of outcrop, sub-crop (sub-soil normally < 1 m) and around karst features, whereas the remainder of the area with shallow sub-soil (1–3 m) could be ranked as 'high'.

Achieveable vulnerability mapping

The production of vulnerability maps for areas where the level of Quaternary geology information is poor is not recommended, as the level of uncertainty with the maps will make them indefensible and is likely to devalue the technique.

In Ireland, the GSI's Quaternary geology map series is designed to meet the requirements for the production of, *inter alia*, groundwater vulnerability maps (see Warren & Horton 1991) in so far as is possible. The major difficulty is in achieving adequate depth-to-bedrock data.

If the available Quaternary geology information is inadequate, it is essential that, prior to preparing vulnerability maps, the information is improved by reconnaissance mapping, trial pitting, drilling and grain size analyses. The following information should be obtained and is being obtained by the GSI when undertaking groundwater protection schemes:

1. areas where Quaternary sediment is generally less than or equal to 1 m thick (outcrop and sub-crop);
2. sand and gravel deposits (> 1 m thick);
3. till deposits with details on texture (>1 m thick) (because in Ireland tills vary rapidly in texture, each till unit will have a probability value attached to its general matrix characteristics, for example a till may be clayey (80%) giving an 80% probability that the till will be clayey at any one point);
4. peat bog;
5. alluvium;
6. lake clays;
7. depth to rock map showing shallow sub-soils (3 m), moderately thick sub-soils (3–10 m) and areas of thick sub-soils (probably > 10 m); however, the 10 m contour can only be attempted where the existing borehole information is adequate;
8. sand and gravel aquifers with a thin (3 m) unsaturated zone;
9. karst features;
10. indicators of sub-soil permeability, such as drainage density, and crop and vegetation types.

This information allows the following vulnerability ratings:

Extreme

1. Outcropping bedrock and sub-crop.
2. Within 30 m of karstic features (including along the area of loss or losing of sinking streams) and within 10 m on either side of losing streams upflow of the area of loss (in certain circumstances, for instance where overland surface run-off is likely, greater distances may be needed.)

Probably Extreme

1. Areas with thin ($\leq$ 3 m) sub-soil over bedrock.
2. Sand and gravel aquifers with a thin ($\leq$ 3 m) unsaturated zone.
3. Alluvium.
4. Blanket bog.

High

1. Areas where high-permeability sand and gravel are > 3 m thick.
2. Areas where intermediate permeability sub-soil such as sandy till is known, from borehole records, to be 3–10 m thick.
3. Areas where low-permeability sub-soil such as clayey till, clay and/or peat is known, from borehole records, to be 3–5 m thick.

Probably High

1. Areas where intermediate-permeability sub-soil such as sandy till is interpreted to be > 3 m and $\leq$ 10 m thick, from sparse borehole records.
2. Areas where low-permeability sub-soil such as clayey till, clay and/or peat is interpreted to be > 3 m and $\leq$ 5 m thick, from sparse borehole records.
3. Unconfined sand and gravel aquifers where the unsaturated zone is > 3 m thick.

Moderate

1. Areas where intermediate-permeability sub-soil such as sandy till is known, from borehole records, to be > 10 m thick.
2. Areas where low-permeability sub-soil such as clayey till, clay and/or peat is known, from borehole records, to be 5–10 m thick.

Probably Moderate

1. Areas where intermediate-permeability sub-soil such as sandy till are interpreted to be > 10 m thick, from sparse borehole records.
2. Areas where low-permeability sub-soil such as clayey till, clay and/or peat is interpreted to be 5–10 m thick, from sparse borehole records.

Low

Areas where low-permeability sub-soil such as clayey till and/or clay and/or peat is known to be > 10 m thick, from borehole records.

Probably Low

Areas where low-permeability sub-soil such as clayey till, clay and/or peat is interpreted to be > 10 m thick, from sparse borehole records.

In order to discourage production of vulnerability maps with inadequate data, the GSI has given guidelines on minimum standard vulnerability mapping. These are not given in this paper.

Map scales

Vulnerability maps are used as part of both source protection and resource protection schemes. As greater detail is needed for planning and contamination prevention purposes in the vicinity of groundwater sources, larger-scale maps are produced for source protection. At present, 1:10 560 (or 10 000 as they become available) scale maps are used for source protection, whereas 1:63 360 (or 1:50 000 as they become available) scale maps are used for resource protection purposes. Smaller-scale maps (e.g. 1:10 0000) are considered inadequate when account is taken of the objectives of the maps and the variability of the sub-soils in many areas over short distances.

Limitations and uncertainty in vulnerability assessments

Uncertainty is inherent in all methods of assessing groundwater vulnerability (Committee on Techniques for Assessing Ground Water Vulnerability 1993). A pragmatic, qualitative approach is taken to this issue. The only situation where the vulnerability rating is considered to be 'certain' is where there are precise detailed data and accurate locations; for instance, bedrock outcrops, karst features and high-quality borehole information. In all other situations, the uncertainty regarding the vulnerability rating is indicated by the use of the qualifying term 'probably', as indicated above under the achievable mapping guidelines. Admittedly, this does not take account of errors in compiling the data, although care is taken to reduce errors. A further effort to deal with this issue is by means of the following disclaimer on vulnerability maps:

> This **vulnerability map** is designed for general information and strategic planning usage. The boundaries are based on the available evidence and local details have been generalized to fit the map scale. Evaluation of specific sites and circumstances will normally require further and more detailed assessments, and will frequently require site investigations to determine the risk to groundwater.

Vulnerability is a term used to represent the intrinsic geological and hydrogeological characteristics that determine the ease with which groundwater may be contaminated by human activities.

The map shows the vulnerability of the first groundwater encountered (in either sand/gravel aquifers or in bedrock) to contaminants released at depths of 1–2 m below the ground surface. Where contaminants are released at significantly different depths, there will be a need to determine groundwater vulnerability using site-specific data. The characteristics of individual contaminants have not been taken into account.

This disclaimer also serves the purpose of outlining certain limitations of the map.

Verification of vulnerability as shown on vulnerability maps

As groundwater vulnerability predictions are made in a relative rather than an absolute sense, and as the area mapped to date in Ireland is relatively small, verification is difficult. However, initial indications are positive. Deakin (1995) found that public supply sources in one county showing signs of contamination were usually located in areas where the groundwater was mapped in the extreme and/or high-vulnerability categories. Also, Doak (1995) found a good correlation between the mapped vulnerability and the vulnerability of a number of springs in the karst lowlands of the west of Ireland, as indicated by the coefficient of variation in electrical conductivity of the spring discharge. While neither of these studies provides a proper evaluation of the validity of the general approach, they nevertheless provide initial supportive evidence.

Site assessments of vulnerability

The use of the vulnerability concept arose from the need to include the likelihood of contamination in the groundwater protection scheme. The vulnerability mapping guidelines evolved to fulfil this need. However, recently the concept has begun to be used as a means of assessing the risk posed by particular developments on specific sites. For instance, the Environmental Protection Agency requires the vulnerability rating for integrated pollution control licensing purposes and may require this information for landfill site selection. This is creating the need for the adaptation of the mapping guidelines to suit site-specific vulnerability determinations. In order to do this, (i) more detail is needed on the release point (or potential release point) of contaminants, and (ii) decisions are needed on the quantitative values of the division between ‘high’ and ‘moderate’ permeabilities and between ‘moderate’ and ‘low’. Information on the former is usually readily available, while decisions on the latter are not possible in the short term owing to the lack of appropriate data.

There is a critical need to determine, through research, the permeabilities of Irish tills. Until this is done, it is not possible to give permeability values at present; the guidelines in Table 2 are recommended for use in the meantime.

Table 2. *Vulnerability guidelines applicable to site assessment*

Vulnerability rating	Hydrogeological requirements *below the point of release of contaminants*				
	Sub-soil permeability (type) and thickness			Unsaturated zone	Recharge type
	high permeability (sand/gravel)	moderate permeability (sandy till)	low permeability (clayey till, clay. peat)	(sand & gravel aquifers *only*)	
Extreme	0–2.0 m	0–1.5 m	0–1.0 m	0–2.0 m	point (< 30 m radius)
High	> 2.0 m	1.5–8.0 m	1.0–3.0 m	> 2.0 m	diffuse
Moderate	N/A	> 8.0 m	3.0–8.0	N/A	diffuse
Low	N/A	N/A	> 8.0 m	N/A	diffuse

N/A = not applicable.
Precise permeability values cannot be given at present.

Integration of vulnerability into the groundwater protection scheme

The scheme combines the four components of risk and risk management: (i) vulnerability; (ii) groundwater value, as indicated by the aquifer category and the proximity to a public groundwater supply; (iii) control measures for (iv) specific hazards. The hydrogeological factors are encompassed on a land-surface zoning map that results from the integration of the three elements of land-surface zoning (vulnerability categories, source protection areas and resource protection areas). In practice, this is achieved by superimposing the vulnerability map on source protection area maps and on the aquifer map, and results in the matrix of groundwater protection zones given in Table 3. Each zone is represented by a code (e.g. Rf/M, which represents areas of *regionally important fissured* aquifers where the groundwater is *moderately* vulnerable to contamination). The vulnerability ratings (i.e. vertical axis) take account of the vertical movement of contaminants from the point of release to the water-table in bedrock or sand/gravel aquifers, while the source protection and resource protection elements (horizontal axis) take account, to a certain degree at least, of the (largely) horizontal movement of groundwater and contaminants in the saturated zone.

The control of groundwater contamination sources is by means of a code of practice which lists the degree of acceptability of potentially polluting activities for each zone and describes the recommended controls for both existing and new activities. It is shown by means of a level of response or restriction, which is applied to each potentially polluting activity or group of activities, with the level of response depending on the different elements of risk: the vulnerability, the value of the groundwater (with public supply sources being more valuable than resources and regionally important aquifers more valuable than locally important and so on) and the contaminant loading.

Four levels of response (R) to the risk of a particular potentially polluting activity are recommended for the Irish situation:

R1	Acceptable subject to normal good practice.
$R2^{a,b,c,...}$	Acceptable in principle, subject to conditions in note a, b, c, etc. (the number and content of the notes will vary for each zone and for each activity).
$R3^{m,n,o,...}$	Not acceptable in principle; some exceptions may be allowed subject to the conditions in note m, n, o, etc.
R4	Not acceptable.

The approach is illustrated for a hypothetical potentially polluting activity in the matrix in Table 4.

The matrix encompasses both the geological/ hydrogeological and the contaminant loading aspects of risk assessment. In general, the arrows (→ ↓) indicate directions of decreasing risk, with the ↓ arrow showing the decreasing likelihood of contamination and the → arrow showing the direction of decreasing consequence. The contaminant loading aspect of risk is indicated by the activity type in the table title.

The response to the risk of groundwater contamination is given by the response category allocated to each zone and by the site investigations and/or controls and/or protective measures described in notes a, b, c, d, m, n and o.

In considering the scheme, it is essential to remember that: (a) the scheme is intended to provide guidelines for the location of developments (the control measures are not statutory); and

Table 3. *Matrix of groundwater protection zones*

Vulnerability rating	Source protection			Resource protection					
				Regionally Imp.		Locally Imp.		Poor Aquifers	
	Site	Inner	Outer	Rk	Rf/Rg	Lm/Lg	Ll	Pl	Pu
Extreme (E)	SS/E	SI/E	SO/E	Rk/E	Rf/E	Lm/E	Ll/E	Pl/E	Pu/E
High (H)	SS/H	SI/H	SO/H	Rk/H	Rf/H	Lm/H	Ll/H	Pl/H	Pu/H
Moderate (M)	SS/M	SI/M	SO/M	Rk/M	Rf/M	Lm/M	Ll/M	Pl/M	Pu/M
Low (L)	SS/L	SI/L	SO/L	Rk/L	Rf/L	Lm/L	Ll/L	Pl/L	Pu/L

After Daly (1995)

Table 4. *Groundwater protection scheme matrix for activity X*

Vulnerability rating	Source protection			Resource protection						
				Regionally Imp.		Locally Imp.		Poor Aquifers		
	Site	Inner	Outer	Rk	Rf/Rg	Lm/Lg	Ll	Pl	Pu	
Extreme (E)	R4	R4	R4	R4	R4	$R3^m$	$R2^d$	$R2^c$	$R2^b$	↓
High (H)	R4	R4	R4	R4	$R3^m$	$R3^n$	$R2^c$	$R2^b$	$R2^a$	↓
Moderate (M)	R4	R4	$R3^m$	$R3^m$	$R2^d$	$R2^c$	$R2^b$	$R2^a$	R1	↓
Low (L)	R4	$R3^m$	$R3^o$	$R2^d$	$R2^c$	$R2^b$	$R2^a$	R1	R1	↓
	→	→	→	→	→	→	→	→	→	

After Daly (1995)

(b) delineation of the groundwater protection zones is dependent on the data available and consequently may be wrong in places. However, the intention is that the statutory authorities should apply the scheme in decision-making on the basis that the best available data are being used. The onus is then on the developer to provide new information which might enable the zonation to be altered and improved and, in certain circumstances, the planning decision to be changed.

Vulnerability maps, as a component of groundwater protection schemes, have been produced for five (out of 29) local authority areas in the Republic of Ireland, and mapping has commenced in three further counties.

Discussion

Criticisms of the application of the vulnerability concept

There are potential difficulties in using vulnerability assessments and maps, and these can lead to the following criticisms and objections.

- The results are subjective and usually lack quantification, particularly regarding permeabilities, travel times and contaminant attenuation capacity. While scoring systems are sometimes used to give a 'number', this frequently gives merely a perception of quantification.
- The concept and its usage are sometimes ill-defined and unclear.
- They do not usually take account of complex geological and hydrogeological situations.
- Most refer to a 'universal contaminant', which may not be appropriate to specific contaminants.
- The concept can be used in too simplified a manner to be of benefit and this situation may have the opposite effect of posing a risk to groundwater.
- Preferential flowpaths are usually not taken into account.
- Planners, who usually have no expertise or experience in hydrogeology, may misuse the maps.
- Uncertainty is inevitable in most instances and devalues the concept. This is particularly the case where vulnerability mapping is attempted in the absence of adequate data.
- Evaluation and verification are difficult.

Answers to the criticisms

Undoubtedly, vulnerability assessments are not a panacea in the planning process for preventing groundwater contamination. However, the approach adopted in Ireland has attempted to overcome the difficulties outlined above.

- While the vulnerability concept used in Ireland lacks quantification, the relative ranking of areas or sites with regard to their vulnerability to groundwater contamination in qualitative terms can:
  - ◆ help bring the groundwater interest to the attention of decision-makers, such as planners;
  - ◆ encourage and enable decision-makers to place less emphasis on engineering control measures and more on the natural, intrinsic characteristics of an area or site;

- provide a 'first-off' warning system before site visits and investigations are made;
- help in risk assessment;
- by identifying areas with less natural protection to contamination, act as a guide in the location of potentially polluting activities;
- aid in rapid and consistent decision-making;
- help in the identification of existing activities posing a risk to groundwater.

- Vulnerability maps are not recommended as an alternative to site-specific investigations and contaminant risk assessments.
- The concept has been clearly set out with regard to the definition of vulnerability, the reference location within the groundwater system, the release point of contaminants, mapping guidelines, and the integration into the groundwater protection scheme.
- Vulnerability is not used as a single tool within the protection strategy. It is the component of the protection scheme that provides information on the likelihood of contamination.
- Sub-soils mapping (Quaternary geology) is now specifically carried out by Quaternary geologists prior to the production of vulnerability maps in areas where the existing data are inadequate.
- The map scales used are large enough to clearly designate zones.
- Account can be taken of specific contaminants, at least to some degree, by appropriate control measures.
- The legend used on the maps reduces the likelihood of misuse.

The vulnerability guidelines used by the GSI have been influenced by discussions with and contributions from Paul Johnston (TCD), Catherine Coxon (TCD), Malcolm Doak (Sligo RTC), Richard Thorn (Sligo RTC), Margaret Keegan (GSI), Geoff Wright (GSI), Natalie Doerfliger (University of Neuchatel, Switzerland), Jean-Pierre Tripet (Swiss Hydrological and Geological Survey) and Brian Adams (British Geological Survey). This paper is published with the permission of Dr. P. McArdle, Director, Geological Survey of Ireland.

References

ALLER, L., BENNETT, T., LEHR, J. H. & PETTY, R. J. 1985. *DRASTIC: a standardised system for evaluating groundwater pollution potential using hydrogeologic settings*. US Environmental Protection Agency.

COMMITTEE ON TECHNIQUES FOR ASSESSING GROUND WATER VULNERABILITY. 1993. *Groundwater vulnerability assessment : predicting relative contamination under conditions of uncertainty*. National Research Council (US), National Academy Press.

DALY, D. 1995. Groundwater protection in Ireland : a scheme for the future. *In*: *Proceedings of 15th Annual Groundwater Seminar, The Role of Groundwater in Sustainable Development*. International Association of Hydrogeologists (Irish Group).

—— & WARREN, W. P. 1994. Mapping groundwater vulnerability to pollution : GSI guidelines. *The GSI Groundwater Newsletter*, **25**, 10–15.

DEAKIN, J. 1995. *Groundwater protection in County Limerick*. MSc thesis, Trinity College, Dublin.

DOAK, M. 1995. *The vulnerability to pollution and hydrochemical variation of eleven springs (catchments) in the karst lowlands of the west of Ireland*. MSc thesis, Sligo Regional Technical College.

FOSTER, S. S. D. 1987. Fundamental concepts in aquifer vulnerability, pollution risk and protection strategy. *In*: VAN DUIJVENBOODEN, W. & VAN WAEGENINGH, H. G. (eds) *Vulnerability of soils and groundwater to pollution*. TNO Committee on Hydrological Research, The Hague, Proceedings and Information No. **38**, 69–86.

GEOLOGICAL SURVEY OF IRELAND. 1962. *Geological Map of Ireland*.

LINDSTROM, R. & SCHARP, C. 1995. *Approaches to groundwater vulnerability assessments – a state of the art report*. Division of Land and Water Resources of the Royal Institute of Technology, Stockholm, Sweden.

STRUCKMEIR, W. F. & MARGAT, J. 1995. *Hydrogeological maps: a guide and standard legend*. International Association of Hydrogeologists. International Contributions to Hydrogeology, Vol. **17**.

WARREN, W. P. & HORTON, A. 1991. Mapping glacial deposits in Britain and Ireland. *In*: EHLERS, J., GIBBARD, P. L. & ROSE, J. (eds) *Glacial Deposits in Great Britain and Ireland*. Balkema, Rotterdam, 377–387.

WRIGHT, G. R., ALDWELL, C. R., DALY, D. & DALY, E. P. 1982. *The Groundwater Resources of the Republic of Ireland*. Commission of the European Communities.

ZAPOROZEC, A. (ed.) 1985. *Groundwater Protection Principles and Alternatives for Rock County*. Wisconsin Geological and Natural History Survey, Madison, Special Report **8**.

Assessment of groundwater vulnerability in England and Wales

R. C. PALMER[1] & M. A. LEWIS[2]

[1]*Soil Survey and Land Research Centre, Cranfield University, The Innovation Centre, York Science Park, Heslington, York, YO1 5DG, UK*

[2]*British Geological Survey, Maclean Building, Crowmarsh Gifford, Wallingford, Oxfordshire OX10 8BB*

Abstract: Groundwater represents around 35% of water distributed for potable supply in England and Wales and is also utilized by industry and for private supplies. It is of high quality and requires little treatment before use. It is important, therefore, that a robust policy of groundwater protection is available to protect the resource from pollution, as rehabilitation of polluted groundwater is expensive, and often may not be technically feasible. One way to protect aquifers is to control activities which pose a threat to groundwater quality, especially in areas where they are naturally poorly protected. Areas where groundwater is at risk of pollution from surface activities are indicated on maps by characterizing the properties of the underlying soils and rocks. The rocks are classified on the basis of their permeability, into Major, Minor and Non-Aquifers. The soils overlying aquifers are then categorized on their physical and chemical properties into leaching potential classes. The Groundwater Protection Policy of the Environment Agency, of which the groundwater vulnerability maps are part, provides guidance on those types of development which require controlling if groundwater quality is to be maintained in the long term. The relevance of these maps to other environmental issues is discussed.

The Groundwater Protection Policy (NRA 1992) of the Environment Agency incorporates a concept of groundwater vulnerability which was developed in conjunction with the Soil Survey and Land Research Centre (SSLRC) and the British Geological Survey (BGS). These two organizations have been commissioned by the Environment Agency to prepare a national series of fifty-three 1:100 000 scale vulnerability maps, to cover all of England and Wales by early 1998. By May 1996 twenty-eight maps had been published (Fig. 1), covering most of the outcrop areas of the Chalk and Sherwood Sandstone Group aquifers. The groundwater vulnerability maps, together with the definition of source protection zones and a set of policy statements on the principal activities posing a threat to groundwater form key elements of the Groundwater Protection Policy (GPP). The GPP document includes a generalized vulnerability map at 1:1 000 000 scale covering the whole of England and Wales. Similar synoptic maps have been prepared for Scotland at 1:625 000 scale (ADRIS 1995) and for Northern Ireland at 1:250 000 scale (DENI 1994).

The concept of vulnerability has been widely discussed by Robins *et al.* (1994) and Foster & Skinner (1995) and alternative methods of mapping the vulnerability of groundwater were initially considered. In particular the DRASTIC System (Aller *et al.* 1987), which evaluates Depth to water-table, net Recharge, Aquifer media, Soil media, Topography, Impact of vadose (unsaturated) zone and hydraulic Conductivity of aquifers, was assessed but, because of its relative complexity, it was deemed to be more suitable for zoning large regional aquifers than the relatively small local aquifers which typically occur throughout England and Wales.

The vulnerability system, designed for use in the GPP maps, zones the soil and geological horizons of the unsaturated zone in terms of their physical and chemical properties in order to assess the ease with which a pollutant released at the surface would be likely to reach the underlying groundwater body (Palmer *et al.* 1995). The maps indicate the vulnerability of groundwater to both diffuse source (e.g. agricultural fertilizers, pesticides and sludge spreading) and point-source (e.g. chemical spillages) pollutants where they are applied at the surface, when the soil is intact and has the potential to ameliorate the effect of the potential pollutant on the underlying groundwater. The maps are designed to be used by planners, developers, consultants and regulatory bodies to provide guidance on possible constraints to potential development.

The GPP indicates the key variables in assessing the vulnerability of groundwaters to surface derived pollutants to be:

- the presence and nature of the overlying soil cover;
- the presence and nature of the Quaternary deposits;
- the nature of the underlying strata;
- the thickness of the unsaturated zone.

PALMER, R. C. & LEWIS, M. A. 1998. Assessment of groundwater vulnerability in England and Wales. *In*: ROBINS, N. S. (ed.) *Groundwater Pollution, Aquifer Recharge and Vulnerability*. Geological Society, London, Special Publications, **130**, 191–198.

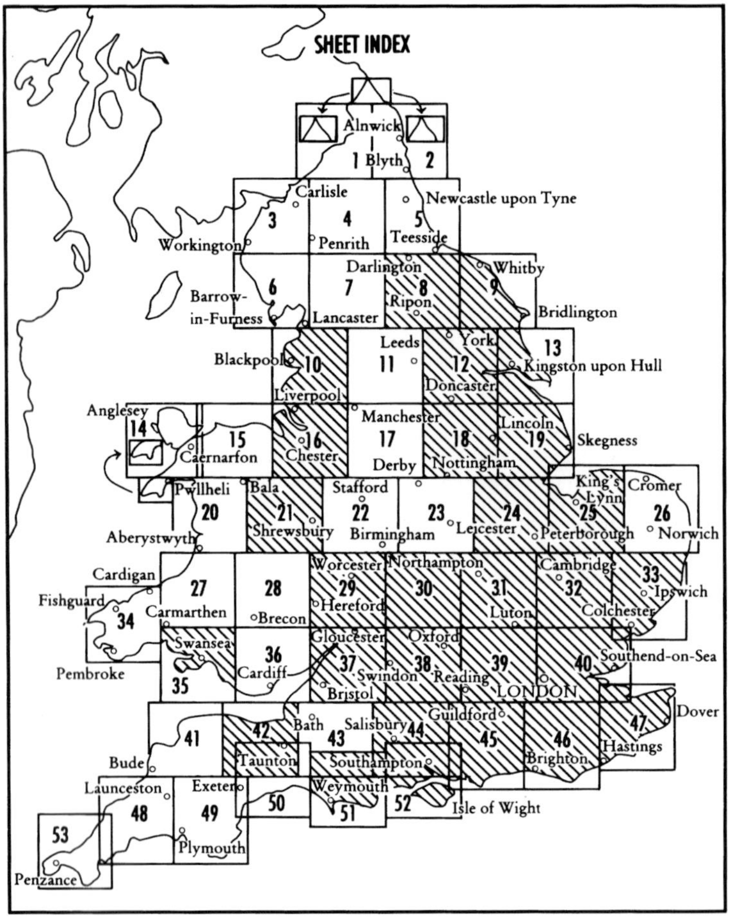

Fig. 1. The groundwater vulnerability map series for England and Wales. Shading indicates maps published by May 1996.

The first three of these parameters are incorporated into the groundwater vulnerability assessments used on the maps. The thickness of the unsaturated zone has not been used in the methodology for several reasons:

- sufficient, good quality data are only available for Major Aquifers;
- depth to water changes seasonally in response to recharge;
- in fissured aquifers (i.e. generally highly productive aquifers in the British context), the thickness of the unsaturated zone does not significantly affect the time of the first arrival of a pollutant at the water-table;
- non-degradable pollutants are unaffected by the thickness of the unsaturated zone.

Geology

Geological strata across England and Wales are classified into Major, Minor and Non-Aquifers according to their permeability: highly, variably and weakly permeable, respectively.

Major Aquifers are highly permeable formations usually with a known or probable presence of significant fracturing. They may be highly productive and able to support large abstractions for public supply and other purposes. They include the Chalk, Sherwood Sandstone Group and Magnesian Limestone aquifers.

Minor Aquifers include fractured or potentially fractured rocks, which do not have a high primary permeability and other formations of variable permeability including unconsolidated deposits. Although these aquifers seldom produce large quantities of water for abstraction, they are important both for local supplies and in supplying baseflow to rivers. They include most of the river terrace deposits, the Crag and the Millstone Grit.

Non-Aquifers are generally regarded as transmitting insignificant quantities of groundwater. However, groundwater flow through such rocks, although imperceptible, does take place and needs to be considered in assessing the risk associated with persistent pollutants. Some Non-Aquifers can yield water in sufficient quantities for domestic use and may also provide some baseflow to rivers. They include the Jurassic clays, Mercia Mudstone Group and most of the Lower Palaeozoic formations.

A summary of the aquifer classification is given in Table 1. This aquifer classification is initially applied to the uppermost pre-Quaternary rocks present beneath the land surface whether they are at outcrop or sub-crop. Where these rocks are a Non-Aquifer, any overlying permeable Pleistocene or Recent deposits present at the ground surface are then shown. These unconsolidated superficial deposits are generally classified as Minor Aquifers. However, locally thick highly permeable river terrace deposits in the Thames valley have been designated a Major Aquifer. For the purposes of this assessment the Norwich Crag and Red Crag, of Pleistocene age, are classified with the pre-Quaternary.

Aquifer boundaries correspond to lines shown on geological maps but aquifers are designated according to lithology rather than chronostratigraphy. Figure 2 is an example of how the aquifer boundaries (lithological) relate to the formation boundaries (stratigraphic) for the Middle Jurassic rocks of Oxfordshire.

The groundwater vulnerability maps subdivide geological formations on the basis of their permeability or the ease with which a pollutant will travel through them. This is usually, but not always, synonymous with their resource potential. The Carboniferous Limestone can transmit potential pollutants very rapidly through its unsaturated zone, but its resource potential may not be very high because the water-bearing fissures are irregularly spaced and difficult to locate.

Lateral changes in the permeability and lithology of formations can lead to changes in aquifer classification so the karstic Carboniferous Limestone of the Peak District, north and south Wales and the Mendips is classified as a Major Aquifer but the Yoredale facies present in the Pennines of northern England is only regarded as a Minor Aquifer.

Quaternary deposits are generally unconsolidated and contain few fractures but their lithology can vary considerably both laterally and

Table 1. *Generalized classification of strata after Palmer et al.* (1995).

Major Aquifers (highly permeable)	Minor Aquifers (variably permeable)	Non-Aquifers (negligibly permeable)
river terrace deposits (Middle Thames Valley)	blown sand	Palaeogene clays
Chalk	alluvium	Gault and other Lower Cretaceous clays
Upper Greensand	dry valley deposits	Jurassic clays
Lower Greensand	river terrace deposits (except Middle Thames Valley)	Penarth Group
Spilsby Sandstone	glaciofluvial sand and gravel deposits	Mercia Mudstone Group
Jurassic limestones	Crag	Permian marls
Upper Lias Sands	Pliocene gravels	Silurian (except limestones)
Dolomitic Conglomerate	Palaeogene sands	Ordovician
Sherwood Sandstone Group	Lower Cretaceous sands (except Greensands and Spilsby Sandstone)	Cambrian
Permian sandstones	Jurassic limestones, sandstones and ironstones (except where Major Aquifers)	Precambrian
Magnesian limestones	Permian breccias and conglomerates (SW England)	intrusive igneous rocks
Carboniferous Limestone	Coal Measures	
	Millstone Grit	
	Lower Carboniferous limestones and shales	
	Old Red Sandstone	
	Silurian limestones	
	volcanic rocks	

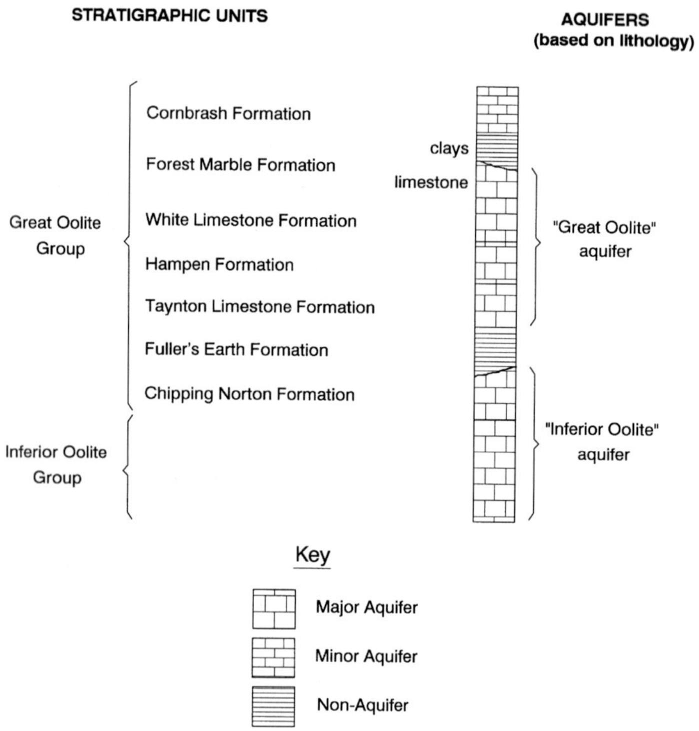

Fig. 2. Relationship of aquifers to stratigraphical divisions of the Middle Jurassic of Oxfordshire.

vertically over relatively short distances. Their unsaturated zone is usually thin and they are often in hydraulic continuity with surface water. For vulnerability mapping purposes superficial deposits are classified into permeable deposits, which are unlikely to limit recharge into underlying aquifers, and low-permeability deposits, which may limit recharge. The coarser-grained permeable deposits have little ability to attenuate pollutants, as they contain only small amounts of clay and organic matter. Where low-permeability drift deposits (Clay-with-Flints, glaciolacustrine clays and tills) are present at the surface and they are underlain by pre-Quaternary Major or Minor Aquifers, their extent is shown by a black stipple ornament on the maps (Fig. 3a).

However, in Holderness, the Vale of York, Teesside and across the Cheshire/Lancashire Plain, where there is known to be a total of more than 5 m of clayey deposits within the Quaternary sequence, it is considered that this provides protection, from surface-derived pollutants, to any underlying groundwater. Aquifers occurring beneath these deposits are therefore not shown as it is assumed they are not vulnerable. The uppermost superficial deposits are zoned instead (Fig. 3b). Elsewhere, and in these areas where there is less than 5 metres of clay in the drift sequence, only the uppermost superficial deposits are considered; underlying layers which may have a significant effect on the movement of pollutants from the land surface to an aquifer are not taken into consideration because of the potentially wide vertical and lateral variation in lithology, knowledge of which is often limited to borehole locations.

Soil

The potential role of the soil zone in adsorbing or attenuating surface-derived pollutants is too often overlooked in groundwater studies. Soil is defined in many ways, differing according to the interests of the user. In UK groundwater vulnerability studies, soil is defined as the weathered zone into which plants will root and is the zone or layers formed from the interaction of climate, soil organisms (including vegetation) and geology. It is generally 1.5–2.0 m thick although in the tropics this zone can exceed 5 m in depth.

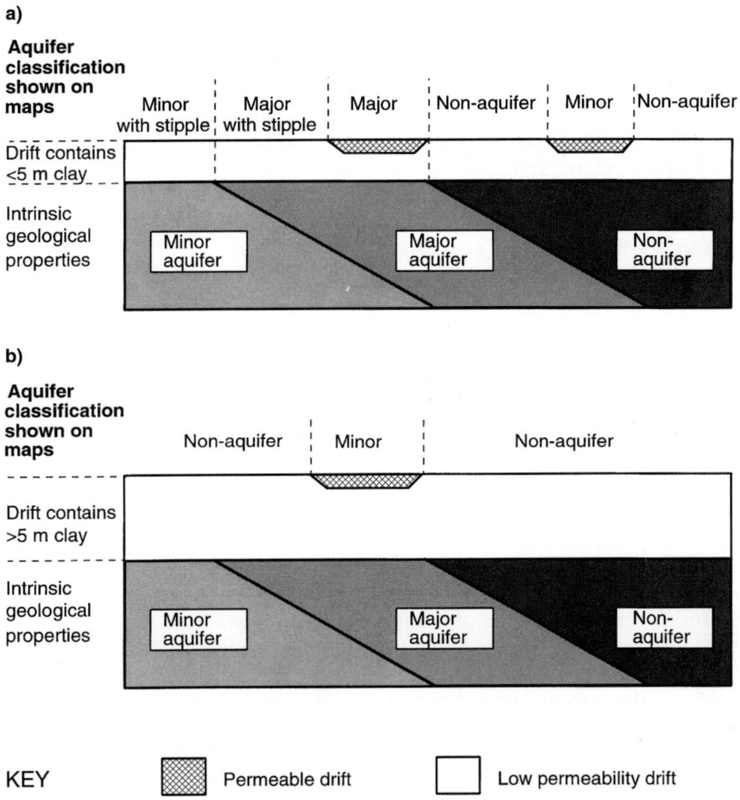

Fig. 3. Assessment of aquifer vulnerability with regard to the thickness of clays in superficial deposits: (**a**) less than 5 m of clay in superficial deposits; (**b**) more than 5 m of clay in superficial deposits.

Soil is important in groundwater vulnerability studies as it is the most chemically active part of the unsaturated zone. It has the highest levels of organic matter which is significant as clay–organic complexes can attenuate the movement of certain heavy metals and pesticides. It also consistently has a higher cation exchange capacity compared with the underlying geological zones. Soil is also very active biologically and usually contains millions of organisms per cubic metre, ranging in size from mammals and earthworms down to the microscopic. Soil can provide natural physical barriers to the vertical movement of water and pollutants in the form of pans and dense, clay-rich, slowly permeable layers. Finally, soils have a major influence on land use especially in conjunction with climate. Locally, the soil–climate interactions can determine which agriculturally derived pollutants will occur within a catchment.

Soil maps in England and Wales show the distribution of soil types (soil series) based largely on three criteria: the texture of the whole soil profile; the soil water regime (depth to and duration of waterlogging in a soil); and the substrate type (the geology into which soils are developed). The 725 soil series defined, described and used on the soil maps of England and Wales (Clayden & Hollis 1984), have been classified, for groundwater vulnerability studies, according to their potential for allowing contaminants, present on the land surface, to be leached into the underlying permeable geological formations (aquifers). The classification is based on those physical and chemical properties which are routinely measured and estimated during soil survey projects. These properties include: texture, stoniness, organic matter content, presence of raw peaty topsoils and slowly permeable layers; and soil water regime.

The leaching potential classification used on the maps is based on an understanding of the physical and chemical properties of the soils of Britain supported by the results of a wide range of fundamental and applied research projects undertaken during the last 20 years. One important element of this research culminated in the production of the HOST classification (Hydrology of Soil Type) which relates soil type to two important hydrological indicators: baseflow index and standard percentage run-off (Boorman *et al.* 1995). Considerable work has also been undertaken in England and Wales to investigate the fate and behaviour of nitrate and both individual biocides and groups of pesticides in a range of soil types.

The leaching potential classification (Table 2) is a simple three-fold classification: high, intermediate and low, with a limited number of subclasses of the high and intermediate categories. Soil distribution in urban areas is often less well-understood than in rural locations and there is also a greater possibility of disturbance and the incorporation of imported soil materials. These changes can alter the permeability and/or thickness of the soil-forming materials of a site. For these reasons urban areas are assumed to have a high leaching potential (HU), unless proven otherwise by site-specific investigation or the availability of good-quality historical records.

Upland blanket or raised peats, where the peat is often still accumulating or humification is inhibited by unfavourable moisture or temperature conditions or very low nutrient status, are classified as having a low leaching potential. Lowland peats, which have been drained and used for agriculture, have relatively well-aerated earthy topsoils. The high organic matter content of these soils provides the capacity to attenuate certain pollutants and these soils are allocated their own subclass in the system (I2). When lowland peats are undrained they are classified as having a high leaching potential (H1) because of their almost permanent saturation at shallow depth.

Discussion

The Environment Agency's series of groundwater vulnerability maps provides a first indication of the soil and hydrogeological setting of a site whenever land-use change is being considered. The maps, by necessity, involve the simplification of complex, often intricate, natural spatial distributions of soils and rocks and can only represent the situation at or near the surface.

The maps allow planners, developers and regulatory bodies to make better informed initial judgements on the location of new developments and so avoid placing potentially polluting activities in high-vulnerability areas. It has always been assumed that any project which might have an impact on groundwater will require site-specific investigations to ascertain its suitability. While site-specific vulnerability assessments should not be made from these maps the basic methodology used for interpreting soil, superficial and pre-Quaternary geology and

Table 2. *SSLRC soil leaching potential classification for groundwater vulnerability studies*

High Soil Leaching Potential

There are four sub-classes with H1 having the highest potential.

H1	Soils with groundwater at shallow depth; Soils with rock, rock-rubble or gravel at shallow depth; Undrained lowland peat soils.
H2	Sandy soils with a low topsoil organic matter content.
H3	Sandy soils with a moderate topsoil organic matter content; Soils with rock, rock-rubble or gravel at relatively shallow depth.
HU	Soils in urban environments and sites of restored mineral workings where H1 leaching potential is assumed until proven otherwise.

Intermediate Soil Leaching Potential

There are two sub-classes, one for mineral soils and one for peaty soils or humose mineral soils.

I1	Deep loamy and clayey mineral soils unaffected by marked seasonal waterlogging;
I2	Drained peat with humified topsoil; Soils with a humose mineral topsoil.

Low Soil Leaching Potential

There are no sub-classes.

L	Soils with a slowly permeable layer restricting downward water movement; Upland soils with a raw peaty topsoil.

aquifer properties for the maps can be adapted for use at the site-specific level.

The maps attempt to simplify the complex technical and scientific issues relating to groundwater vulnerability in order that groundwater vulnerability concepts can be readily used by non-specialists. The requirement for a clear, understandable end-product necessitates simplification and this has occurred in four main instances.

- Depth to water has not been incorporated. This information is not always available and is too complex to show informatively on regional-scale maps especially in areas of multi-layered aquifers. However, this parameter is important to vulnerability assessment and should be investigated in any site-specific interpretation.
- The thickness, properties and stratigraphy of Quaternary deposits are only partially incorporated in the maps. The sequences are locally extremely complex and often imperfectly understood. The nature of the uppermost deposit is incorporated into the published maps by showing low-permeability deposits with a stipple shading and indicating permeable deposits as Minor Aquifers where they are underlain by pre-Quaternary Non-Aquifers.
- Groundwater quality is not addressed by the maps. Following a seminar on groundwater in the Hague in 1991 (RIVM/RIZA 1991), attended by EC Environment Ministers, an action programme was adopted which recognizes the need for the management and protection of groundwater on a sustainable basis. This requires that all groundwaters be protected and is the basis on which the vulnerability maps are being produced.
- The vulnerability maps always incorporate the best soil and geological information currently available. However, national coverage of high-quality detailed soil and geological mapping information is not complete and therefore the vulnerability assessment may need updating as and when improved data become available. Locally, the variability in the geology and the complexity of soil distribution is too great to be adequately represented on maps at a scale of 1:100 000. In these cases some simplification of the available data is required in order to prepare appropriate regional-scale maps. In some cases, therefore, detailed information not shown on the published vulnerability map is available from published soil and geology maps should it be required for site-specific interpretation.

These vulnerability maps, based on spatial soil and geological information, are specifically designed to be applicable to the zoning of groundwaters with regard to diffuse-source pollutants and liquid discharges deposited on the land surface. These include the widespread use of nitrogenous fertilizers, pesticides and sludge spreading . The hazards associated with certain point-source spillages of potential groundwater pollutants can also be assessed using the maps, so long as the spill occurs on the land surface and the sorption/degradation properties of the pollutant are compatible with the soil leaching potential classification (Palmer *et al.* 1995).

The movement of non-aqueous phase liquids (NAPL), both dense (DNAPL) and light (LNAPL), depends on their individual properties such as density and viscosity. The soil leaching potential classification used on the Environment Agency maps cannot be used directly for these pollutants and class boundaries may require some adjustment in order to represent the movement of these pollutants through soil layers. Both DNAPLs and LNAPLs behave similarly within the unsaturated zone, travelling through the air-filled pores in preference to water-filled pores. Where the aquifer is fractured they have the potential to travel quickly through the unsaturated zone but in unsaturated intergranular aquifers a significant part of the liquid will be retained as it displaces air in the pore spaces. Once LNAPLs reach the water-table they spread out across the surface and may only present a limited problem to groundwater quality because they usually remain floating on the groundwater surface, even when pumping causes a cone of depression to form. On the other hand, once DNAPLs have penetrated to the water-table they can move rapidly down through the aquifer until an aquiclude is encountered. In fractured aquifers, dissolved-phase DNAPLs may be drawn rapidly into supply sources. Further research is required to study the movement of representative DNAPLs and LNAPLs in both soil and the underlying unsaturated zone before groundwater vulnerability systems can be devised to cater for these pollutants.

Point-source pollution can occur beneath the soil zone and in these cases the threat to groundwater quality will usually be significantly higher than that indicated on the maps. A provisional scheme for amending the current leaching potential classification system for pollutants spilled or injected into subsoils or soil substrates has been presented by Palmer *et al.* (1995). Further work is required to refine this kind of assessment in order that the risks posed by septic-tank drainage fields, underground storage tanks and facilities, and the subsoil injection of wastes may be adequately assessed.

A full site investigation is obviously always required for any type of landfill activity. However, the vulnerability maps provide a broad indication of the permeability of underlying rocks and hence indicate areas that may or may not be suitable for further investigation for siting purposes and give some assistance with the assessment of the pollution risks associated with current landfills.

Conclusion

Groundwater vulnerability as assessed in England and Wales provides a robust methodology for informing both specialists and non-specialists of the likely distribution pattern and the natural physical and chemical properties of the soils and rocks underlying a district. It allows all participants in the planning process to make informed initial judgements regarding the threats to groundwater quality from the siting of potentially polluting activities. Groundwater vulnerability maps provide a key element in the regional strategy to protect Britain's valuable groundwater resources as long as they are not interrogated for site-specific detail. Detailed assessments of groundwater vulnerability require site-specific information although the vulnerability assessment methodology devised for these maps can be readily used as a model into which locally collected data may be fed. It should, however, be stressed that the maps were designed to be used for diffuse and certain point-source spillages occurring at the land surface. Where the soil is partially or completely removed, this assessment methodology will not be applicable.

The concepts presented in this paper were developed for the National Rivers Authority under the guidance of Dr A. C. Skinner and the National Groundwater Centre, Solihull. This paper is published with the permission of the Environment Agency (formerly the National Rivers Authority) and the Directors of the Soil Survey and Land Research Centre (Cranfield University) and the British Geological Survey (NERC).

References

ADRIS. 1995. *Groundwater Protection Strategy for Scotland.* Association of Directors and River Inspectors of Scotland, Perth.

Aller, L., Bennett, T., Lehr, J. H., Petty, R. J. & Hackett, G. 1987. *DRASTIC: A standardised system for evaluating groundwater pollution potential using hydrogeologic settings.* US Environment Protection Agency, Ada, Oklahoma, **EPA/600/2–87/035**.

Boorman, D. B., Hollis, J. M. & Lilly, A. 1995. *Hydrology of soil types: a hydrologically-based classification of the soils of the United Kingdom.* Institute of Hydrology, Wallingford, Report No. **126**.

Clayden, B. & Hollis, J. M. 1984. *Criteria for differentiating soil series.* Soil Survey Technical Monograph **17**, Harpenden.

DENI. 1994. *Groundwater Vulnerability Map of Northern Ireland, scale 1:250 000.* Department of the Environment, Northern Ireland, Belfast.

Foster, S. S. D. & Skinner, A. C. 1995. Groundwater protection: the science and practice of land surface zoning. *IAHS Conference on Groundwater Quality: Remediation and Protection*, Prague, Czech Republic, 15–18 May 1995, 471–482.

NRA. 1992. *Policy and Practice for the Protection of Groundwater.* National Rivers Authority, Bristol.

Palmer, R. C., Holman, I. P., Robins, N. S. & Lewis, M. A. 1995. *Guide to Groundwater Vulnerability Mapping in England and Wales.* HMSO, London.

RIVM/RIZA. 1991. *Sustainable use of groundwater: problems and threats in the European Communities.* Rijksinstitut voor Volksgezondheid en Milieuhygiene, Bilthoven, The Netherlands, Report No. **600025001**.

Robins, N. S., Adams, B. Foster, S. S. D. & Palmer, R. C. 1994. Groundwater vulnerability mapping: the British perspective. *Hydrogéologie*, **3**, 35–42

Methods used to delineate groundwater source protection zones in England and Wales

D. B. BURGESS[1] & S. W. FLETCHER[2]

[1] *The Environment Agency, Anglian Region, Kingfisher House, Orton Goldhay, Peterborough PE2 5ZR*

[2] *The Environment Agency, Midlands Region, 10 Warwick Road, Olton, Solihull B92 7HZ*

Abstract: Groundwater source protection zones (GPZs) are an integral part of land surface zoning within the Environment Agency's Policy and Practice for the Protection of Groundwater. The methods and experience gained in delineating GPZs around 1000 groundwater sources in England and Wales are reviewed. The methods include : manual calculations, hydrogeological mapping, analytical and semi- analytical modelling and numerical modelling. The large numbers of sources within a diverse number of hydrogeological settings has led to a structured approach to selecting the GPZ method appropriate to each groundwater source. This includes consideration of the protected yield, data acquisition and formulation of a conceptual flow model. The GPZ methods also include consideration of parameter uncertainty. Further developments in GPZ delineation methods are required to cope with fissured/fractured groundwater flow and to exploit the results of transient groundwater models.

The delineation of groundwater source protection zones (GPZs) around strategically important groundwater supplies forms an important part of the Environment Agency's *Policy and Practice for the Protection of Groundwater* (National Rivers Authority 1992). The policy has three key elements:

- a classification of groundwater vulnerability and maps that describe the susceptibility of the underlying groundwater resource to threats from surface development or land-use proposals;
- a series of groundwater source protection zones around existing potable groundwater supplies requiring immediate and specific protection;
- a series of statements and acceptability matrices for each zone and vulnerability area that relate to land use or development that may pose a threat to the sustainability of groundwater resources and existing groundwater users.

An overview and experience in the use of the policy is described in Skinner & Foster (1995) while the method of vulnerability assessment is given by Palmer & Lewis (1998).

The term 'groundwater source' in the context of this paper refers to the point of abstraction of groundwater, e.g. a well, a borehole or boreholes in close proximity.

It is estimated that there are over 78 000 groundwater sources in England and Wales where the use requires the water quality to be of a potable standard and where land use in the vicinity of the source needs to be controlled to maintain the groundwater quantity and quality. Of these, some 75 000 are small, domestic boreholes and wells pumping less than 20 $m^3 day^{-1}$ and in isolated rural areas often supplying only one household. By contrast, there are nearly 2000 larger public and commercial water supply boreholes pumping from 500 to over 10 000 $m^3 day^{-1}$ supplying urban areas or the more densely populated rural areas. Therefore, the priorities for delineating GPZs vary with the amount of groundwater abstracted. For the numerous smaller sources, an arbitrary fixed-radius protection zone of 50 m is currently in use, with more detailed GPZs being computed as the need arises and in the face of local development pressures (National Rivers Authority 1995*a*). For those potable sources pumping over 500 $m^3 day^{-1}$ of groundwater, a more proactive approach has been adopted and a programme for delineating GPZs around these started in 1992 and is expected to be completed by 1998. This paper reviews the methods and approach used in delineating GPZs for over 1000 large, potable sources within this programme.

Definition of groundwater source protection zones

GPZs encapsulate the area around a particular groundwater source within which, if a stable contaminant were to be introduced into the satu-

BURGESS, D. B. & FLETCHER, S. W. 1998. Methods used to delineate groundwater source protection zones in England and Wales. *In*: ROBINS, N. S. (ed.) *Groundwater Pollution, Aquifer Recharge and Vulnerability*. Geological Society, London, Special Publications, **130**, 199–210.

rated groundwater flow, it would eventually be pumped from that groundwater source. Using this definition, three different zones are recognized around each source:

1. Inner Zone I, defined as the 50-day travel time from any point beneath the groundwater table to the source. This zone has a minimum radius of 50 m.
2. Outer Zone II , defined by either a 400-day travel time from a point beneath the water-table to the source or the minimum recharge area required to support 25% of the mean rate of annual abstraction from the source, whichever is the larger.
3. Source Catchment Zone III, defined as the area within which all groundwater would be eventually pumped at the source. Under steady-state conditions this zone is synonymous with the total capture zone (EPA 1987). In confined aquifers, the source catchment may be displaced some distance from the source.

A schematic representation of how the three zones vary under different aquifer conditions is shown in Fig. 1. It should be noted that the two travel-time GPZs, the Inner and Outer Protection Zones, although dominantly based on hydraulic considerations, are also underpinned by arbitrary and pragmatic *de minimus* values of distance and area. A full discussion of the basis and application of each zone was given by Adams & Foster (1992) and within the original groundwater policy document (National Rivers Authority 1992).

The area of each hydraulically derived GPZ is directly proportional to the abstraction rate. The abstraction rate on which each GPZ is based is termed the protected yield. This is defined as the maximum authorized and sustainable rate of groundwater pumping that can take place from the source in question. Generally, the protected yields are taken as those quantities stated within the abstraction licence as issued under the Water Resources Acts, 1963 and 1991. For each licence, the maximum daily quantity is used to derive the inner or 50-day zone, while the annual licensed rate is used to derive the outer 400-day zone and the total source catchment zone (National Rivers Authority 1995*b*).

History of methods used to delineate groundwater protection zones

The complexity of the methods used to delineate groundwater protection zones has varied with the growth in the use of groundwater and the increased perceived risk to groundwater quality from surface contamination. Initially, simple distance criteria were adopted for the protection of groundwater sources. For instance, a 3 km radius circle was adopted widely by the UK water industry following the recommendations of a Ministry of Health Memorandum in 1948 (Ministry of Health 1948). By the 1970s protection zones were more firmly based on hydraulic considerations following the European experience in adopting groundwater catchments and travel time zones for land use and development controls. The Severn Trent Water Authority chose

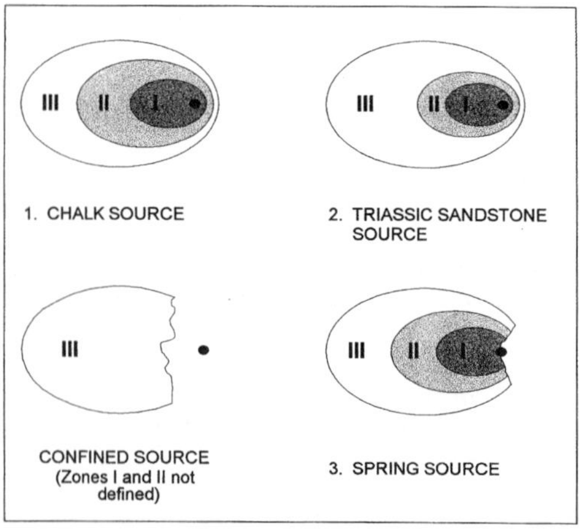

Fig. 1. Schematic diagrams of the relationship between source and protection zones (after National Rivers Authority 1992).

to protect a 1 km radius circle around public water supply boreholes. This decision was largely based on the 1 km radius circle approximating to between a 50- and 100-day saturated flow travel time within Severn Trent's major aquifer, the Triassic sandstones (Selby & Skinner 1979). Southern Water (1985) developed this approach further for the Chalk aquifer of southern England by deriving 50-day travel time, egg-shaped protection zones based on analytical equations and following a technique first published by Bear & Jacobs (1965).

Methods used by the Environment Agency

The methods adopted within the GPZ delineation programme vary in complexity from the declaration of simple arbitrary fixed radii circular zones, through empirical methods based on hydrogeological mapping, to methods based on numerical groundwater modelling. These methods are grouped and described under the following headings:

- manual calculations;
- hydrogeological mapping;
- analytical and semi-analytical modelling;
- numerical modelling.

The parameters represented, and advantages and disadvantages of each set of methods are summarized in Table 1.

Manual calculations

Manual calculations are used in areas where data are sparse and the time available to collect any further data is limited. The most widely used manual calculation is the construction of circular zones, the extent of which represent the area of

Table 1. *Methods used by the Environment Agency to delineate groundwater source protection zones (GPZs)*

Capture zone methodology	Parameters which can be represented	Advantages	Disadvantages
Manual calculations	• Recharge • Time of travel • Abstraction rate	• Low cost • Easy and quick to implement with minimal data requirements • Some technical basis	• Does not have regard to local hydrogeological conditions • Over-simplified
Hydrogeological mapping	• System boundaries and approximated divides	• Should be combined with all other methods • Good for karst and fractured aquifers with strong geological control	• Poor in areas with distinct boundaries • Not quantitative • Does not represent time of travel
Analytical and semi-analytical modelling – WHPA and Bear & Jacobs (1965)	• Hydraulic gradient • Hydraulic conductivity • Aquifer thickness • Effective porosity • Recharge • Simple recharge/ barrier boundaries • Some limited parameter variability possible	• Capture zones based on idealized representation of local aquifer conditions • Can represent a simple system • Quantitative • Simple and uniform boundaries and recharge allowed for	• Isotropic conditions assumed • Modest data requirements • Assumes infinite aquifer • Does not allow complex boundary and recharge effects to be considered
Numerical modelling – FLOWPATH, MODFLOW	• All hydrogeological parameters • Much flexibility in varying parameters	• Can represent most geological and hydrogeological boundaries	• Conceptual model of local hydrogeology required • Significant data requirements • Isotropic conditions assumed • Practical difficulties of obtaining adequate resolution of the flow field

recharge required to balance the rate of abstraction.

The area A_s (m^2) of a source catchment zone in a region subject to a mean annual recharge R ($m\,a^{-1}$) may be calculated from the simple water balance relationship:

$$A_s = \frac{Q}{R} \quad (1)$$

where Q is the protected yield of the source ($m^3\,a^{-1}$). The mean annual recharge is usually estimated from several years of effective rainfall from the Meteorological Office's MORECS dataset (Thompson *et al.* 1981) with adjustments made for surface run-off. The radius of the circular source catchment zone is defined as:

$$r = \sqrt{(A_s/\pi)} \quad (2)$$

where r is the radius of the boundary of the source catchment in metres (m).

For inner and outer protection zones an estimate of the area A_t (m^2) of the t_d (days) time-of-travel zone may be made using a volumetric approach so that:

$$A_t = \frac{Qt_d}{bn} \quad (3)$$

where t_d is either 400 days for the outer zone or 50 days for the inner zone, b (m) is the aquifer thickness and n is the effective porosity.

Often the aquifer characteristics around older public water supply sources have to be estimated from the results of periods of short test pumping that took place when the borehole or well was constructed. The data from such tests are usually limited to water levels within the pumped borehole and immediate vicinity only. Even so, sufficient information may be available to estimate the aquifer transmissivity T ($m^2\,day^{-1}$) and infer the hydraulic gradient i. In such circumstances, and with a number of limiting assumptions, two equations adapted from Todd (1959) have been used to estimate the source catchment area. These define the distance to the null point or catchment divide directly downstream of the pumping borehole x (m) as:

$$x = -\frac{Q}{2\pi Ti} \quad (4)$$

and the flow front or maximum width of the catchment orthogonal to the flowlines, y (m) as:

$$y = \frac{Q}{Ti} \quad (5)$$

The above equations should be used with caution as they apply only to an isotropic aquifer of infinite extent and with steady groundwater flow. As such, it should be stressed that these manual calculations represent an over-idealized approach to zone delineation and their use on their own can only be justified when hydrogeological information is limited. Nevertheless, these relatively simple manual calculations continue to be used as an overall check and quality control on the results of GPZs delineated using numerical modelling techniques (Aspinwall & Co. 1993).

Hydrogeological mapping

Hydrogeological mapping is used to delineate a GPZ when the hydrogeological regime is complex and there is a paucity of information on aquifer characteristics or water levels. Hydrogeological mapping uses a combination of local knowledge and published information on hydrology, topography and soils. For unconfined aquifers, the total catchment area may often be drawn by reference to topographic maps, from which the surface drainage may be assessed together with geological maps indicating the extent of the permeable strata (Headworth & Skinner 1986). For sources pumping from confined aquifers, the source catchment area may be displaced several kilometres away from where the aquifer outcrops, as in the case of the Lincolnshire Limestone aquifer of eastern England.

Where groundwater gradients are thought to be shallow, hydrogeological mapping may be used in conjunction with the manual methods above to define source catchment and time-of-travel zones. Figure 2 gives an example of this approach for a Triassic sandstone source near Bearstone, Shropshire in which the derived source catchment area is termed the best-estimate zone (BEZ). The Sherwood Sandstone dips to the west with the main aquifer comprising the Wilmslow Sandstone unit. The source catchment zone is defined by a circle using Eqs (1) and (2) above. However, the aquifer is overlain to the west by the poorly permeable Upper Helsby Sandstone making it locally confined as far west as the Norton and Wem faults which are taken as the effective western catchment boundary. To the east the catchment boundary is drawn where the aquifer thins and unconformably overlies the Keele Beds which are mainly impermeable mudstones.

Hydrogeological mapping techniques are also extensively used to define GPZs for sources pumping groundwater from karstic limestone aquifers. Karstic aquifers pose two problems for the calculation of GPZs. Firstly, the surface flow in such areas may be poorly integrated, being disrupted by sinkholes and springs, making the extent of catchment areas difficult to infer from topographic maps. Secondly, the

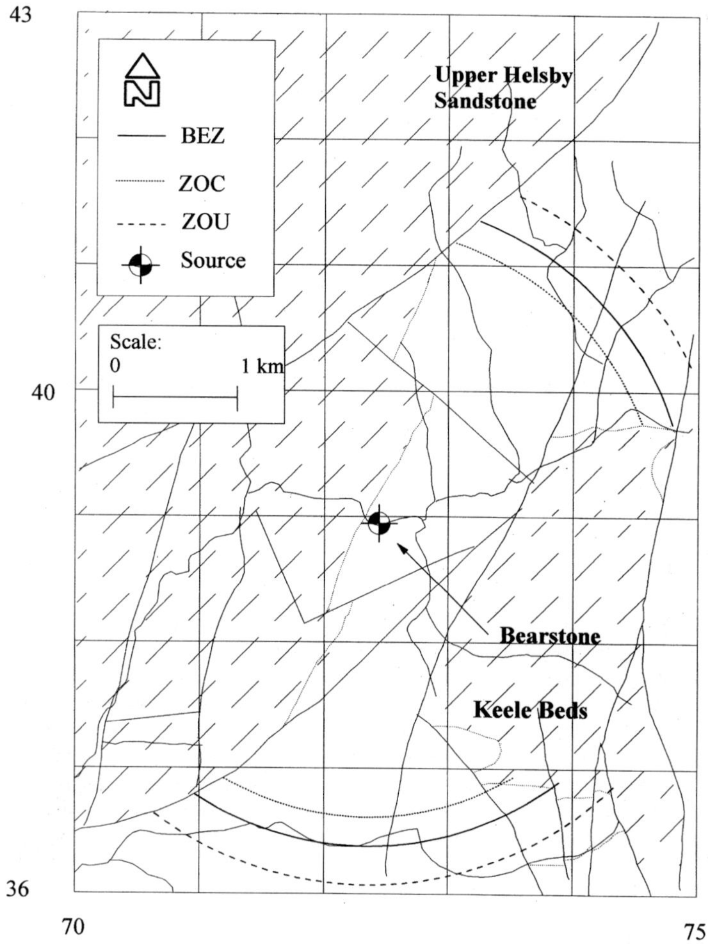

Fig. 2. Example of GPZ drawn using manual calculations and hydrogeological mapping: Bearstone Pumping Station, Shropshire.

concentration of groundwater flow within well-defined channels and conduits represents a significant departure from the assumptions of groundwater flow within the analytical and numerical models conventionally used to calculate GPZs (Huntoon 1995). For these reasons, the accurate definition of GPZs on karstic aquifers may require the results from detailed and time-consuming field mapping and tracer techniques (e.g. Atkinson & Drew 1974).

Analytical and semi-analytical models

Analytical and semi-analytical models have been used both prior to and within the programme for GPZ delineation. The assumptions behind such models are restrictive, idealizing groundwater flow to a uniform direction and gradient within an isotropic, homogeneous aquifer of infinite extent. Despite these restrictions, the models have been used in three contexts: firstly, to provide rapidly and with the minimum of effort, an initial or 'first pass' set of protection zones in advance of the use of more time consuming and complex techniques; secondly, as a rapid means of providing protection zones around small sources in the face of potentially threatening nearby development; and finally, analytical models have been used to verify the codes of numerical solutions such as FLOWPATH and MODFLOW/MODPATH (Barker *et al.* 1995).

Most analytical models are an adaptation of the equations derived by Bear & Jacobs (1965) which define the boundary line and travel times for water injected into a borehole. These equations can be used to compute travel times to a borehole from any point in the surrounding flow field. However, the inverse problem of delineating a specific travel time line such as for 50 or 400 days normally requires the use of numerical methods to find lines of equal time of travel within the analytically derived velocity field. Such methods are termed semi-analytical models. For instance, Lerner (1992) describes a semi-analytical model which incorporates spatially varying recharge and aquifer boundaries.

The semi-analytical models within the US Environmental Protection Agency's Well Head Protection Area (WHPA) computer package have been adopted for use in the GPZ programme (EPA 1993). The two semi-analytical models within WHPA that have been most widely used to date are RESSQC and MWCAP. RESSQC delineates time related capture zones around several interfering sources in a homogeneous aquifer of infinite extent and uniform groundwater flow. The RESSQC model has been adapted to rapidly delineate 'first pass' GPZs for several groundwater sources pumping from the Chalk aquifer of East Anglia. The MWCAP model delineates either steady-state, time-related, or hybrid capture zones for sources in homogeneous aquifers with steady and uniform ambient groundwater flow. The aquifer may be considered to be infinite in extent or it may be bounded by parallel stream or barrier boundaries. The MWCAP model provides one of the methods recommended for use in delineating GPZs around small sources (National Rivers Authority 1995*a*).

Numerical models

For the range and complexity of groundwater flow and boundary conditions in England and Wales, numerical models have proved the most robust for deriving GPZs. To date most of the models used have been based on the two-dimensional and steady-state assumptions. Although these assumptions may be considered to over-idealize more complex situations, in practice the use of numerical models has provided an acceptable balance between cost and accuracy. Of the GPZs completed to date, two-dimensional, steady-state models have been used to define over 80% of them. These represent a large sample of the groundwater flow conditions and at a variety of scales ranging from a few square kilometres around an isolated source to several sources pumping from an aquifer unit of over 1000 km^2.

The use of numerical models to delineate GPZs is based on the well-established approach of representing the groundwater flow system within the grid or mesh of an accepted model code. The model is then calibrated with groundwater pumping at historic rates by adjusting the parameters within the model, within acceptable ranges, to simulate measured water levels or flows. The calibrated groundwater model is then used to predict the groundwater heads that would result from pumping at the protected yield for the source in question. These simulated heads, together with a particle tracking code, are used to trace out the pathlines or flowpaths either for the source catchment area or the 50- or 400-day time-of-travel zones. The theory and practice involved in the application of numerical groundwater flow models and of particle tracking routines are described in Anderson & Woessner (1992).

The numerical model most widely used within the GPZ programme has been FLOWPATH. FLOWPATH is a well-integrated package of computer programs that includes a two-dimensional, steady-state flow model and particle tracking routines. The groundwater flow code is similar to the well-proven PLASM model modified for mesh centred boundary conditions. The particle tracking routine uses reverse distance interpolation with Euler integration to move particles along flow lines (Guiger & Franz 1991). The model was chosen for its ease of use and ability to simulate groundwater flow over the wide range of aquifer conditions encountered in England and Wales (Barker *et al.* 1991).

The input of data to, and output of results from the model are carried out interactively using computer graphical methods that allow the rapid construction of model grids against a background of the principal geological and hydrological features. The computer program allows for the rapid input and display of maps showing recharge, internal flow boundaries, groundwater contours, aquifer characteristics and finally output and display of travel time or catchment pathlines. These are illustrated in Fig. 3 which illustrates the use of the FLOWPATH to generate pathlines for the Edgmond Bridge borehole which pumps groundwater from the Sherwood Sandstone to the north of Telford, Shropshire.

In complex hydrogeological conditions such as multilayered aquifers, or where confined and

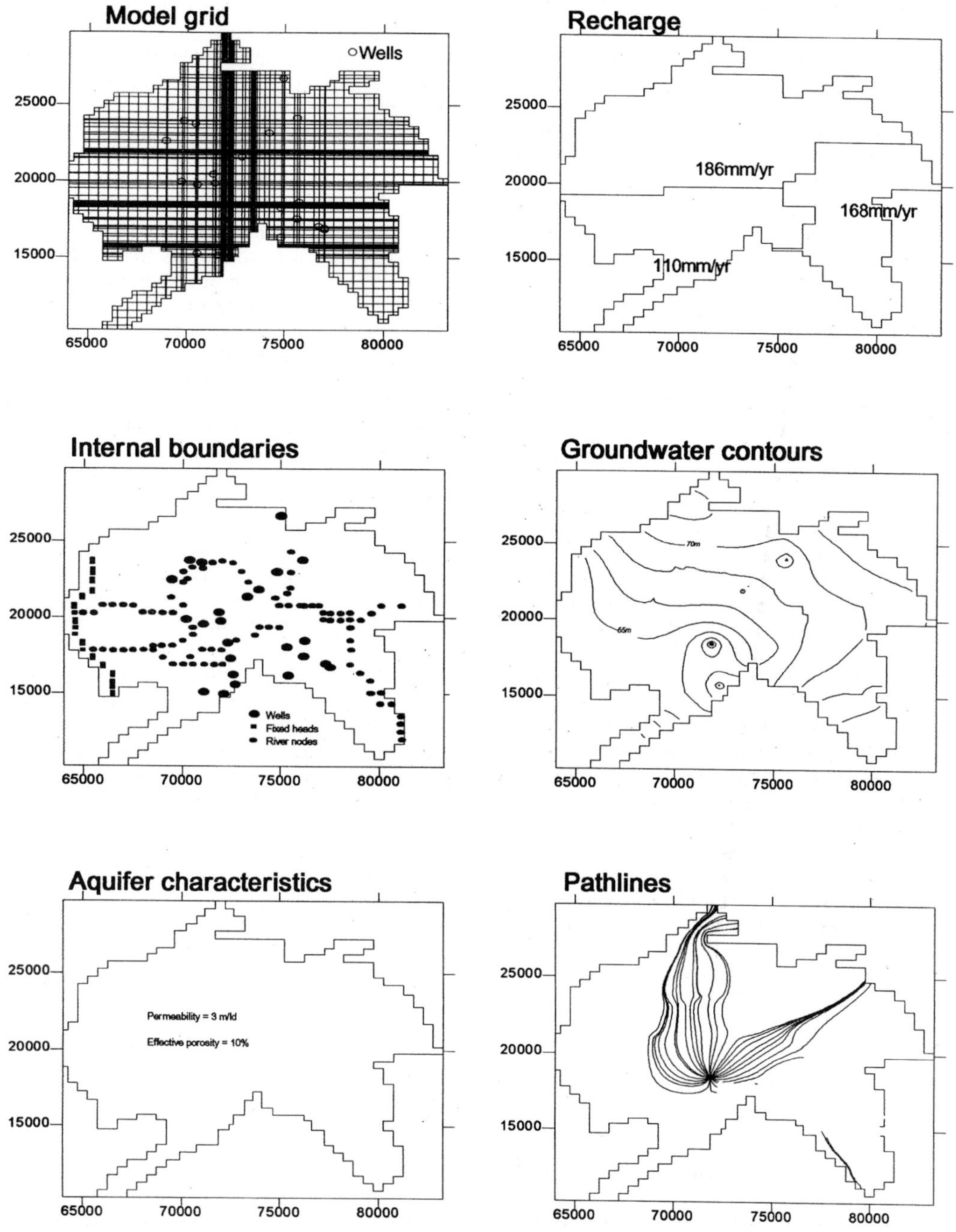

Fig. 3. Example of FLOWPATH maps showing model grid, model parameters and generated pathlines: Edgmond Bridge Pumping Station, Shropshire (after National Rivers Authority 1995*b*).

unconfined conditions exist within the same area to be modelled, the US Geological Survey models MODFLOW and MODPATH have been used to delineate GPZs. The MODFLOW model is an internationally accepted and well-proven groundwater flow model (McDonald & Harburgh 1988). MODFLOW is a three-dimensional model that is capable of dealing with a wider range of aquifer and boundary conditions than FLOWPATH. It can also be used to simulate transient or groundwater flow conditions. MODPATH is a three-dimensional particle tracking code that is designed to be used with MODFLOW. However, the current version is not designed for use under transient groundwater flow conditions (Pollock 1988).

The MODFLOW/MODPATH codes were adopted for the GPZ programme following a detailed evaluation and comparison with FLOWPATH (Barker *et al.* 1995). MODFLOW/MODPATH have been used to model layered aquifers such as the Chalk/Greensand aquifer of southern England. Unlike FLOWPATH, MODFLOW/MODPATH can be used in situations where unconfined, leaky and confined conditions exist within the same model. The packages have also been used in aquifer units where steady-state modelling for GPZ delineation is regarded as a prelude to more complex regional transient models used to evaluate the groundwater resource and simulate the effects of future groundwater management options.

The output from both FLOWPATH and MODPATH is in the form of pathlines running from the source to the outer limit of the source catchment or time-of-travel zone as seen in the last diagram of Fig. 3. These are expressed as zonal lines by joining the outer limits by eye to produce a continuous zone boundary usually using a computer aided design program. In order to minimize the subjectivity of this process and to ensure consistency between different zones, guidelines and rules have been produced (Environment Agency 1996).

Selection of GPZ method for each source

The above methods have been used to delineate GPZs for a wide range of hydrogeological conditions within England and Wales. This, together with the large number of sources, has necessitated a well-documented and considered approach to data acquisition and choice of method for delineating GPZs within the Environment Agency programme. The steps in the approach include:

- definition of source and protected yield
- data acquisition
- formulation of a conceptual groundwater flow model
- selection of method to delineate GPZs

and are represented diagramatically in Fig. 4. Detailed guidelines on the application of this approach are given in the Environment Agency's manual of zone delineation (Environment Agency 1996).

The definition of the pumping rate or protected yield of a particular source is the most important parameter controlling the size of the derived GPZs. Typically, a groundwater abstraction licence for public water supply will allow the use of several boreholes within the overall licensed quantities. Therefore, the first step in zone delineation includes a definition of the protected yields of individual boreholes or each sourceworks that is consistent with the licence conditions and operational practice.

The hydrogeological data required to delineate each set of GPZs are drawn from a wide variety of data sources and reports, both inside and outside the Environment Agency. These range in detail and in scale from borehole construction details, provided by the source operator, through to regional geological maps provided by the British Geological Survey, to groundwater-level maps from the Environment Agency's own groundwater-level observation network. Around some sources, data may be sparse or non-existent. Around others, a fully calibrated groundwater model may already exist together with detailed hydrogeological reports. Such a diversity of information has called for a well-structured approach to data acquisition to a common, well-defined format, on pro-forma sheets. In this way a high degree of consistency and quality control can be applied to the data acquisition process. Wherever possible, hydrogeological data are gathered for an area that corresponds to the aquifer resource unit for the source. However, where this is difficult to define, an arbitrary 5 km radius of search may be adopted around the source.

The data collected are used to form a conceptual model of the groundwater flow system surrounding the source. This is usually done by use of hydrogeological cross-sections and diagrams that are used to summarize what is known about the aquifer properties and flow processes of the area in question. Where data are sparse, the conceptual model may simply be a statement of the hydraulic gradient and assumed aquifer parameters. However, for the principal and well-developed aquifers of England

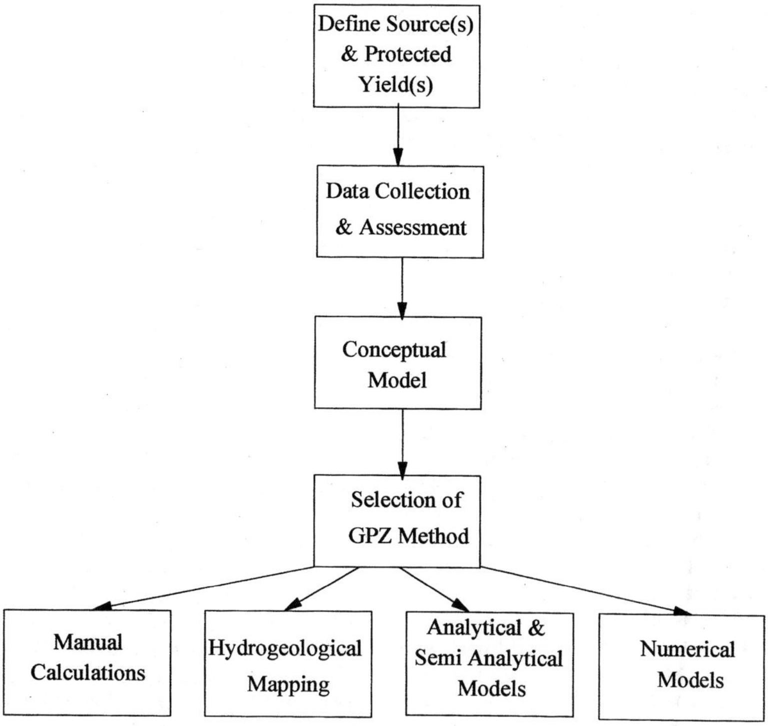

Fig. 4. Diagram showing the steps in selecting the method used to delineate GPZs around each source.

and Wales the conceptual model will include a full definition of the hydrostratigraphical units, a preliminary water budget and a definition of the flow system including boundaries.

The availability of data and the complexity of the hydrogeological setting as defined by the conceptual model combine to limit the choice of the methods available to delineate a particular set of GPZs. These may be further limited by time and resources. For instance, if a set of GPZs is required quickly for a small source in an isolated gravel aquifer with little or no groundwater-level information, the GPZs may take the form of arbitrary fixed-radius circles. By contrast, a large public water supply source drawing from the Chalk or Triassic sandstone aquifer, with relatively detailed hydrogeological information, would justify the use of the more complex and time-consuming numerical modelling techniques to delineate the GPZs. In complex aquifers such as the Carboniferous Limestone, relatively simple analytical methods may be used to draw GPZs in the short term, to be followed by more detailed field and tracer investigations as time and money allow.

Precision and uncertainty in delineating GPZs

The variable precision of either original or derived hydrogeological data used to calibrate groundwater flow models is well recognized (National Research Council 1990). Hydrogeological parameters such as hydraulic conductivity or recharge may often be expressed as a range of values over particular areas of aquifer. In addition, when numerical models are used to delineate GPZs, different combinations of parameters may produce what seem equally valid calibrated models and hence generate a number of equally valid GPZs (Evers & Lerner 1995).

Within the GPZ programme, uncertainties in both the measurement and selection of parameters are used to derive a boundary envelope either side of each GPZ. The outer or larger boundary line is termed the zone of uncertainty and the inner or smaller boundary line the zone of confidence. The methods used to derive these extremes are largely qualitative and may include consideration of the range of either one or more different aquifer parameters on the size and shape of each GPZ. The number of different parameter ranges considered varies with the method used to derive the GPZ. For example, GPZs delineated using manual calculations may use an uncertainty analysis that is limited to consideration of the maximum and minimum plausible values of groundwater recharge on the size and extent of the GPZ. This method was used to define the range of the circular source catchment zone for the Bearstone Pumping Station source in Fig. 2. In this diagram, the zone of confidence (ZOC) and the zone of uncertainty (ZOU) lie either side of the original GPZ boundary, termed the best-estimate zone (BEZ).

By contrast, the uncertainty method attached to GPZs delineated using numerical models uses ranges of values for three parameters: recharge, hydraulic conductivity and effective porosity. The range of value for each parameter is derived from field values and by a sensitivity analysis on the parameters within the original calibrated groundwater flow model. These values are used in different combinations to generate different flow models and pathlines. The different pathlines are then superimposed and the area of mutual overlap is taken as the zone of confidence. The area that encloses the outermost pathlines generated by all the different model combinations is taken to be the zone of uncertainty. A full description of the uncertainty method used with the FLOWPATH model is given by Fermor *et al.* (1995). This method was used to derive the zones of confidence and of uncertainty for both the source catchment and outer 400-day travel time zone around the Edgmond Bridge Pumping Station source (see Fig. 5).

The concept of zones of confidence and uncertainty has been used when delineating Nitrate Sensitive Areas and Nitrate Vulnerable Zones as part of the UK Government's and European Community's programmes for limiting the leaching of nitrates to groundwater. For example, the zones of confidence for source catchment areas have been used to help draw the boundaries of the Nitrate Vulnerable Zones designated under the EC Nitrate Directive (Council of European Communities 1991). Such a high degree of certainty has been considered necessary in the face of the statutory status and the agricultural restrictions operating within Nitrate Vulnerable Zones (Skinner & Foster 1995).

Conclusions

The adoption of the above methods and approach has evolved and been developed in the face of experience gained in delineating GPZs around the larger groundwater sources included within the Environment Agency's programme for England and Wales. As such, the different methods used represent the outcome of a pragmatic approach, tempered by the need that the procedure used to delineate each GPZ should be well-documented and technically robust.

There remain several areas where the methods used to delineate GPZs require further development. In particular, a more robust methodology is required for delineating GPZs around sources drawing from fractured or heavily fissured aqui-

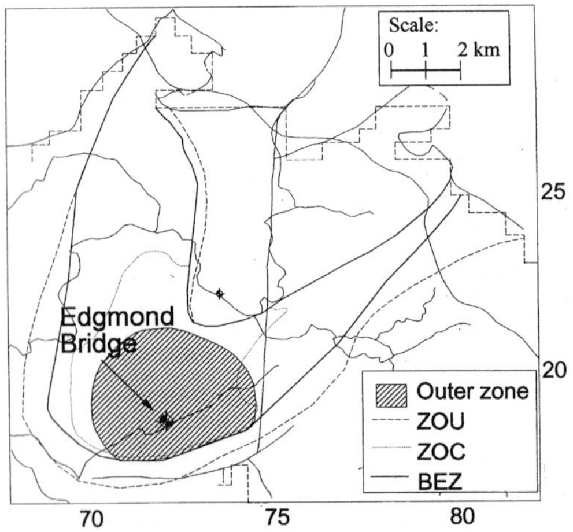

Fig. 5. Example of final GPZs drawn using FLOWPATH model output and computer-aided design program: Edgmond Bridge Pumping Station, Shropshire.

fers. Groundwater flow under such conditions is dominated by flow in interconnected conduits, a feature not allowed for within the assumptions on which the methods described in this paper are based.

In addition, guidelines are required for the use of regional, time-variant or transient groundwater models in delineating GPZs. To date, transient flow models have been used to derive 'snapshot' protection zones under extreme seasonal, quasi-equilibrium conditions of high and low water level (Salmon *et al.* 1996). However, there remains no accepted methodology for representing GPZs under transient groundwater flow conditions. A method is required that defines the position of a particular zone on a frequency basis such as a source catchment boundary that would be expected to be equalled or exceeded for say 95% of the time.

GPZs are now a well-established tool in limiting and controlling the risk of groundwater contamination in the face of continued development of the land surface. They have also proved to be a useful tool in defining areas to limit the continued deterioration of groundwater quality from diffuse pollutants such as nitrates.

The methods and concepts described in this paper are the result of over five years' experience and work by the staff of the then National Rivers Authority (NRA) and Consultants working for the NRA Groundwater Centre, Solihull, and under the overall direction of Dr A. C. Skinner. Nevertheless, the opinions expressed in this paper are those of the authors and may not necessarily reflect the current policy of the Environment Agency.

References

ADAMS, B. & Foster, S. S. D. 1992. Land surface zoning for groundwater protection. *Journal of Institute of Water & Environmental Management*, **6**, 312–320.

ANDERSON, M. P. & WOESSNER, W. W. 1992. *Applied Groundwater Modelling: simulation of flow and advective transport*. Academic, San Diego.

ASPINWALL & CO. 1993. *NRA Protection Policy: Training Manual for the Determination of GPZs*.

ATKINSON, T. C. & DREW, D. P. 1974. Underground drainage of limestone catchments in the Mendip Hills. *In*: GREGORY, K. J. & WALLING, D. E. (eds) *Fluvial Processes in Instrumented Watersheds*. Institute of British Geographers, London, Special Publication No. **6**.

BARKER, J. A., MACDONALD, D. M. J. & ADAMS, B. 1991. *Computer codes for the construction of source protection areas*. British Geological Survey, Technical Report **WD/91/41R**.

——, WILLIAMS, A. T., COLEBY, L. M. & WAGSTAFF, S. J. 1995. *NRA National Groundwater Protection Zone Programme: comparison of two model codes for phase III*. British Geological Survey, Technical Report **WD/95/8R**.

BEAR, J. & JACOBS, M. 1965. On the movement of water bodies injected into aquifers. *Journal of Hydrology*, **3**, 37–57.

COUNCIL OF EUROPEAN COMMUNITIES. 1991. Directive concerning the protection of waters against pollution caused by nitrates from agricultural sources (91/676/EEC). *Official Journal*, **L375**, December 1991.

ENVIRONMENT AGENCY. 1996. *Groundwater Protection Zones: Manual of Standard Zone Delineation Methodologies*. Environment Agency, Bristol.

EPA. 1987. *Guidelines for Delineation of Wellhead Protection Areas*. US Environment Protection Agency, Washington DC.

—— 1993. *WHPA Well Head Protection Area Delineation Code*. US Environment Protection Agency, Washington DC.

EVERS, S. & LERNER, D. N. 1995. Zones of certainty and uncertainty in estimating borehole catchments. *In*: *Groundwater Quality, Remediation and Protection*. IAHS Publication **225**, Prague, 483–491.

FERMOR, M. M., MORRIS B. L. & FLETCHER, S. W. 1995. Modelling groundwater source catchments in the UK: Implementation and review of a qualitative method of representing uncertainty. *In*: RITCHEY, J. D. & RUMBAUGH, J. O. (eds), *Subsurface Fluid-Flow Modelling*. American Society for Testing and Materials, Philadelphia.

GUIGER, N. & FRANZ, T. 1991. Development and application of a wellhead protection area delineation computer program. *Water Science & Technology*, **24**, 51–62.

HEADWORTH, H. G. & SKINNER, A. C. 1986. *Hydrogeological Investigations*. *In*: BRANDON, T. W. (ed.) *Groundwater: Occurrence, Development and Protection*. Institute of Water Engineers and Scientists, London, Water Practice Manual No. **5**, Ch. 15.

HUNTOON, R. W. 1995. Is it appropriate to apply porous media groundwater circulation models to Karstic aquifers? *In*: EL-KADI, A. I. (ed.) *Groundwater Models for Resources Analysis and Management*. CRC, Florida, 339–360.

LERNER, D. N. 1992. A semi-analytical model for borehole catchments and time-of-travel zones which incorporates recharge and aquifer boundaries. *Quarterly Journal of Engineering Geology*, **25**, 137–144

MCDONALD, M. G. & HARBAUGH, A. W. 1988. A modular three-dimensional finite difference model. *Techniques of Water Resources Investigations*, Book 6, Chapter A1. US Geological Survey.

MINISTRY OF HEALTH, 1948. *Safeguards to be adopted in the day to day administration of water undertakings*. Memorandum **221**, HMSO, London.

NATIONAL RESEARCH COUNCIL 1990. *Ground Water Models: Scientific and Regulatory Applications*. National Academy Press, Washington DC.

NATIONAL RIVERS AUTHORITY. 1992. *Policy and Practice for the Protection of Groundwater*. National Rivers Authority, Bristol.

—— 1995*a*. *Groundwater Protection Zones for Small Sources* (3 volumes). National Rivers Authority, Bristol.

—— 1995*b*. *Guide to Groundwater Protection Zones in England and Wales*. HMSO, London.

PALMER, R. C. & LEWIS, M. A. 1998. Assessment of groundwater vulnerability in England and Wales. *This volume*.

POLLOCK, D. W. 1988. Semi-analytical computation of path lines for finite difference models. *Ground Water*, **26**, 743–750.

SALMON, S. D., CHADA, D. & SMITH, D. 1996. Development of a groundwater resource model for the Yorkshire Chalk. *Water and Environmental Management*, **10**, 413–422.

SELBY, K. H. & SKINNER, A. C. 1979. Aquifer protection in the Severn-Trent Region: Policy and Practice. *Water Pollution Control*, **78**, 254–269

SKINNER, A. C. & FOSTER, S. S. D. 1995. Managing land to protect water: the British experience in groundwater protection. *26th Congress* of *the International Association of Hydrogeologists*, Edmonton, Canada

SOUTHERN WATER. 1985. *Aquifer Protection Policy*, Southern Water Authority, Worthing.

THOMPSON, N., BARRIE, I. A. & AYLES, M. 1981. *The Meteorological Office Rainfall and Evaporation System: MORECS*. Meteorological Office, Bracknell, Hydrological Memorandum No. **45**.

TODD, D. K. 1959. *Ground Water Hydrology*. Wiley, New York.

Is all groundwater worth protecting? The example of the Kellaways Sand

JOHN MATHER[1], DAWN HALLIDAY[1] & JEREMY JOSEPH[2]

[1]*Dept of Geology, Royal Holloway, Egham, Surrey TW20 0EX*
[2]*Shanks and McEwan, Woburn Sands, Bucks MK17 8TA*

Abstract: Many formations treated as Minor Aquifers for the purposes of resource protection do not meet accepted definitions of aquifers. The Kellaways Sand is a typical example, and modern waste-disposal licences require the installation of an engineered containment system to protect groundwater within it. However, the formation is clay-rich and only 4 to 5 m in thickness, and has a transmissivity of around 1.3×10^{-1} $m^2 day^{-1}$. Contained water quality is poor with salinities well above maximum admissible concentrations for drinking water. As the Kellaways Sand does not yield water to pumping wells, the only rational reason for protection is that the formation might contribute to baseflow in the River Ouse to the northeast. However, hydraulic gradients are such that seepage rates to the river will not exceed more than a few cubic metres per year over the length of outcrop adjacent to the river. It is concluded that the installation of engineered mineral liners at the base of brick-pits in the Lower Oxford Clay do nothing to reduce the risks associated with waste disposal. Resources spent on protecting the Kellaways Sand could be better used in other areas of environmental protection.

The requirement that all groundwater resources must be protected derives from the European Community Groundwater Directive (80/68/EEC) and generally leads to the installation of some sort of engineered containment at the base of new landfill developments. The Groundwater Protection Policy developed by the National Rivers Authority (now part of the Environment Agency) states that the disposal of domestic, commercial and industrial wastes is only acceptable on Major or Minor Aquifers with such containment (NRA 1992). Even on Non-Aquifers it is indicated that engineering methods may be necessary to protect surface waters, to which groundwater baseflow may be a major contributor.

Aquifers can be considered as layers of rock sufficiently porous to store water and permeable enough to allow water to flow through them in economic quantities (Price 1996). Major Aquifers are defined in the Protection Policy as highly productive strata of regional importance which are often used for large potable abstractions, and Minor Aquifers as formations of variable permeability which support locally important abstractions (NRA 1992). Non-Aquifers have negligible permeability and support only very minor abstractions, if any. There are few arguments over the designation of Major Aquifers but the designation of Minor Aquifers is a more difficult area. The examples of Minor Aquifers in the Groundwater Protection Policy (NRA 1992) include the Coal Measures, Devonian, river gravels and the Palaeogene sands. Non-Aquifers include all clays, shales, marls and siltstones.

Experience shows that the NRA classifications tend to be applied unquestioningly by waste regulators. Thus Coal Measures sites are required to have engineered containment despite the fact that local groundwater may already be affected by acid mine drainage. Any geological formation with the word sand in its title is taken to be a Minor Aquifer despite the fact that its groundwater may be saline with no potential for utilization as a public, or other, water supply source either now or in the future.

The need to provide protection in the form of engineered containment impacts on the waste management industry in two ways. The first is the cost of installation of the liner itself. A liner specified in a site licence to protect groundwater beneath a landfill on the Coal Measures in Yorkshire includes a 1 m layer of colliery spoil with a hydraulic conductivity of 1×10^{-8} $m\,s^{-1}$ overlain by a more permeable detection layer with drains installed. This is overlain by another 1m layer of colliery spoil engineered to a hydraulic conductivity of 1×10^{-9} $m\,s^{-1}$. This in turn is covered by a 300 mm stone leachate drainage blanket and a 300 mm stone protection layer. Even if the colliery spoil is available free of charge, this results in a cost in excess of £30/m^2. At a site in Wales the laying of a plastic geomembrane with overlying drainage blankets and protection has resulted in a cost of £13/m^2 and the construction of a double membrane composite liner with incorporated drainage increases this by at least £20/m^2. These costs assume that any clay required is available on site; if it has to be brought in at a price of £7 to £9/m^3, this will add further to unit costs.

MATHER, J. D., HALLIDAY, D. & JOSEPH, J. B. 1998. Is all groundwater worth protecting? The example of the Kellaways Sand. *In*: ROBINS, N. S. (ed.) *Groundwater Pollution, Aquifer Recharge and Vulnerability*. Geological Society, London, Special Publications, **130**, 211–217.

It can be argued that with a void of say 30 m depth, such costs are not significant and that a liner cost of £30/m^2 results in an increase in the gate price for waste disposal of only £1/tonne, if it is assumed that 1 tonne of waste occupies a volume of 1 m^3. However, there is a second way in which engineered containment affects landfill economics. If material for liners, drainage layers, etc., has to be imported into a site it uses up valuable void space. Thus, an engineered containment system 1m thick installed at the base of a large open-cast site or brickpit 70 ha in area will occupy a volume of 700 000 m^3, even if no allowance is made for drainage falls. At a gate price of £10/m^3 this results in a loss in turnover as a result of liner installation of some £7M. Put another way, with a void 30 m in depth the turnover and hence potential profit over the lifetime of the site is reduced by some 3.3% and the capacity of the resource is reduced. Hence it is pertinent to question whether or not such liner systems are always necessary.

In this paper the protection of the Kellaways Sand, which underlies the Oxford Clay through Bedfordshire and Buckinghamshire and beyond, will be taken as an example. This nominal sand is treated as a Minor Aquifer by the regulatory authorities. Modern waste-disposal licences require the installation of an engineered containment system with a minimum thickness of 1.5m of natural and engineered clay above the Kellaways Sand, of which not less than 0.5 m is native clay and not less than 1 m is engineered to a hydraulic conductivity equal to or lower than 1×10^{-9} m s^{-1} (Fig. 1).

Landfill takes place in many disused brickpits predominantly within the Lower Oxford Clay, which has ideal properties for the manufacture of bricks by the Fletton process. The weathered clay or 'callow' was unsuitable and stockpiled; it is this callow which is now often used to form the engineered containment. A number of these disused brickpits extend from Calvert in the southwest to Elstow in the northeast (Fig. 2). The large voids and excellent communications by road and/or rail mean that the pits are much in demand as landfill sites for domestic, commercial and industrial wastes. Most of the waste management operations are run by Shanks and McEwan (Southern) Ltd, who took over London Brick Landfill in 1986.

Geology of the Kellaways Sand

The Middle Jurassic Kellaways Formation lies immediately below the Lower Oxford Clay (Fig. 3). Almost everywhere it can be sub-divided into a lower mudstone member (the Kellaways Clay) and an upper relatively more sandy member (the Kellaways Sand). Although recorded thicknesses of the two members can vary significantly, they are laterally persistent from the Dorset coast to North Yorkshire (Cope *et al.* 1980). In the Marston Vale, which lies between the M1 motorway and Bedford (Fig. 2), boreholes have proved thicknesses of the Kellaways Sand from 3.5 to 5.2 m (Barron *et al.* 1992). To the west at Bletchley (Fig. 2) thicknesses of 3 to 5 m have been proved in boreholes drilled recently by Shanks and McEwan. The transitional nature of the boundary with the underlying Kellaways Clay means that some of the variation in thickness may result from misidentification of the junction between the two members.

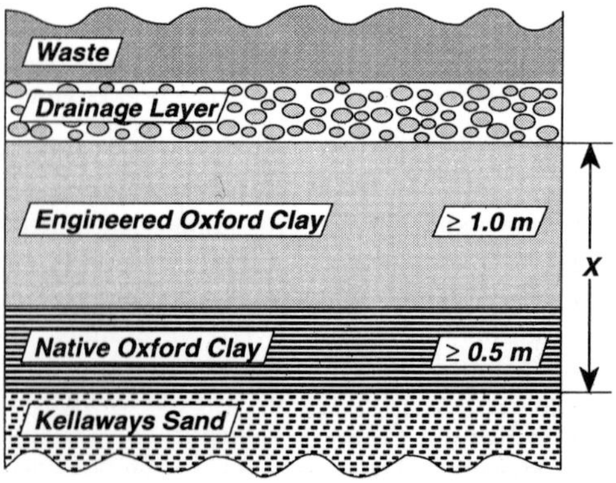

Fig. 1. Schematic cross-section through the base of an Oxford Clay landfill site, showing the presence of the engineered clay liner required by the regulatory authorities. In practice the thickness of cell X may range from 1.5 to 5.0 m.

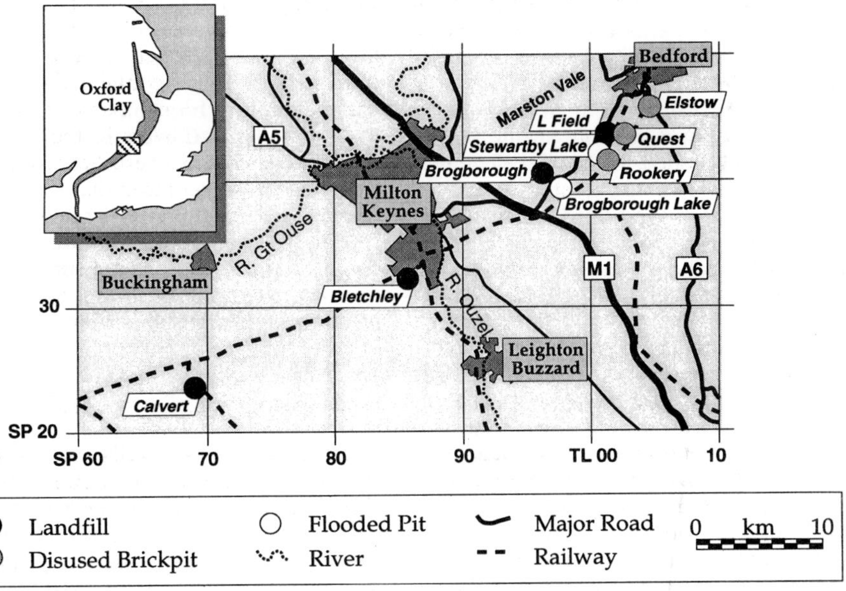

Fig. 2. Distribution of the major sites between Calvert and Elstow. Inset map shows the outcrop of the Oxford Clay in southern England.

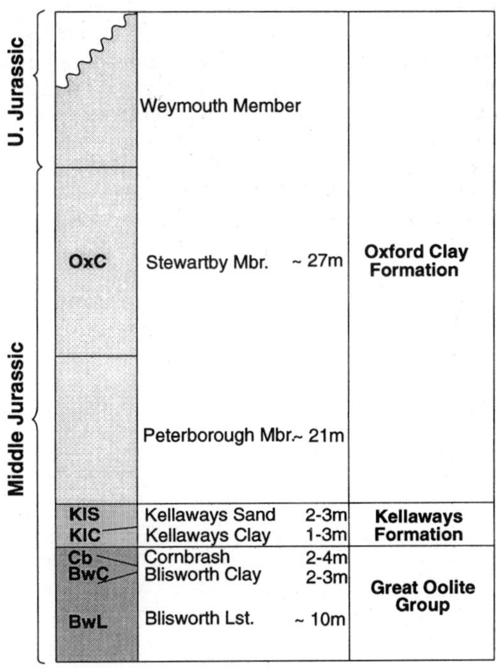

Fig. 3. Generalized geological section (modified from Barron *et al.* 1992).

The Kellaways Sand consists of mid-greenish-grey, fine- to very-fine-grained, clayey sand, silt and sandy silty clay, wispily bedded in places (Barron *et al.* 1992). The fraction of material below sand grade varies from around 20 to 40%. The formation is variably shelly, barren in parts, but with shell beds of belemnites, bivalves and gastropods, and is intensely bioturbated with well-defined sand-filled burrows. Carbonaceous and pyritized plant material is common and localized calcite cementation forms 'doggers', hard tabular masses of sandstone up to 3 m or so across. The underlying Kellaways Clay is about 0.5 to 2 m in thickness and is a mid- to dark-grey smooth pyritic mudstone. In the Marston Vale the base of the Kellaways Sand is taken as the first appearance of sandy beds or laminae within this mudstone (Barron *et al.* 1992), but at Bletchley the junction is more clearly defined.

Petrographic analysis of the Kellaways Sand from boreholes at Harwell, in Oxfordshire to the southwest, shows a close-packed fabric of sub-rounded detrital quartz with subordinate potash-feldspar, muscovite and organic fragments. The clay matrix consists of detrital illite and kaolinite with trace amounts of smectite and mixed-layer illite-smectite. Pyrite is abundant in the clay matrix and represents the major authigenic phase (Milodowski & George 1985).

The Kellaways Formation is underlain by the Great Oolite Group which consists of a sequence of limestones and mudstones (Fig. 3). The Cornbrash is predominantly limestone with minor mudstones which generally comprise 17 to 27% and exceptionally 43% of the formation (Barron *et al.* 1992). The Blisworth Clay is characteristically a smooth-textured mudstone and the underlying Blisworth Limestone comprises limestones with a small proportion of marls and minor thin mudstones.

Hydrogeology

Over the past few years Shanks and McEwan have drilled a large number of investigatory and monitoring boreholes in the Marston Vale and around the Calvert and Bletchley sites to the southwest (Fig. 2). In addition, UK Nirex Ltd drilled a network of boreholes around Elstow as part of their now-abandoned plans to investigate the Elstow Storage Depot, 5 km south of Bedford, as a potential repository for the shallow land burial of low- and intermediate-level radioactive wastes. Many of these boreholes penetrated the Kellaways Sand and some have been continued into the Blisworth Limestone.

Quoted values for the hydraulic conductivity of the Kellaways Sand are low, reflecting their clay and silt-rich nature. Available data from field tests in the area between Calvert and Elstow (Fig. 2) are tabulated in Table 1. The field tests used were mainly rising-head tests in which a slug of water is removed from the borehole and recovery monitored. A plot of all the data points gives a median value of 3.10×10^{-7} m s^{-1} with 25% of the values below 1×10^{-7} m s^{-1} and 25% above 1×10^{-6} m s^{-1}. These values can be compared to those given by Domenico & Schwartz (1990) where 10^{-7} m s^{-1} is said to be representative of fine-sand to silt-grade material. Laboratory determinations of both vertical and horizontal hydraulic conductivity quoted by Williams (1985) are generally one or two orders of magnitude less than those calculated from the field tests. Nine rising-head tests in the Blisworth Limestone, recently carried out at Bletchley by the authors, gave values of hydraulic conductivity ranging between 10^{-7} and 10^{-8} m s^{-1}, comparable to those of the Kellaways Sand.

It is clear from the above discussion that with a thickness of below 5 m and transmissivities of the

Table 1. *Hydraulic conductivity data for the Kellaways Sand*

Area	Source	No. of tests	Type of test	Hydraulic conductivity range (m s^{-1})
Marston Vale	Klinck (1992)	6	Rising head	$1.41 \times 10^{-6} - 2.03 \times 10^{-8}$
Elstow	Klinck & Noy (1992) (includes some data abstracted from confidential reports)	14	Mainly rising head	$7.10 \times 10^{-6} - 1.30 \times 10^{-8}$
'L Field' (Stewartby)	Parry (1972)	2	Falling head	$1.26 \times 10^{-7} - 3.00 \times 10^{-8}$
Bletchley	Halliday (unpublished)	5	Rising head	$2.40 \times 10^{-6} - 1.15 \times 10^{-8}$

order of 1.5×10^{-6} $m^2 s^{-1}$ or 1.3×10^{-1} $m^2 day^{-1}$, the Kellaways Sand will not yield water to pumping wells at any useful rate. This is confirmed by experience at Shanks and McEwan's L Field site (Fig. 2) where investigatory boreholes can be pumped dry by a small pump in a few minutes but take around 24 h to refill.

Piezometric levels have been recorded in the Kellaways Sand from the network of monitoring boreholes drilled in the Marston Vale and to the southwest. On a regional basis piezometric levels indicate movement within the Sands from the southwest towards the northeast (Fig. 4). The piezometric surface is relatively flat-lying in the west but steepens towards the outcrop along the western side of the River Ouse Valley. In the main part of the Marston Vale the gradient is about 1 in 350 or 0.003.

On a regional scale, heads in the Kellaways Sand are generally above those in the underlying Blisworth Limestone, indicating downward flow from the Sand into the Limestone. However, immediately down the hydraulic gradient at many of the pits this situation is reversed as a result of local perturbations resulting from pit excavation. At many of the sites, drainage trenches were excavated into the Kellaways Sand and pumped to remove surface water run-off from working pits and this has had an impact on local groundwater levels. Dry-weather flows have not been recorded in these drainage channels. This again illustrates the impermeable nature of the Sand, which itself made very little water and as far as the operation of the site was concerned was not obviously more permeable than the overlying Oxford Clay.

The final consideration with any groundwater which is to be used for water supplies is its quality. Woodward & Thompson (1909) reviewed previous work and stated that 'The Kellaways Beds yield water, but not as a rule of good quality. It is particularly saline in the area to the south and south-west of Bedford . . .' For some wells they concluded that 'This water is simply an impossible one to contemplate using as a regular supply, either for drinking or general domestic use. The amount of salt is enormous' (Woodward & Thompson 1909).

A review of the available historic analyses shows that, in all cases, chloride exceeds the current maximum admissible concentration in drinking water of 400 $mg\,l^{-1}$. These analyses were all carried out prior to brickclay extraction or landfill operations and highlight the naturally poor quality of the Kellaways Sand groundwaters. Modern water analyses are plotted in Fig. 5 where the high Na, SO_4 and Cl demonstrate the poor quality of these groundwaters in comparison to a typical Chalk groundwater used for public supply.

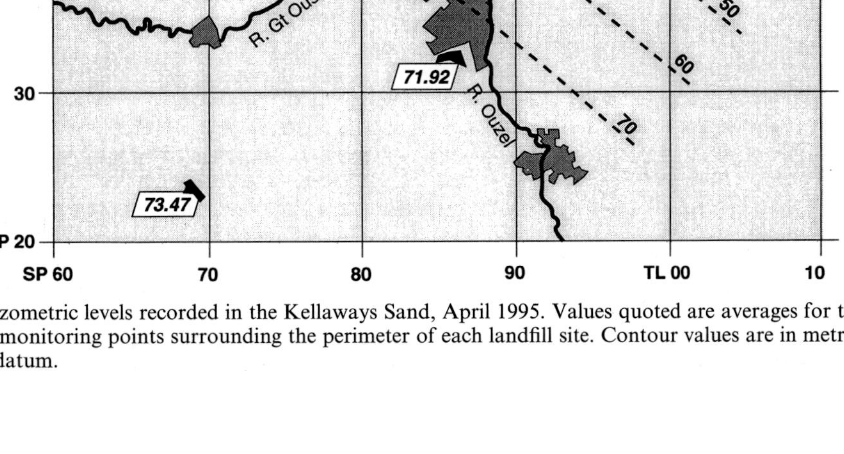

Fig. 4. Piezometric levels recorded in the Kellaways Sand, April 1995. Values quoted are averages for the numerous monitoring points surrounding the perimeter of each landfill site. Contour values are in metres above ordnance datum.

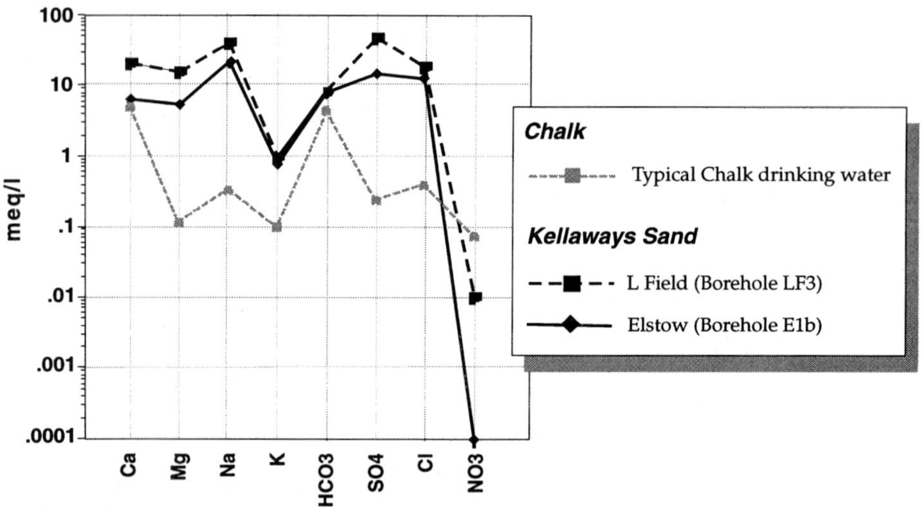

Fig. 5. A Schoeller plot to show the poorer quality of the Kellaways Sand groundwater compared with a typical Chalk drinking water.

Discussion

The fieldwork carried out over the last few years together with an analysis of the existing data confirm that the Kellaways Sand is a thin, fine-grained, clayey sand. It has a low hydraulic conductivity with a transmissivity of around 1.3×10^{-1} $m^2 day^{-1}$. Hydraulic gradients are low and water quality is poor. It will not yield water to pumping wells and cannot in any way be regarded as a Minor Aquifer. In fact it corresponds well to the definition of a Non-Aquifer given in the Groundwater Protection Policy (NRA 1992) i.e. 'Formations with negligible permeability. Only support very minor abstractions, if any'.

If it is accepted that the Kellaways Sand is a Non-Aquifer, the only reason for insisting on engineering containment to protect it is that it may contribute to baseflow in adjacent rivers, or groundwater from it may flow into an adjacent aquifer. It has already been established that discharge from the Kellaways Sand is towards the River Ouse in the northeast (Fig. 4). A simple calculation based on the hydraulic gradient and transmissivity values obtained indicates that seepage rates to the River Ouse will not exceed more than a few cubic metres per year distributed along an outcrop length of some kilometres. There are no springs issuing from the Kellaways Sand outcrop and a site visit failed to find any evidence of discrete discharges.

There is also an argument that water from the Kellaways Sand can enter the Blisworth Limestone, part of the Great Oolite formation. However, slug tests in the Blisworth Limestone have shown that hydraulic conductivities are comparable with those in the Kellaways Sand and water quality is generally as poor.

So what is the risk to groundwater from waste disposal into old brickpits in the Lower Oxford Clay which have no engineered containment system? From the data reviewed in this paper it is clear that the containment system provides negligible additional security. Whatever the risks associated with the disposal of domestic and industrial waste to these old brickpits, they are not reduced by the installation of engineered mineral liners. Such liners cost considerable sums of money and fill valuable void space but have insignificant impact on the risks associated with the waste-disposal operation.

Conclusions

The introduction to this paper questioned the necessity of installing engineered containment systems, which are often complex, to line many voids used for waste disposal. Such systems are expensive to install and use up valuable void space. The case against unnecessary containment took as an example the Kellaways Sand, which is a thin, clayey sand underlying the Oxford Clay. However, there are many other cases where resources are invested indiscriminately without any proper evaluation of the reduction in risk which is achieved. Examples include the protection of many thin and discontinuous Coal Measures Sandstones and protection in estuarine environments where groundwater is already contaminated by saline water.

Risk arises when an activity such as landfill takes place at a particular location. The risk to groundwater can be assessed by taking into account the nature of the waste to be disposed of and the leachate which will be generated from it, the natural vulnerability of the groundwater and the scale of preventive measures proposed. It is suggested that in the case of the Kellaways Sand and similar formations, the preventive measures which are installed are out of all proportion to any rational assessment of groundwater vulnerability arising from waste disposal.

Groundwater resources in the United Kingdom would be better served if money which goes on unnecessary protection were to be invested in research into the real problem areas. These include the remediation of groundwater contaminated by both light and dense non-aqueous phase liquids, and reducing the impact of nitrate and pesticides used in agriculture. Assessment of the cost of protection against the benefit to be gained will be a challenge for the new Environment Agency as it takes over responsibility for both groundwater protection and waste regulation.

The authors are grateful to Shanks and McEwan for permission to publish this paper and for the support given to the underlying research. The views expressed are those of the authors and do not necessarily represent company policy.

References

BARRON, A. J. M., HORTON, A. & SEN, M. A. 1992. *Geology of the Marston Vale*. British Geological Survey, Technical Report **WA/92/23C**.

COPE, J. W. C., DUFF, K. L., PARSONS, C. F., TORRENS, H. S., WIMBLEDON, W. A. & WRIGHT, J. K. 1980. *A Correlation of Jurassic Rocks in the British Isles. Part 2: Middle and Upper Jurassic*. Geological Society, London, Special Report No. **15**.

DOMENICO, P. A. & SCHWARTZ, F. W. 1990. *Physical and Chemical Hydrogeology*. Wiley, New York.

KLINCK, B. A. 1992. *Marston Vale groundwater study, Bedfordshire: results of rising head slug tests*. BGS Fluid Processes Group, Technical Report **WE/92/19C**.

—— & NOY, D. J. 1992. *A groundwater flow model for the Marston Vale*. British Geological Survey, Technical Report **WE/92/21C**.

MILODOWSKI, A. E. & GEORGE, I. A. 1985. *The petrography of the Jurassic core from the Harwell Research Site. Part 2: Kellaways Beds, Great Oolite Group and Inferior Oolite Group*. British Geological Society Report **FLPU 85–8**.

NATIONAL RIVERS AUTHORITY (NRA) 1992. *Policy and Practice for the Protection of Groundwater*. National Rivers Authority, Bristol.

PARRY, R. H. G. 1972. Some properties of heavily over-consolidated Oxford Clay at a site near Bedford. *Geotechnique*, **22**, 485–507.

PRICE, M. 1996. *Introducing Groundwater*, 2nd edn. Chapman and Hall, London.

WILLIAMS, G. M. 1985. *Preliminary assessment of the geology and hydrogeology for the proposed site investigation at Elstow Storage Depot, Bedfordshire*. Nirex Report No. **22**.

WOODWARD, H. B. & THOMPSON, B. 1909. *The water supply of Bedfordshire and Northamptonshire from underground sources with records of sinkings and borings*. Memoir of the Geological Survey.

Index

References in *italics* are to Figures or Tables